Global Environmental
Forest Policies

Global Environmental Forest Policies

An International Comparison

*Constance L. McDermott, Benjamin Cashore
and Peter Kanowski*

publishing for a sustainable future

London • Washington, DC

First published in 2010 by Earthscan

Earthscan Ltd, Dunstan House, 14a St Cross Street, London EC1N 8XA, UK
Earthscan LLC, 1616 P Street, NW, Washington, DC 20036, USA
Earthscan publishes in association with the International Institute for Environment and Development

For more information on Earthscan publications, see www.earthscan.co.uk or write to earthinfo@earthscan.co.uk

ISBN: 978-1-84407-590-4 hardback

Typeset by MapSet Ltd, Gateshead, UK
Cover design by Susanne Harris

A catalogue record for this book is available from the British Library

Library of Congress Cataloging-in-Publication Data

McDermott, Constance
  Global environmental forest policies : an international comparison / Constance McDermott, Benjamin Cashore, and Peter Kanowski.
     p. cm.
  Includes bibliographical references and index.
  ISBN 978-1-84407-590-4 (hardback)
  1. Forest policy–Environmental aspects. 2. Forest management–Environmental aspects. 3. Forests and forestry–Environmental aspects. 4, Global environmental change. I. Cashore, Benjamin William, 1964– II. Kanowski, Peter. III. Title.
  SD561.M33 2010
  333.75–dc22

                                                                                    2009051709

At Earthscan we strive to minimize our environmental impacts and carbon footprint through reducing waste, recycling and offsetting our CO_2 emissions, including those created through publication of this book. For more details of our environmental policy, see www.earthscan.co.uk.

Printed and bound in the UK by CPI Antony Rowe.
The paper used is FSC certified.

Contents

PART I SETTING THE SCENE

PART II REGIONAL ANALYSES

PART III SUMMARY AND CONCLUSIONS

List of Figures, Tables and Boxes

Figures

Tables

Boxes

In memory of
Mark N. McDermott
and for
Lillian
Walter, Theresa and Joseph
Maureen and Peter

Acknowledgements

We thank the many colleagues worldwide who provided invaluable assistance to our research, from its early stages to its metamorphosis as a published book. These include student interns at the Yale School of Forestry and Environmental Studies, as well as academics and practitioners from all 20 case study countries and elsewhere.

While it is impossible to name all those who have so importantly influenced our thinking and approach, we would like to specifically acknowledge those individuals who were directly involved in reviewing and case study research. We thank those who reviewed draft chapters of the book, including Auro Almeida, Tony Bartlett, Nuno Borralho, Chris Burchmore, Andrzej Czech, Glen Kile, Robert Hendricks, Bruce Manley, Andrew Mathews, Ilze Silmakele, John R. Vitello and Graham Wilkinson. For specific chapter contributions we thank Michael Blyth, Pingyang Liu, Terhi Koskela, Melanie H. McDermott, Brian Milakovsky, Ulrich Schraml and Takuya Takahashi.

We are also grateful to a range of faculty and students who helped us better understand regulatory approaches in a number of the countries under review, including: Gabriela Alonso, Laura Bozzi, Marisa Camargo, Elizabeth Egan, Monika Kumar, Arvind Nagrajan, Emily Noah, Camille Rebelo, Steve Rhee, Krishna Roka, Lisa Smith, Corrina Steward, Anna Tikina, Meredith Trainor, and Lisa Curran and Lloyd Irland.

Finally, we thank the range of funders and supporting institutions who made this book possible, including The American Forest and Paper Association; The Australian National University; The Department of Sustainability and Environment, Victoria; The Forests and Forest Industry Council of Tasmania; The Forest Products Association of Canada; Forest and Wood Products Australia; The James Martin 21st Century School, University of Oxford; Natural Resources Canada; The Canadian Forest Service; The Oxford Centre for Tropical Forests; The United States Department of Agriculture, Forest Service; The University of New Brunswick; The Weyerhaeuser Foundation; and The Yale School of Forestry and Environmental Studies.

List of Acronyms and Abbreviations

AAC	annual allowable cut
ABARE	The Australian Bureau of Agricultural and Resource Economics
ACPWP	Advisory Committee on Paper and Wood Products
ACT	Australian Capital Territory
AF&PA	American Forest and Paper Association
AFS	Australian Forestry Standard
AITPN	Asian Indigenous Tribal Peoples Network
ANCSA	Alaska Native Claims Settlement Act
APEC	Atlantic Provinces Economic Council
ASQ	Annual Sale Quantity
ASRD	Alberta Sustainable Resource Department
ATFS	American Tree Farm System
BaySF	Bayerische Staatsforsten
BC MFR	British Columbia Ministry of Forests and Range
BC MWLAP	British Columbia Ministry of Water, Land and Air Protection
BIA	Bureau of Indian Affairs
BMP	best management practice
BMVEL	Ministry of Consumer Protection, Food and Agriculture
BSAP	Biodiversity Strategy and Action Plan
BSLF	Bayerisches Staatsministerium für Landwirtsch aft und Forsten
CARPE	Central African Regional Program for the Environment
CARTS	Conservation Areas Reporting and Tracking System
CBD	Convention on Biodiversity
CBP	Customs and Border Patrol
CCAEC	Canada–Chile Agreement on Environmental Cooperation
CCEA	Canadian Council on Ecological Areas
CCFM	Canadian Council of Forest Ministers
CDM	Clean Development Mechanism
C&E	Compliance and Enforcement
CIESIN	Center for International Earth Science Information Network
CIFOR	Center for International Forestry Research
CIRAD	Centre de coopération internationale en recherche agronomique pour le développement (France)
CITES	Convention on International Trade in Endangered Species
CMS	Convention on Migratory Species

CoC	Chain of Custody
CONADI	National Corporation of Indigenous Development (Chile)
CONAF	Chilean National Forestry Corporation
CONAFLOR	Coordinating Commission for the National Forestry Programme (Brazil)
CONAMA	National Environment Commission (Chile)
COSEWIC	Committee on the Status of Endangered Wildlife in Canada
CPET	The Central Point of Expertise on Timber Procurement
CSA	Canadian Standards Association
DAFF	Department of Agriculture, Fisheries and Forestry (Australia)
DBH	diameter at base height
DEWHA	Department of the Environment, Water, Heritage and the Arts (Australia)
DIPE	Department of Infrastructure, Planning and Environment (Australia)
DRC	Democratic Republic of Congo
DWAF	Department of Water Affairs and Forestry (South Africa)
EECONET	Convention on the Conservation of European Wildlife and Natural Habitats (The Netherlands)
EFC	European Forestry Commission
ENF	National Strategy for Forests
ENGOs	Environment Non-Government Organizations (Australia)
EPA	Environmental Protection Agency
ESA	Endangered Species Act
ESCC	Endangered Species Conservation Committee
FEMAT	Forest Ecosystem Management Assessment Team
FESA	Forest Engineering Southern Africa
FFA	Finnish Forest Association
FIEC	Environmental Guidelines for Plantation Forestry in South Africa
FLEGT	Forest Law Enforcement Governance and Trade
FMAs	Forest Management Associations
FPA	Forest Practices Authority (Tasmania)
FPA	Forest Practices Plan
FPB	Forest Practices Board
FPC	Forest Products Commission (Australia)
FPG	Forest Practice Guidelines
FSC	Forest Stewardship Council
FUNAI	Indian National Foundation (Brazil)
FWIGF	Forest Watch Indonesia Global Forest Watch
GAP	Gap Analysis Program
GDP	gross domestic product
GOI	Government of India
HDI	Human Development Index
IBAMA	Brazilian Institute of Environment and Renewable Resources
IBGE	Instituto Brasileiro de Geografia e Estatística

ICCN	Congolese Institute for the Protection of Nature
IELRC	International Environmental Law Research Centre
IFCO	International Forestry Cooperation Office
IIASA	International Institute for Applied Systems Analysis
IIED	International Institute for Environment and Development
IMFP	International Model Forests Program
INAC	Indian and Northern Affairs Canada
INE	Instituto Nacional de Estatística
INPE	Instituto Nacional de Pesquisas Espaciais
INR	Institute of Natural Resources (South Africa)
ISO	International Organization for Standardization
ITTO	International Tropical Timber Organization
IUCN	International Union for the Conservation of Nature
IUFRO	The Global Network for Forest Science Cooperation
IUPHHK	Commercial Timber Utilization Permits
IWMS	Identified Wildlife Management Strategy
JCP	Joint Certification Protocol
JFM	Joint Forest Management
JOFCA	Japan Overseas Forestry Consultants Association
LEDC	Latvian Environment Data Center
LEI	Lembaga Ekolabel Indonesia
LFD	Land and Forest Division (Alberta)
LRMPs	Land and Resource Management Plans
LVM Ltd	Latvia's State Forests
MAF	Ministry of Agriculture and Forestry
MAR-SFM	Monitoring, Assessment and Reporting on Sustainable Forest Management
MCPFE	Ministerial Conference on the Protection of Forests in Europe
MFR	Ministry of Forests and Range
MMA	Ministry of Environment (Brazil)
MNRWP	Ministry of Natural Resources, Wildlife (MNRWP: English; MRNFP, French)
MOUs	Memoranda of Understanding
MP	Madhya Pradesh
MPIG	Montreal Process Implementation Group
MSP	Maximum Sustained Production
MTK	Central Union of Agricultural Producers and Forest Owners (Finland)
NAFA	National Aboriginal Forestry Association
NAFI	National Association of Forest Industries (Australia)
NAFTA	North American Free Trade Agreement
NEPA	National Environmental Policy Act
NERA	New England Regional Assessment
NGO	non-governmental organization
NMFS	National Marine Fisheries Service

NOAA	National Oceanographic and Atmospheric Administration
NPS	National Park Service
NRC	Natural Resources Canada
NSMD	non-state market driven
NTFP	Non-Timber Forest Products
NZFOA	New Zealand Forest Owners Association
ODI	Overseas Development Institute
OEMA	Órgão estadual do meio ambiente
OMNR	Ontario Ministry of Natural Resources
OWEB	Oregon Watershed Enhancement Board
PAD-US	Protected Area Database Partnership
PAs	Protected Areas
PCI	Principles, Criteria and Indicators
PEFC	Programme for the Endorsement of Forest Certification
PGF	Forest Management Plan (Portugal)
PNF	National Forest Programme (Brazil)
PROFEPA	Federal Office for Environmental Protection (Mexico)
REDD	Reduced Emissions from Deforestation and Degradation
RENACE	Red Nacional de Acción Ecologista – Argentina/Chile
RFA	Regional Forest Agreement
RONV	Range of Natural Variation
RRI	Rights and Resources Initiative
SAC	Special Areas of Conservation
SARA	Species at Risk Act
SCDHEC	South Carolina Department of Health and Environmental Control
SFA	Swedish Forest Agency
SFI	State Forest Inventory (Russian Federation)
SFI	Sustainable Forestry Initiative
SFIF	Swedish forest industries federation
SFL	sustainable forest licence
SFRA	Southern Forest Resource Assessment
SFM	Sustainable Forest Management Series
SFS	State Forest Service
SGEC	Sustainable Green Ecosystem Council (Japan)
SILNA	South Island Landless Natives Act
SMZs	Special Management Zones
SNUC	Sistema Nacional de Unidades de Conservação da Natureza
SPAs	Special Protection Areas
sph	stems per hectare
TPTI	Tebang Pilih Tanam Indonesia
UNEP-WCMP	United Nations Environment Programme – World Conservation Monitoring Centre
USDA	United States Department of Agriculture (USA)
USFS	US Forest Service
USFS FIA	US Forest Service Forest Inventory and Analysis

USFWS	US Fish and Wildlife Service
USGS	US Geological Survey
VPA	Voluntary Partnership Agreement
WDPA	World Database on Protected Areas
WTO	World Trade Organization
WWF	World Wide Fund for Nature
Y2Y	Yellowstone-to-Yukon

PART I

Setting the Scene

Chapter 1

Introduction

Constance L. McDermott, Benjamin Cashore and Peter Kanowski

INTRODUCTION

The world's forests play a fundamental role in shaping and sustaining both the global environment and human society. Forests are the world's most biologically diverse terrestrial ecosystems and, despite unprecedented rates of loss over the past century, still cover some 30 per cent of the world's land surface. They provide critical ecosystem services in climate regulation and the protection of soil and water resources. Both wood and non-wood forest products are essential to the livelihoods of more than a billion of the world's poorest people. Overall, per capita wood consumption increases with wealth, supporting a forest products industry that accounts for 1 per cent of global gross domestic product (GDP) (World Bank, 2004; FAO, 2007).

The world's forests are also under enormous pressure: from conversion to agricultural and other uses, from illegal and unsustainable harvesting of forest products and from climate change. Some of these pressures reflect increasing human populations and resource consumption; some reflect conscious societal choices about land use and development strategies; some reflect poor forest policies and governance (Woodwell and Ullsten, 2001; World Bank, 2004; Millennium Ecosystem Assessment, 2005). There is general consensus that – despite progress on some issues and in many places – the scale and rate of forest loss and degradation remain untenable if the needs and interests of future generations are not to be ignored.

This book represents an effort to understand and assess one critical component of this complex story: environmental policies and regulations that have emerged to address commercial timber harvesting practices. Amongst the many pressures on forests, commercial harvesting merits attention because it has been of increasing significance globally over the past century. Its impacts have accelerated over the past

50 years, as growing demand and national development strategies encouraged, and new technologies enabled, the expansion of access into frontier forests accompanied by large-scale harvesting in each of the boreal, temperate and tropical zones. Wood production remains a designated objective of half, and the primary objective of one-third, of the world's forests (FAO, 2007, p67). Consequently, policies that regulate forest harvesting for wood production remain central to the implementation of sustainable forest management, and thus to the conservation and responsible use of the world's forests.

Despite an extensive body of literature on global forest governance, and many country-level case studies, there has been little comparative analysis to assess country-level forest policy measures intended to improve the environmental outcomes of forest harvesting. This book helps fill this gap by conducting a systematic, large-scale comparison of the regulation of commercial wood harvest in 20 case study countries that are significant in the extent of their forest cover and/or role in international wood products trade. Our analysis of each case study country begins with a description of the context of forestry and forest practices, as the basis for a standardized comparison of forest practice policies. This contextualized analysis allows us to situate policies within jurisdictions, to compare them on a transparent basis and to facilitate policy learning for more sustainable forestry globally.

This book is organized in three parts. The first, comprising Chapters 1 and 2, introduces our topic and methodologies and sets the overall context of the study. The second, comprising Chapters 3 through 9, outlines key contexts for each case study country and assesses policies which regulate forest harvesting for wood production within the analytical framework explained in Part 1. The third part, Chapters 10 and 11, presents the overall findings and conclusions of our work.

THE IMPORTANCE OF GLOBAL FOREST POLICY COMPARISONS

Forest policies have now been a focus of global concern for a quarter of a century, since the loss and degradation of tropical forests first began to command the attention of the international community. During this period, the emphasis of international policy initiatives has evolved: from the original focus on only tropical forests to all of the world's forests; from principally biodiversity to the broader suite of environmental services; from measures intended to enhance forest conservation and sustainable forest management to those focused on forest governance and illegal logging. Most recently, realization of the significance of climate change impacts of greenhouse gas emissions from deforestation and forest degradation has brought renewed impetus to efforts to conserve and better manage forests globally.

Despite these shared concerns, the international community has not been able to reach any binding global agreement on forest conservation and management (Humphreys, 2006). Consequently, the policies that govern how forests are conserved and managed remain those defined by national and sub-national governments for the forests under their jurisdiction. It is through these policies that fundamental forest policy challenges – of sustaining both livelihoods and forests, of balancing environmental protection with economic development and of reconciling competing interests in forests, at scales from local to global – are addressed. Because, as discussed above, 50 per cent of the world's forests are managed for some level of wood production, the issue of how national and sub-national policies protect the full range of forest values in forests from which wood is harvested is central to the realization of sustainable forest management.

The policies governing forest management, and the forest practices which they allow and require, have changed – sometimes dramatically – in most countries over the past quarter century. Typically, they have become more restrictive and more demanding, as governments have given progressively greater weight to the environmental values of forests. These policies must also accommodate a diversity of forest tenure and production types from which wood is harvested – not only the natural forests under state ownership, which have been the focus of much international attention, but also natural forests under private and various forms of community management important in many countries, and plantation forests, which now provide one-third of the world's industrial wood.

Although the policies governing forest practices have become, almost universally, more environmentally oriented, there remains major debate over the definition of 'sustainable' forest practices. This debate is not just between the 'browns' and the 'greens', but within the environmental community itself, reflecting differences between those who seek to engage major market actors in incremental improvements and those focused on transforming global markets in support of low intensity, more locally based, models of wood production.

How ought those committed to sustainable management of the world's forests assess such contradictory arguments? How do countries actually compare in their forest practices? In this book, we address these questions by developing a framework for comparing policies that govern forest practices and by applying the framework to a representative sample of countries.

This research also informs related questions that we visit in the concluding chapters, such as how the evolution of forest policies in one country impacts those in another. Might the gains made in forest practices in one country be undermined by 'leakage' to countries with poorer practices, or do they lead to policy standardization as more governments, firms and non-governmental organizations (NGOs) commit to similar sets of 'sustainable forest management' principles and practices? Overall, do forest practice policies in an era of increasing globalization reflect improvements in the balancing of environmental conservation and social benefits with economic development, or do they resemble a 'race to the bottom'? Even partial answers to these questions can help policy-makers, and other forest stakeholders, reflect on the benefits and costs of particular policy approaches.

Our approach is timely and important for a number of reasons. First, there is widespread concern and frustration globally that, while forest loss and degradation continue at what the FAO (2007, p64) calls an 'alarming' rate, only limited progress has been made in developing an effective global response. As noted above, this means that national and sub-national policies continue to be critical to realizing forest policy objectives. A related issue concerns the adoption and effectiveness of forest certification, a market-based policy instrument promoted initially by NGOs as a strategy to harness the power of market forces and bypass what were viewed as overly slow-moving inter-governmental efforts (see Cashore et al, 2004). However, it is now recognized that certification and state-based governance are intricately intertwined. For example, existing government regulations may influence both the performance requirements of certification standards (McDermott et al, 2008) and producer willingness to pursue certification, which may lead to variable impacts of certification in different regions (Cashore et al, 2007; Auld et al, 2008).

Meanwhile, forest certification has been very slow to evolve in developing countries, where illegal logging frequently circumvents even the most basic government requirements. Recognition of this has led many environmental groups to return to work with governments, for example through government procurement policies, trade prohibitions on illegally produced products[1] or the European Union's 'Forest Law Enforcement Governance and Trade (FLEGT)' process (European Commission, 2008). The FLEGT process promises improved market access for countries with high rates of illegal logging in exchange for systems that distinguish wood that is legally produced from wood of unknown or illegal origin. Among the questions that arise in such processes are: what are the rules that define 'legal' forest harvest in different countries and how do they compare to forest laws elsewhere. The approach we adopt is also relevant to these questions.

Second, recognition that forest loss and degradation is responsible for nearly 20 per cent of global greenhouse gas emissions (IPCC, 2007) has defined a new 'Reduced Emissions from Deforestation and Degradation' (REDD) agenda as important to both future international climate negotiations and to future sustainable forest management regimes (e.g. Kanninen et al, 2007). The effectiveness of REDD initiatives will depend not only on reducing deforestation and degradation in the areas on which such initiatives focus, but also on minimizing leakage (i.e. unintended adverse consequences in the form of increased emissions from consequent activities elsewhere);[2] this, in turn, will require an understanding of forest practice requirements globally. The approach we adopt, and the analyses we report in this book, provide the basis for such an understanding.

Third, both practitioner and academic communities would like to understand better the comparative *effectiveness* of alternative policy approaches to sustainable forest management. As we discuss further below, effectiveness is a complex issue and beyond the scope of the analyses we conduct here. However, the capacity to assess effectiveness of forest policies has been limited by lack of systematic analyses of what the policies actually require. In addressing the latter issue, we provide a platform for addressing the former.

Finally, the analytical framework we develop and present facilitates a new generation of systematic research into forest policy questions. While our empirical analysis, focused on understanding and comparing forest practices policies, informs those issues, it also illustrates the value of a new direction for policy studies in general, and global forest policy development in particular. One outcome may be to promote a more common global approach to the fundamentals of sustainable forest practices; such an approach can only help facilitate broader learning and knowledge generation within both practitioner and scholarly communities that is a prerequisite to addressing the continuing loss and degradation of the world's forests.

OUR METHODOLOGIES

This study comprises three methodological elements, which we introduce in the following order. First, in the next section in Chapter 1, we present analytical frameworks that permit comparison and classification of forest practice policies across national and sub-national jurisdictions. Second, we identify the policy variables that we will compare. Third, in Chapter 2, we identify and apply a methodology for selecting our case study countries and situate these countries in global context.

FRAMEWORKS FOR COMPARATIVE POLICY ANALYSIS

A policy taxonomy

Cashore and Howlett (2007) and Howlett and Cashore (2007) have drawn on and adapted existing work within policy studies (see especially Hall, 1993), to develop a taxonomy that identifies six levels of policy located within a three by two matrix (Table 1.1). The vertical axis distinguishes policy 'ends' (the ultimate aim of policy) from 'means' (instruments), while the horizontal axis distinguishes degrees of abstraction. At the 'ends' level, they identify abstract 'goals' (such as environmental protection or economic development), measurable 'objectives' (such as maintaining species diversity, protected areas, or employment) and actual policy 'settings' that dictate what is specifically required (such as how far from a stream harvesting operations can occur, or the level of harvesting that is permitted in a given year). At the 'means' level, they identify 'logic' that permeates decisions about policy instruments (such as preferences for coercion versus suasion), actual 'instrument' choice (such as 'command and control' regulations versus tax incentives or subsidies), and precise 'calibrations' of these instruments (such as how does the actual tax incentive work, or what are the levels of subsidies?).

This study is focused largely on policy settings (the top right cell in Table 1.1) – the specific, on-the-ground, requirements of policy. Policy scholars have often ignored this issue, while both scholars and practitioners often lack a strong international perspective on it. In part because of the complexity of policy, systematic cross-country characterizations of policy have usually taken one of three directions:

Table 1.1 *A modified taxonomy of policy components following Hall*

	High level abstraction	Policy content programme level operationalization	Specific on-the-ground measures
Policy ends or aims	**GOALS** **What general types of ideas govern policy development?** (e.g. environmental protection, economic development)	**OBJECTIVES** **What does policy formally aim to address?** (e.g. saving wilderness or species habitat, increasing harvesting levels to create processing jobs)	**SETTINGS** **What are the specific on-the-ground requirements of policy** (e.g. considerations about the optimal size of designated stream-side riparian zones, or sustainable levels of harvesting)
Policy focus			
Policy means or tools	**INSTRUMENT LOGIC** **What general norms guide implementation preferences?** (e.g. preferences for the use of coercive instruments, or moral suasion)	**MECHANISMS** **What specific types of instruments are utilized?** (e.g. the use of different tools such as tax incentives, or public enterprises or 'direct regulation' (command and control)	**CALIBRATIONS** **What are the specific ways in which the instrument is applied?** (e.g. what are the specific subsidy levels or tax incentives? How are 'direct regulation' policies developed? Regulation? Legislation? Directives?)

Note: Cells contain examples of each measure.
Source: Howlett and Cashore (2007), modified from Cashore and Howlett (2007)

they have been abstracted to broad national-level generalizations (Rayner and Howlett, 2003), focused on a very limited number of selected comparisons (Hoberg, 1993b; Cashore, 1997), or have been single case studies (Tollefson, 1998; Cashore, 2001; Hoberg, 2001). In this study, we fill the gap in previous work by focusing on understanding and comparing what forest policies governing harvesting actually require. Focusing on this gap will advance understanding among domestic and global forest policy communities, as well as contribute to scholarly literature explaining policy development.

Policy comparisons, whether they be small or large scale, general or specific, are ultimately only a first step in answering the larger question of policy 'effectiveness'. Our analysis, by itself, does not answer the key question of how policy affects 'on-the-ground' behavioural change nor the results of this change in addressing policy goals. However, it does take a necessary first step by identifying exactly what is required of forest managers. Only by first identifying and contrasting policy approaches and content can we begin assess how different policies might differently shape on-the-ground outcomes.

Classifying policy settings

Our focus on policy settings still poses significant challenges for our comparative task. Most contemporary forest policies are complex, with detailed policy settings that may run to many volumes. We address this complexity by developing a clear and carefully delineated protocol to identify and describe the forest policies that exist in each jurisdiction. We then develop an analytical framework for classifying the type of regulatory approach employed by different governments.

Analytic classification system

The analytic framework for comparing policy settings we have designed for this study draws on Cashore (1997) and Cashore and McDermott's (Cashore and McDermott, 2004; McDermott and Cashore, 2007) earlier work. Cashore's original policy schematic (1997) classified forest policy according to:

1   Structure: whether policies are worded in a discretionary versus a non-discretionary manner; and
2   Method: whether policies emphasize procedures or plans (such policies are often referred to as 'systems-based') or are 'substantive' (often referred to as 'performance-based'), specifying on-the-ground behaviour.

Structure
A key question under this classification system is whether rules are voluntary (discretionary) or mandatory (non-discretionary). Major policy differences can result from subtle differences in wording – the uses of words like 'must' and 'shall' limits discretion, in contrast to the use of words like 'may' or 'where appropriate' which expands discretion. There may even be more subtle distinctions, where words like 'must' are used, but where the wording of the requirement allows broad room for interpretation, or where government officials are able to grant exemptions.

Method
Policies can be further distinguished by whether they focus exclusively on procedural rules, which cover such things as the requirement for written plans and procedures, but which do not detail specific forest management practices. Forest practice policies have often been more reliant on procedures than on substantive requirements. For example, requiring that certain management objectives, such as riparian protection for example, be addressed in management plans is different from prescribing specific on-the-ground practices, such as the establishment of a 50-metre buffer zone. It is thus important to distinguish procedural rules, whose effect on on-the-ground forest management is both indirect and uncertain, from non-discretionary rules prescribing actual forest management practices. Classifying forest practice policies in this way does not imply that a procedural approach is superior or inferior to a rule-based approach; as we discuss below, there are arguments in favour of each. However, it does recognize that they are fundamentally different.

Table 1.2 *Forest policy classification system*

Structure	Approach
1 Voluntary	Rules encourage, but don't require, a course of action
2 Mandatory	Rules require a specific course of action

Method	
1 Substantive	Rules address on-the-ground changes
2 Planning/procedural	Rules address management systems, rather than on-the-ground actions

Source: Adapted from Cashore (1997)

These distinctions lead to the four 'styles' of forest policy regulation identified in Table 1.3. *Procedural voluntary* policies are voluntary and involve the development of processes or plans, rather than prescriptions for on-the-ground practices. Such flexible approaches *could* lead to significant change in forestry practices, but it is difficult to predict their effect without a case-by-case analysis of their implementation. *Procedural mandatory* policies involve requirements for the development of plans or procedures. An example of such an approach is the US National Environmental Policy Act's (NEPA) requirements that federal projects undergo an environmental assessment. Procedural non-discretionary policies help to guarantee that planning has taken place, but provide little certainty about the forest practices that result. *Substantive voluntary* policies involve those cases where specific forest practice rules or guidelines exist, but are voluntary in nature. Finally, the *substantive mandatory*, or *prescriptive*[3] policy category refers to mandatory, on-the-ground requirements or restrictions, such as a rule that no timber harvest may occur within x metres of a river of y width.

   These different policy approaches can be viewed along a continuum of policy 'prescriptiveness'. Voluntary policies are the least prescriptive, in that they afford the forest manager the maximum flexibility. Mandatory substantive policies are the most prescriptive, particularly when they involve quantitative performance 'thresholds' that prescribe the precise actions required of all forest operators. Highlighting these distinctions is not to argue that one approach is necessarily better than the other, since each has advantages and disadvantages. Many environmental advocacy organizations tend to support highly 'prescriptive' policies, i.e. mandatory substantive requirements involving environmental thresholds, due to a lack of trust in forest managers to adequately exercise their own discretion (McDermott, 2003). Among these and other non-producer stakeholders there is concern that procedural and/or voluntary approaches (Porter and van der Linde, 1995) may permit intransigent forest managers to simply maintain and/or undercut *status quo* forest practices (Sharma, 1996). Prescriptive approaches, if enforced, guarantee a minimum threshold of environmental performance. However, they also leave little room for adaptation, are generally not receptive to local knowledge or local concerns and may discourage innovation or creative 'win–win' solutions on the part of forest managers, forest industries and other stakeholders.

Table 1.3 *Matrix of four policy styles ('subspecies')*

	Voluntary	Mandatory
Procedural (systems-based)	Suggested planning processes	Planning requirements
Substantive (performance-based)	Suggested performance measures	Prescribed performance requirements

Source: Adapted from Cashore (1997)

Many forest policy conflicts have focused on policy 'means', regarding which policy instruments are most appropriate and/or effective. For instance, conflicts over endangered species protection on federal lands in the United States have often centred as much on the appropriateness of different policy approaches to achieving habitat protection as they have on the end goal of conserving biodiversity (Hoberg, 1993b, 1997; Cashore, 1997, 1999). Different views on the merits of 'command and control' approaches frequently underlie this debate. To continue the US example, the 'command and control' approach is recognized to have been effective in improving natural habitat (Kohm 1991), reducing harvesting on US national forestlands (Hoberg, 1993a, 1993b; Cashore, 1997, 1999), and in greatly impacting the change in forest policy settings (Cashore and Howlett, 2007). However, it is also highly bureaucratic and fosters adversarial relations between governmental and industry officials that have led to 'bomb proofing' at the expense of long range planning (United States Congress Office of Technology Assessment, 1992, p65). Likewise 'command and control' has been blamed for reducing industry innovations that could lead to increased environmental protection (Cashore and Vertinsky, 2000), and for burdening industry with increased costs (Northwest Forest Resources Council, Association of O&C Counties, 1991; Lippke and Oliver, 1993; Evergreen, 1994; Flick, 1994; Flick et al, 1995). Outside the forests sector, these limitations of 'command and control' approaches have led to the development of 'new generation' instruments for environmental protection (e.g. Gunningham and Sinclair, 2002).

Perceptions of appropriate policy approaches may vary depending on the type of land ownership in question – a theme to which we will return in concluding chapters. Some argue that a command and control approach is less useful for private lands, given the dangers of creating perverse landowner incentives. For example, prescriptive rules governing endangered species protection could induce forest managers to destroy forested habitats in order to prevent endangered species from inhabiting their property (Polasky, 1998; Zhang, 2001). These concerns, and the resistance of private landowners to strong regulation, have often led to a more flexible regulatory approach to private land management.

This book does not seek to resolve controversies such as those discussed above, but it does seek to encourage policy learning by identifying the different policy styles and settings adopted to regulate forest practices. As discussed previously, understanding these is a prerequisite to examining policy effectiveness in different contexts.

RESEARCH DESIGN: THE CONTEXTUAL AND POLICY VARIABLES

This section overviews the contextual information we present in Chapter 2 and in each of the country chapters (3–9) and identifies and discusses the specific policy setting variables we compare for the sample of case study countries and jurisdictions.

The global, regional and national-level forest context

As many authors have observed (Westoby, 1983; Cubbage et al, 1993; Romm, 1993; Mayers and Bass, 1999), forest policy is an articulation of a society's values for its forests and of its intent as to how those values should be realized. Consequently, forest policy – in terms of all cells of the matrix presented in Table 1.1 – can only be fully understood in the context of the particular society in which it has been developed. For this reason we devote Chapter 2, and sections of each regional chapter, to outlining the context for forest policy in the case study countries. In Chapter 2, we present standardized data for each country on a suite of relevant parameters – including key socio-economic and governance indicators, as well as indicators relating to forest area and management. Chapters 3 through 9 are organized by world region. In the introductory sections of each, we describe key features of that country relevant to forest practices policies. For each, we describe the general status of its native and plantation forests,[4] general structures of land ownership and forest governance and overview forest harvesting, production and trade.

We are aware of the implicit reflexivity of this study, as with any other, in that our approach to policy assessment also contributes to shaping the policy discourse. As will be clear from our standardized comparison, some aspects of forest practice policies – typically those which are more prescriptive – lend themselves more easily to systematic global-scale comparisons than do less prescriptive policy settings. Thus, we reiterate that the full outcomes and impacts of any particular policy can only be fully understood within the regional and local contexts we present.

Biodiversity conservation

Biodiversity conservation has been established as a core goal of sustainable forest management in numerous government policies and inter-governmental environmental initiatives (Tarasofsky, 1999; UNFF, 2004; McDermott et al, 2007). Progress towards this goal, however, is notoriously difficult both to measure and achieve, given the complexity of forest ecosystems and the tremendous number and diversity of forest-dependent species, as well as the array of human activities that affect them. In this book, therefore, we highlight this topic as a key contextual issue worthy of special attention, but one that does not lend itself easily to standardized policy analysis.

Specifically we highlight and compare two key policy measures of central contextual importance:

1   the extent of protected areas, and
2   the identification and protection of species at risk and their habitat.

In so doing, our focus is on core measures that countries have taken specifically to protect biodiversity. We take note, for example, if a country's forest harvesting activities occur amidst a large network of protected areas, or if no such nature reserves exist. Likewise, we consider policies protecting endangered species and habitats as a means to set limits on the extent of human impacts on forest biodiversity. There are of course many other types of policy instruments that may contain biodiversity objectives, including all of the environmental forest practice criteria covered in this book. Our intent in focusing on protected areas and species at risk legislation is not to be comprehensive, but rather to examine biodiversity policies of major relevance to all forest ecosystems, including those designated both for production and protection and involving governance often independent of forest agencies and departments.[5]

Protected areas

Protected areas may be designated to fulfil a range of functions, from providing habitats with minimal human presence to providing places of recreation and to supporting traditional cultures and livelihoods. Whatever the stated purpose, the part or total exclusion of extractive activities within designated natural areas plays a number of key roles in biodiversity conservation. These include maintaining species and ecosystems that require natural or near-natural conditions for survival, providing an 'ark' for threatened species whose surrounding habitats have been heavily disturbed and providing research opportunities for scientists and conservationists to learn lessons about ecosystems that can be used to promote biodiversity conservation elsewhere.

The presence of legal protected area status is important for achieving the above functions, even in countries that lack full capacity to enforce protected area boundaries. A 2001 study of 200 protected areas in 34 countries found that lands with legal protected area status were in better ecological and biodiversity condition than those without legal status. This pattern held true across areas of the development frontier (Bruner et al, 2001). Other authors have also found measurably lower deforestation and/or degradation rates across a variety of legally protected areas (see for example Nepstad et al, 2006; Andam et al, 2008; Clark et al, 2008), although protected areas even in close geographical proximity may vary significantly in their effectiveness (for example Gaveau et al, 2007).

Although biologists have long agreed that protected areas alone will be insufficient to conserve biodiversity (Hansen et al, 1991; Hansen and DeFries, 2007), many also argue that protected areas are the cornerstone of biodiversity conservation strategies (WWF, 2004; Hoekstra et al, 2005; Brooks et al, 2006). Consistent with this understanding, in 2004 the United Nations Convention on Biodiversity (CBD) established a target to protect 'at least 10% of each of the world's ecological regions' by 2010 (Decision VII/28, CBD 2004). As of 2008, about 12 per cent of the Earth's surface (including land and water) was under protected area status (United Nations,

2009). However, the level and distribution of protected areas is so widely variable that the percentage of many terrestrial and marine ecosystem types protected remains well below the CBD 10 per cent target (Hoekstra et al, 2005; Brooks et al, 2006; Schmitt et al, 2008; Coad et al, 2009). While politically negotiated targets such as the CBD 10 per cent goal may be useful in spurring international action, many biologists and activists argue that additional, more nuanced and empirically based approaches are needed to stem rapid and continued loss of species and ecosystems worldwide (Rodrigues et al, 2004).

Our comparison of protected areas across case studies examines the percentage of land area under formal protection, with an emphasis on the degree to which the level of protection limits commercial timber harvest. The standardized portion of our analysis relies on a Geographic Information System (GIS) overlay analysis of 2008 data from the United Nations Environment Programme World Conservation Monitoring Centre (Coad et al, 2009). Ideally, we would also have examined the extent of forests represented in protected areas, but such data are currently available for only a few countries. To partially address this gap, the summary of results in Chapter 10 includes figures from a global forest gap analysis (Schmitt et al, 2008) that provides regional level data on protected area coverage across a range of forest types.

Protecting species at risk

There are essentially two levels of species at risk policies. At the international level, agreements such as CITES limit the trade of threatened and endangered species and multinational agreements such as the Convention on Migratory Species (CMS) facilitate cross-border cooperation. National-level policies, which are the most closely examined in this book, range from a simple prohibition on the killing of endangered species to additional requirements to protect their habitat. For many species at risk, loss of habitat is the single largest threat to their survival (Brooks et al, 2002; Dobson et al, 2006).

Our comparison of species at risk policies identifies the presence of species at risk legislation and the extent to which it is focused on the identification and conservation of species at risk and/or requirements for protecting the habitat of species at risk.

Environmental forest practice policies

Policies governing practices in natural forests

After providing a regional and country-level background for each of our case study jurisdictions, Part II of each regional chapter presents a standardized assessment of forest practice policies based on our analytic classification framework. The following sections introduce and discuss the policy criteria used for this assessment and the indicators used to classify the policy approach.

The primary emphasis of these analyses is on policies governing natural forest management. However, in those countries where plantations play a key role in forest production, our assessment of natural forest policies is followed by a parallel assessment of policies for plantations.[6]

Riparian zone management (Indicator: Riparian streamside buffer zone rules)

Riparian zones – land that adjoins, directly influences, or is influenced by a stream or other body of water – are important for a number of reasons (Robbins, 2002). Riparian zones typically support a different array, and often a greater diversity, of flora and fauna than non-riparian areas. Riparian vegetation is often taller, denser and more structurally complex than adjacent vegetation, with a more humid micro-climate. Riparian zones can also provide important wildlife corridors in otherwise fragmented landscapes. The integrity of the riparian zone is critical to protecting water quality and to maintaining aquatic habitat. In some environments, riparian vegetation can also play a significant role in regulating water yield into streams.

The same high productivity of riparian areas that makes them important for habitat diversity also makes them greatly valued for timber production and can lead to conflicts between production- and more conservation-oriented goals. Where aquatic biodiversity includes commercially or culturally important elements, such as salmon on the Pacific Coast of Canada and the US, riparian zone management has become a major focus of forest management.

Numerous studies from around the world have investigated the impacts of forest management on riparian ecosystems (Iwata et al, 2003; Parkyn et al, 2003; Semlitsch and Bodie, 2003; Olson et al, 2007). These studies have variously demonstrated the ecological importance of shade, coarse woody debris, nutrient levels and other parameters associated with forests and/or other vegetation along stream bank channels.

The protection of trees in buffer zones has been shown to play an important role in moderating stream temperatures, reducing siltation and stabilizing stream channels as well as influencing in-stream nutrient cycling (Nilsson and Svedmark, 2002). In a study of nine managed riparian zones in New Zealand, Parkyn et al (2003) found that water temperatures were the most significant determinant of macro-invertebrate diversity. These authors concluded that 'restoration of in-stream communities would only be achieved after canopy closure, with long buffer lengths, and protection of headwater tributaries'.

While forest practices clearly influence the health of aquatic ecosystems, the precise mechanisms by which they do so, and the best means to mitigate the impacts of forestry activities, remain poorly understood in many environments (Parkyn et al, 2003). Consequently, while there is considerable evidence of the need to protect riparian zones, the appropriate degree of protection, and the means necessary to provide that protection, remain more open questions. The difficulties begin with defining the extent of the riparian zone: while its physical structure is defined by vegetation species composition and structural diversity, its functional attributes depend on the integration of the environmental setting with the biotic community (Loftin et al, 2001). The issue is further complicated in environments in which streams and water bodies are intermittent and ephemeral.

Typically, policy-makers have responded to the importance of riparian zones, their variability and uncertainties about the requirements for protection, by requiring or advising the establishment and maintenance of riparian buffer zones. These are often of a specified width, which usually varies with the size of the watercourse or water body and/or with other features of the landscape and topography. Harvesting is often precluded or significantly restricted in the buffer zone.

An alternative approach to the establishment of standardized buffer zone widths is one which is more 'results-based'. A 'results-based' approach would focus on the impacts of forest management activities on the riparian environment. This would require the identification of key indicators of stream health and the establishment of performance thresholds based on these indicators, rather than on prescribed buffer zone widths. Under this approach, any forest management practice would be acceptable so long as the measurement of indicators revealed no damage to key riparian functions. Given the complexity of riparian ecosystems, and the importance of riparian management to stakeholders, however, reaching agreement on appropriate variables and thresholds presents a formidable challenge.

There is as yet little agreement on appropriate results-based measures and thus riparian zone protection in most countries is based on defining appropriate forest practices within buffer zone areas. Consequently, our comparative analysis examines the presence or absence of requirements to establish buffer zones in streamside riparian areas. It then compares specifications for buffer widths and the restriction of harvesting and harvest-related activities within those buffers. We distinguish between 'no-harvest' zones that prohibit commercial harvest and 'special management zones' that place special limits on harvest activities within the zone. In some jurisdictions, streamside buffers may be required but no exact width is specified. This would be classified as a 'mixed' approach.

Our analysis focuses on prescriptions that are specific to streamside riparian buffers. It does not consider other policies that may impact on the riparian zone and its function from outside the defined zone – for example, silvicultural systems and harvesting systems abutting the riparian zones (for example, clearcutting or single tree selection, road building or helicopter logging) and other factors which are believed to impact on the zone, such as the seasonal timing of harvest, terrain variability, climate and the spatial and temporal distribution of forest buffers along stream channels and throughout entire watersheds and landscapes (Naiman et al, 2000; Parkyn et al, 2003).

Roads (Indicators: Culvert size at stream crossings, road decommissioning)
Road building has been described as one of 'the main causes [of] the environmental degradation of most forest regions' (Spinelli and Marchi, 1996; see also Hay, 1994). This is so for two reasons. The first is the access that roads provide. In many previously unroaded tropical forest regions, such as in the Amazon or Congo Basins, roads serve to open forests to immigrant farmers, to poachers and to illegal loggers, leading to further forest degradation (Laurance et al, 2002). In other situations, the impacts may be less obvious and severe, but nevertheless significant – for example, in Idaho, USA, the introduction of roads has been closely correlated with high grizzly bear mortality, due to such factors as hunting, road kill and on-foot human–bear encounters (Boyce and Waller, 2003).

The second set of reasons is associated with the direct impacts of poorly built roads on soil and slope stability, water quality and landscape productivity. The larger the road network, and the poorer construction and maintenance standards, the greater the risk of decreased permeability of soil, erosion and slope failures and siltation of waterways.

The environmental impact of roads can be addressed by policies governing, for example, roading density, construction and maintenance standards and decommission requirements. Our analysis is limited to two of these factors, which we use as a proxy for road construction and maintenance standards more generally. The chosen indicators are culvert size at stream crossings, reflecting the environmental significance of road–stream interactions (Lane and McDonald, 2002) and road decommissioning, reflecting the potential impacts of improperly maintained roads (Lugo and Gucinski, 2000).

CULVERT SIZE AT STREAM CROSSINGS

Our first road building indicator, culvert size at stream crossings, is a critical factor influencing aquatic biodiversity through its effects on fish passage and soil stability in riparian zones. A key issue in this context is whether culverts are designed to accommodate stochastic flood events. Regulations and guidelines, therefore, often set the level of acceptable risk by specifying the 'peak flow' levels that culverts must accommodate. Peak flow refers to the maximum flood level likely to occur over a defined period of time. For example, culverts designed for 50-year peak flow would be built to withstand the maximum flood level expected over a 50-year period. In addition to, or in place of, peak flow specifications, some jurisdictions may establish standardized minimum culvert diameters.

ROAD DECOMMISSIONING

Typically, some roads are left in place once harvesting is complete, for future forest management or other activities, and are maintained as part of the permanent road infrastructure. Other, usually more minor roads, are not required beyond the immediate harvesting operation. Failure to decommission roads no longer in use (close them and undertake necessary remediation work) is likely to lead to erosion and stream sedimentation and facilitate poaching and other human disturbance.

The decision of whether or not to decommission a road, however, depends on many factors, such as the requirements of future forest management activities and the utility of the road to legitimate users such as local communities or tourism operators. Our study, therefore, does not assess the ways in which different jurisdictions determine which roads should be permanently removed from the road network. Instead, we identify whether or not there are standards governing the treatment of roads that will be decommissioned. We assess, furthermore, the specificity of decommissioning rules, distinguishing between general requirements to 'close' and 'stabilize' the road from specific management prescriptions. Examples of the latter include 'remove all water bars and drainage structures', 're-contour slopes' and monitor stream sedimentation.

Clearcutting (Indicator: Clearcut size limits or other relevant cutting rules)
Clearcutting is defined as the complete clearing of all trees, other than seedlings and

occasional saplings, in a harvesting coupe in a single harvesting operation (Kimmins, 1992; Forestry Tasmania, 2005). Clearcutting is perhaps the most controversial forest harvesting practice (American Forest and Paper Association, 1994), because of its actual and perceived impacts on other forest values, including ecological function and processes (Franklin et al, 1999) and landscape amenity (Wood, 1971; Biswas and Sankar, 2002). Clearcutting has been the most common forest harvesting method in temperate and boreal forests worldwide (Kimmins, 1992: Chapter 6, pp73, 76) in part because it facilitates regeneration of the light-demanding species that dominate these forests (e.g. Kimmins, 1992) and in part because it is operationally efficient and commercially attractive (e.g. Binkley, 1999). However, the criticism clearcutting has attracted – from ecologists (see for example Franklin et al, 1999; Spence, 2001; Lindenmayer and Franklin, 2003), socially (Bliss, 2000) and in the marketplace (Stanbury et al, 1995; Bernstein and Cashore, 1999) – has led to a search for, and the implementation of, alternative silvicultural systems (Robertson 1992; CFS 2000; Forestry Tasmania 2005) .

Earlier industry and government exploration of alternatives to clearcutting, spurred in part by regulations which restricted the size of clearcuts, focused on using smaller, more dispersed clearcuts known as 'checker boarding'. However, research has shown that checker boarding can lead to increased road building (Franklin and Forman, 1987) and excessive amounts of forest edge (Chen et al, 1992), which further exacerbate clearcutting's ecological impacts (Franklin and Forman, 1987).

More recently, the concept of the 'range of natural variation' has taken broader hold in some policy circles (Landres et al, 1999; CSA, 2002). This approach calls for designing cut blocks to more closely mimic natural disturbance regimes. For example, in areas where the dominant natural disturbance is small clearings caused by windthrow of individual trees, single tree selection may be appropriate. In contrast, relatively large clearcuts may be the preferred method in ecosystems adapted to large-scale, relatively intense, disturbances (such as some fire-adapted forests). The concept of natural variation relates not only to overall patch size, but also to the design and location of cut-over areas. Within larger clearcuts, there has been increased focus on the retention of patches to maintain structural diversity and biological legacies within the harvest coupe (see for example Forestry Tasmania, 2005). Likewise more attention has been paid to the arrangement of clearcut blocks in relation to slope contours and landscape features.

Many policies governing clearcutting also address factors besides maximum patch size. For example, they often include 'adjacency' requirements that restrict harvest in areas adjacent to clearcuts for specified time frames, or until tree regrowth in the clearcut area has reached specified heights. It is also now common to require the retention of individual living and/or dead trees ('snags'), or groups of snags and seed trees, within a cutblock. This latter practice can in fact lead to conflicting data on the extent of clearcutting, since there is as yet no agreement on the level of tree retention that divides 'clearcutting' from 'uneven-aged management'.

Given that the importance and complexity of harvesting systems, and the significance of their environmental impacts, we use clearcut size limits as a relatively easily measured *indicator* of the policy approach of our case study governments to regulating forest practices. In so doing, we also recognized that no single such forest practice indicator will capture the full range of policies relevant to the mitigation of environmental impacts associated with harvesting.

We make an exception to our use of clearcutting as an indicator of policy approach, however, in the case of the tropical case study countries or others with forest types in which clearcutting is not used. Clearcutting is not a common harvesting system in tropical forests, or in some temperate mixed forest types, reflecting both the ecology of these forests and variation in the economic worth of different species. As a result, prescriptions based around selective harvesting are typical of the regulatory context for managing these forests. For example, in southeast Asia's tropical rainforests, 'minimum diameter cutting limits'[7] are a common form of substantive regulation (Sist et al, 2003), as is the single tree or small group selection typical of Australia's mixed eucalypt forests (Florence, 1996). For these jurisdictions, we assess the policy approach in terms of the presence or absence, and the substantive or procedural nature, of 'cutting rules' that determine harvesting regimes.

Reforestation (Indicator: Requirements for reforestation, including specified time frames and stocking levels)
Effective reforestation following harvesting is fundamental to any concept of sustainable forest management. Methods of reforestation vary with the objectives of management, forest type and harvesting and silvicultural systems. For example, replanting with seedlings of specified geographic and genetic origin is common following clearcut harvesting in many temperate coniferous forests; reforestation practices in western Canada and the USA typify this approach (see for example BC Market Outreach Network 2004). In contrast, in temperate eucalypt forests managed under similar harvesting and silvicultural systems, regeneration from natural or artificial seeding is the preferred reforestation method, although supplementary replanting may also be used in some cases (Florence, 1996). Most silvicultural systems applied in tropical forests rely primarily on natural regeneration (see for example Sist et al, 2003).

Consequently, the explicit requirement for reforestation following harvesting is one of the forest practices criteria we consider. We assess this criterion in terms of two indicators: whether minimum stocking levels (seedlings/stems per hectare) are specified, and whether specific time frames are nominated in which to achieve these targets.

Annual allowable cut (Indicator: Cut limits based on sustained yield)
The annual allowable cut (AAC) is the volume of timber that may, or must, be harvested each year from a specified area. In forests intended to sustain wood production over the long term, the AAC cannot exceed the 'sustained yield' of wood the forest is capable of producing. Consequently, the concept of sustained yield – which can be traced to what has been commonly referred to as the 'German school', first promoted in the late 18th century in Prussia and Saxony, of 'rational' and 'scientific' forest management – generally underlies the concept of AAC (Aplet et al, 1993; Johnson, 1993; Scott, 1998).

The appropriate means to achieve sustained yield has been a subject of enormous debate since it was first espoused in the 18th century (see LeMaster et al, 1982; Walker, 1990), around issues ranging from the ethical – such as what forest attributes (e.g. timber, wildlife, visual quality) should be sustained (e.g. Romm, 1993) – to the practical, such as how sustained yields from old-growth forests should be defined (e.g. Parry et al, 1983). Similarly, the translation of sustained yield to AAC offers considerable scope for debate (Johnson, 1993). For example, for the USA's national forests, as Parry et al (1983) recount, specification of AAC usually included considerations of forest-based development and forest-dependent communities, as well as of the productive capacity of the forest. More recently, broader interpretation of the meaning of the concept of sustainable forest management has led to a rethinking of approaches (e.g. Sample and Anderson, 2008).

Nevertheless, the concept of AAC remains a fundamentally important one, because it is the articulation of the principle of sustaining forest productivity in perpetuity. The AAC calculation may be based on maximizing wood production from a given area of forest, or maintaining a minimum level of production over the long term. When applied to mature or old-growth forests, the former approach leads to substantial harvest volumes in the short term, as standing forests are harvested, followed by a 'fall down' in production before growth rates are ultimately maximized through the full establishment of younger, faster-growing stands. For this reason, it has been termed the 'liquidation-conversion' model of forest management (Wilson, 1998). In contrast, the latter approach – usually termed 'non-declining even flow' – avoids the fall down effect by reducing harvest volumes in the short term to those that can be steadily sustained over the long term, thereby forgoing potential short-term returns and maximized volume production. There is also the issue of the forests to which AAC requirements apply: typically, AAC has applied only to public lands, but it could include private lands, or AAC may not be required at all.

The assessment of AAC in this study considers whether case study jurisdictions specify *any* limit on timber harvest and, if so, what general factors forest managers are required to consider in calculating those limits. If the balance of economic, social and environmental factors is left to the discretion of government agencies, we call this a 'procedural' approach. If it is required that AAC be capped by sustained yield, but with no reference to the time frames over which sustained outputs would be calculated, we classify this as a 'mixed' policy. An AAC requirement for non-declining even flow, or the equivalent, is classified as a mandatory substantive policy.

Policies governing practices in plantation forests

Natural forest policies are the primary focus of this book. However, given the growing importance of plantation forests to the world wood and fibre supply, we also examine what differences, if any, exist for policies governing plantation production.

We do this by selecting and comparing the countries which rank in the top ten of our case studies in terms of the total area classified by FAO (2007) as 'productive plantation', i.e., those with the largest plantation areas established and managed primarily for commercial wood production.

Enforcement and compliance

Enforcement and compliance[8] are critical elements of policy implementation (Dovers, 2005); the best-crafted forest practice regulations are irrelevant if they are not observed in practice. Closing the gap between law and on-ground outcomes is one of the main challenges in the forest sector (Christy et al, 2007) and so issues of enforcement and compliance are amongst the most important arenas of policy analysis. They are also amongst the most complex, reflecting the range of perspectives, disciplines and practicalities that shape the design and implementation of regulatory regimes (see for example Gunningham and Grabosky, 1998; Dovers, 2005).

At the most fundamental level, effective enforcement and compliance is predicated on good forest governance, which the World Bank (World Bank, 2008, p151) describes, in the context of state-based governance, as:

> *characterized by predictable, open and informed policy making based on transparent processes, a bureaucracy imbued with a professional ethos, an executive arm of government accountable for its actions, and a strong civil society participating in decisions related to sector management and other public affairs – and all behaving under the rule of law.*

Achieving these goals can be difficult in any country and is particularly challenging in many of our developing case study countries. In Chapter 2, we present a number of governance indicators used to rank countries' performance in these terms.

We inform our case study reviews of enforcement and compliance by a general overview of forest governance and policy and of forest practices systems, in each case study jurisdiction. Where relevant, we discuss evidence of illegal logging – a term that encompasses such disparate practices as timber theft, tax evasion and regulatory violations – as a crude proxy for the effectiveness of enforcement and level of compliance associated with forest harvesting (see for example Tacconi, 2007: Chapter 1). Where possible, we also outline the specific mechanisms in place for auditing compliance with forest practice regulations. We note two different types of auditing activity: the first relates to verifying compliance with forest practice requirements and the second to evaluating the effectiveness of these practices in achieving desired outcomes. The appropriate balance between these two components in forest practices systems is itself a matter of debate (see Wilkinson, 1999). As with our comparative analysis of forest policies, we do not address the relative effectiveness of different approaches. To do so would require more in-depth, field-based research such as that described by Ellefson et al, 2007. Our case-by-case analysis instead sets the stage for such further research by addressing the kinds of implementation or compliance monitoring performed. Specifically, we identify the agencies responsible for enforcement, review how forest practices are audited and assess the degree to which forest audit records and decisions are made public. For pragmatic reasons related to the ease of access to available data, this analysis is conducted in greatest detail for the US, Canadian and Australian case study jurisdictions.

Non-state regulatory approaches: Forest certification

Any comparison of global forest policy in the current era would be incomplete without addressing forest certification. Hence our empirical chapters each include a section on the development of forest certification within the case study countries.

Forest certification first emerged in the 1990s as an innovative non-state market driven policy instrument with which to promote sustainable forestry. It involves the third-party assessment of forest practices according to a set of environmental, social and economic standards developed through multi-stakeholder processes. Forest operations that meet these standards may use their certification as proof of good forest stewardship, thereby distinguishing themselves from their market competitors.

The roots of forest certification are both local and global in scale, and there is continued debate as to the appropriate level to set certification standards (McDermott, 2003; McDermott and Hoberg, 2003). While various local initiatives were developing in the late 1980s, forest certification first gained the widespread attention of domestic and global forest policy communities in 1993, following the creation of the Forest Stewardship Council (FSC) certification programme (Meidinger, 1997; Domask, 2003). The FSC was founded by an array of internationally focused environmental groups, led by the World Wide Fund for Nature (WWF), social allies and some retailers and other businesses. The creation of the FSC presented the international stage with an innovative institutional design with which to address global forest deterioration. Forest certification was innovative because it turned to the market's supply chain, rather than governments, for policy-making authority and because it represented a change for environmental advocacy groups from traditional 'stick' approaches such as boycott campaigns, to a focus on 'carrots' in the form of *rewards* to companies who practice corporate environmental and social stewardship (Bass, 1997).

The creation of the FSC sparked a number of reactions and trends. First, while environmental and social stakeholders were generally supportive of international standards as a common baseline, many governmental agencies and forest producers argued for national-level standard setting, citing rights of sovereignty. Second, there were objections to the FSC's decision-making structures, which did not permit direct government involvement and were designed to ensure that industry was not able to dominate policy-making processes. This governance approach was central to the FSC's legitimacy for many environmental groups, but was viewed with scepticism and caution on the part of forest owners (Auld et al, 2001; Newsom et al, 2003; Vlosky and Granskog, 2003). Initial opposition by most industry and forest owners to FSC forest certification split in two directions: some owners came to support the FSC, while others helped create alternatives to the FSC (Boström, 2003; Cashore et al, 2004).

The major FSC alternatives to date consist of producer-backed national certification schemes. In the US, the American Forest and Paper Association created the Sustainable Forestry Initiative (SFI), now applicable in both the US and Canada. Within Canada, industry and government interests spearheaded Sustainable Forest

Management certification under the Canadian Standards Association (CSA). Many countries in Europe have also formed national schemes, as have Australia, Japan and an increasing number of developing countries: notably Brazil, Chile and Malaysia, and a number of African countries, with support from an African Timber Organization / International Tropical Timber Organization Initiative (UNCTAD, 2009). Almost all of the developed country national-level schemes, as well as some developing country initiatives, have now joined together under an umbrella system, which first emerged in Europe, and is now known as the Programme for the Endorsement of Forest Certification schemes (PEFC, 2009). Among the exceptions is the Indonesian Lembaga Ekolabel Indonesia (LEI), which has instead signed a joint certification protocol with the FSC (JCP Consortium, 2000).

In addition to forest certification, there are other certification schemes of some relevance that are not exclusive to the forest sector. The long-standing international standards consortium, the International Organization for Standardization (ISO), has created an international 'Environmental Management System' certification programme, entitled ISO 14001. ISO 14001 is applicable to a wide range of natural resource-based industries, including, but not limited to, forestry. According to our comparative policy framework, the ISO certification approach is procedural, or 'systems-based', in that it does not set the standards for appropriate resource management but rather assesses whether or not companies have themselves established and effectively implemented their own environmental management systems.

We do not directly address forest industry adoption of ISO 14001 in this book. However, it is important to note that a number of the forest certification systems discussed have incorporated elements of ISO's 'systems-based' (procedural) approach. In other words, the push and pull between procedural versus substantive ('performance-based') approaches is a policy dynamic as relevant to non-governmental certification systems as it is to governmental regulations.

Certification policies and approaches are highly complex and dynamic and a complete analysis would require another book as lengthy as this one (e.g. Cashore et al, 2004). The discussion of certification in this book, therefore, is limited to an overview of key trends and differences among forest certification schemes and a comparison of the areas certified under each scheme in each case study country.

For the same reasons we outline above in relation to forest practice regulations, our review does not assess the impact of certification on forest practices. To do so would require the measurement of direct effects through on-the-ground assessment, as well as indirect effects at both landscape and normative levels (Auld et al, 2008). The focus on area certified also emphasizes certification of large-scale industries rather than smaller-scale landowners, although smaller-scale and alternative forest actors may play an important role in the development of certification (McDermott, 2003; McDermott and Hoberg, 2003). We return in the conclusions to discuss further research to inform this topic.

Box 1.1 Template for organization of case study country chapters

For each region:

   A. Overview of forests and forestry

For each case study country:

   B. Country overview
      i. Native forests
      ii. Planted forests and plantations
      iii. Forest governance
         a. Forest practice systems
      iv. Forest production and trade
      v. Indigenous and community forestry, where relevant

For each country or sub-national jurisdiction:

   C. Biodiversity conservation measures
      i. Protected areas
      ii. Protection of species at risk

   D. Forest practice regulations – native forests
      i. Riparian zone management
         a. Buffer zone width and management restrictions in buffer zones
      ii. Road building
         a. Stream crossings
         b. Road decommissioning
      iii. Clearcutting
         a. Maximum clearcut sizes
      iv. Reforestation
         a. Specification of time frames and stocking levels
      v. Annual allowable cut rules
         a. Limits on annual harvest levels

   E. Forest practice regulations – plantations (for case studies where there is significant plantation development)
      i. Similarities and differences in rules governing plantation management
      ii. Use of exotic species
      iii. Conversion of natural forests to plantation

   F. Enforcement and compliance
      i. System of enforcement
      ii. Levels of compliance

   G. Forest Certification

   H. Regional summary of findings

STRUCTURE OF THE BOOK

Chapter 2 concludes Part I by addressing our case study selection and placing our country cases in a global context. Part II (Chapters 3 through 9) provides regional and national context and contains our empirical policy analysis. These chapters are organized within the following regions (according to relative value of international forest trade): North America (Chapter 3), Western Europe (Chapter 4), Asia (Chapter 5), Eastern Europe (Chapter 6), Latin America (Chapter 7), Australasia (Chapter 8) and Africa (Chapter 9). Each of these chapters is structured against a common template (Box 1.1). Part III contains our summary and conclusions. It begins with Chapter 10, which provides a criterion-by-criterion summary of our comparative policy results. Chapter 11 then merges our contextual analysis with our analysis of policy to identify key global trends and assess the degree to which those trends are reflected in policy differences. It then considers the implications of this book's findings for the development of policy theory, including the ways in which globalization may influence trends in environmental policy and concludes with suggestions for further research.

NOTES

1   For example, the Convention on International Trade in Endangered Species of Wild Fauna and Flora (CITES), or the US Lacey Act, which prohibits trade in illegally procured (i.e. procured in violation of the laws of the country of origin) animals and animal products, and was recently amended (in 2008) to cover – with limited exceptions – plants and plant products including timber (US CBP 2009).
2   Formally, 'leakage' has been defined by the UNFCCC (UNFCCC, 2003 #494) only for Clean Development Mechanism (CDM) projects under the Kyoto Protocol, as the net change in anthropogenic greenhouse gas emissions by sources of greenhouse gases, which occurs outside the project boundary and which is measureable and attributable to the CDM project activity; see Kanninen et al (2007).
3   For a similar, but slightly different treatment, see Hoberg 2003. Hoberg, drawing on Coglianese and Lazer (Coglianese and Lazer 2003), classifies policies according to guide-lines that 'can be used to identify recommended practices', 'technology- or practices-regulations' that 'specify particular forest practices that must be used in certain circumstances' and 'performance- or results-based regulations' that 'specify an outcome to be achieved rather than a specific practice.'
4   Throughout this book, unless otherwise noted, the term 'native' forest is used interchangeably with 'natural' forest. The definitions of 'natural' and 'plantation' forests are those established for the FAO 2005 Forest Resources Assessment (FAO 2004).
5   The issue of natural forest conversion (the conversion of forests to non-forest land use) is another crosscutting issue of central importance to biodiversity. We address forest conversion in our general discussions of forest context, as well as in our assessment of plantation policies (in this latter case the conversion of natural forests to tree plantations). However,

the complexity of policies affecting conversion of forest to non-forest prevented a standardized cross-country comparison.

6   The definitions of 'natural' and 'plantation' forests are those established for the FAO 2005 Forest Resources Assessment (FAO 2004).

7   Minimum diameter cutting limits are prohibitions on the harvest of trees below a pre-determined stem size.

8   Dovers (2005) discusses how 'enforcement' is usually used only in the context of direct regulatory instruments, and 'compliance' more widely, to encompass a wider range of approaches.

REFERENCES

American Forest and Paper Association (1994) 'Closer look: An on-ground investigation of the Sierra Club's Book, Clearcut', Washington, DC: American Forest and Paper Association

Andam, Kwaw S., Paul J. Ferraro, Alexander Pfaff, G. Arturo Sanchez-Azofeifa and Juan A. Robalino (2008) 'Measuring the effectiveness of protected area networks in reducing deforestation', *PNAS*, 105, 42, pp16089–16094

Aplet, Greg, Nels Johnson, Jeffrey T. Olson and V. Alaric Sample (eds) (1993) *Defining Sustainable Forestry*, Washington, DC: Island Press

Auld, Graeme, Benjamin Cashore and Deanna Newsom (2001) 'A look at forest certification through the eyes of United States wood and paper producers', paper presented to the Auburn Forest Policy Center's conference on Globalization and Private Forestry, 25–27 March, Atlanta, GA

Auld, Graeme, Lars H. Gulbrandsen and Constance L. McDermott (2008) 'Certification schemes and the impacts on forests and forestry', in G. Matson (ed) *Annual Review of Environment and Resources*, Palo Alto: Annual Reviews

Bass, Stephen (1997) 'Introducing forest certification – A report prepared by the Forest Certification Advisory Group (FCAG) for DGVIII of the European Commission', Torikatu, Finland: European Forest Institute

BC Market Outreach Network (2004) *BC Forest Facts: Maintaining British Columbia's Natural Forest Diversity*, BC Market Outreach Network, Vancouver, available at www.bcforestinformation.com, last accessed February 2008

Bernstein, S. and B. Cashore (1999) 'World trends and Canadian forest policy: Exploring the influence of consumers, environmental group activity, international trade rules and world forestry negotiations', *Forestry Chronicle*, 75, 1, pp34–38

Binkley, Clark S. (1999) 'MacBlo deal brightens BC forestry future', *National Post*, 12 July

Biswas, S. and K. Sankar (2002) 'Prey abundance and food habit of tigers (*Panthera tigris tigris*) in Pench National Park, Madhya Pradesh, India', *Journal of Zoology*, 256, pp411–420

Bliss, John C. (2000) 'Public perceptions of clearcutting', *Journal of Forestry*, 98, pp4–9

Boström, Magnus (2003) 'How state-dependent is a non-state-driven rule-making project? The case of forest certification in Sweden', *Journal of Environmental Policy & Planning*, 5, 2, June, pp165–180

Boyce, M. S. and J. S. Waller (2003) 'Grizzly bears for the Bitterroot: Predicting potential abundance and distribution', *Wildlife Society Bulletin*, 31, 3, pp670–683

Brooks, Thomas M., Russell A. Mittermeier, Cristina G. Mittermeier, Gustavo A. B. da Fonseca, Anthony B. Rylands, William R. Konstant, Penny Flick, John Pilgrim, Sara Oldfield, Georgina Magin and Craig Hilton-Taylor (2002) 'Habitat loss and extinction in the hotspots of biodiversity', *Conservation Biology*, 16, 4, pp909–923

Brooks, T. M., R. A. Mittermeier, G. A. B. da Fonseca, J. Gerlach, M. Hoffmann, J. F. Lamoreux, C. G. Mittermeier, J. D. Pilgrim and A. S. L. Rodrigues (2006) 'Global biodiversity conservation priorities', *Science*, 313, 58, pp58–61

Bruner, A. G., R. E. Gullison, R. E. Rice and G. A. da Fonseca (2001) 'Effectiveness of parks in protecting tropical diversity', *Science*, 291, p105

Cashore, Benjamin (1997) 'Governing forestry: Environmental group influence in British Columbia and the US Pacific Northwest', PhD, Political Science, Toronto: University of Toronto

Cashore, Benjamin (1999) 'US Pacific Northwest', in B. Wilson, K. V. Kooten, I. Vertinsky and L. Arthur (eds) *Forest Policy: International Case Studies*, Wallingford: CABI Publications

Cashore, Benjamin (2001) 'Understanding the British Columbia environmental forest policy record in comparative perspective', Auburn, AL: Auburn University Forest Policy Center

Cashore, Benjamin and Michael Howlett (2007) 'Punctuating which equilibrium? Understanding thermostatic policy dynamics in Pacific Northwest forestry', *American Journal of Political Science*, 51, 3, pp532–551

Cashore, Benjamin and Constance L. McDermott (2004) 'Global Environmental Forest Policy: Canada as a constant case comparison of select forest practice regulations', Victoria, BC: International Forest Resources

Cashore, Benjamin and Ilan Vertinsky (2000) 'Policy networks and firm behaviours: Governance systems and firm responses to external demands for sustainable forest management', *Policy Sciences*, 33, March, pp1–30

Cashore, Benjamin, Graeme Auld and Deanna Newsom (2004) *Governing Through Markets: Forest Certification and the Emergence of Non-State Authority*, New Haven, CT: Yale University Press

Cashore, Benjamin, Graeme Auld, Steven Bernstein and Constance McDermott (2007) 'Can non-state governance "ratchet up" global environmental standards? Lessons from the forest sector', *Review of European Community and International Environmental Law*, 16, 2, pp158–172

CFS (2000) 'The state of Canada's forests: 2000', report, Ottawa: Canadian Forest Service, 120pp

Chen, Jiquan, Jerry Franklin and Thomas Spies (1992) 'Vegetation responses to edge environments in old-growth Douglas-fir forests', *Ecological Applications*, 2, 4, pp387–396

Christy, L. C., C. E. Di Leva, J. M Lindesay and P. T. Tokoukam (2007) 'Forests law and sustainable development: Addressing contemporary challenges through legal reform', in T. W. Bank (ed) *Law, Justice and Development Series*, Washington, DC: The World Bank

Clark, Sarah, Katharine Bolt and Alison Campbell (2008) 'Protected areas: An effective tool to reduce emissions from deforestation and forest degradation in developing countries?', Cambridge: United Nations Environment Programme World Conservation Monitoring Centre (UNEP-WCMC)

Coad, Lauren, Neil Burgess, Lucy Fish, Corinna Ravillious, Colleen Corrigan, Helena Pavese, Arianna Granziera and Charles Besançon (2009) 'Progress towards the Convention on Biological Diversity terrestrial 2010 and marine 2012 targets for protected area coverage', *Parks*, 17, 2, pp35–42

Coglianese, C. and D. Lazer (2003) 'Management-based regulation: Prescribing private management to achieve public goals', *Law & Society Review*, 37, 4, pp691–730

CSA (2002) 'CAN/CSA Z809-02 Sustainable Forest Management Requirements and Guidance', Mississauga, Ontario: Canadian Standards Association

Cubbage, Frederick, Jay O'Laughlin and Charles S. Bullock III (1993) *Forest Resource Policy*, New York: John Wiley & Sons, Inc

Dobson, Andrew, David Lodge, Jackie Alder, Graeme S. Cumming, Juan Keymer, Jacquie McGlade, Hal Mooney, James A. Rusak, Osvaldo Sala, Volkmar Wolters, Diana Wall, Rachel Winfree and Marguerite A. Xenopoulos (2006) 'Habitat loss, trophic collapse, and the decline of ecosystem services', *Ecology*, 87, 8, pp1915–1924

Domask, Joe (2003) 'From boycotts to partnership: NGOs, the private sector, and the world's forests', in J. P. Doh and H. Teegen (eds) *Globalization and NGOs: Transforming Business, Governments, and Society*, New York: Praeger

Dovers, S. (2005) *Environment and Sustainability Policy*, Sydney: Federation Press

Ellefson, Paul V., Michael A. Kilgore and James E. Granskog (2007) 'Government regulation of forestry practices on private forest land in the United States: An assessment of state government responsibilities and program performance', *Forest Policy and Economics*, 9, pp620–632

European Commission (2008) http://ec.europa.eu/environment/forests/flegt.htm, accessed May 2009

Evergreen (1994) 'The hidden danger of moral persuasion: The Clinton Plan laid bare', *Evergreen*, July, pp46–47

FAO (2004) 'Global forest resources assessment update 2005: Terms and definitions (final version)', Rome: Food and Agriculture Organization of the United Nations

FAO (2007) 'State of the world''s forests – 2007', Rome: United Nations, Food and Agriculture Organization

Flick, Warren A. (1994) 'Changing time: Forest owners and the law', *Journal of Forestry*, 92, 5, May, pp28–29

Flick, Warren A., Allen Barnes and Robert A. Tufts (1995) 'Public purpose and private property: The evolution of regulatory taking', *Journal of Forestry*, 93, 6, June, pp21–24

Florence, R. G. (1996) *Ecology and Silviculture of Eucalypt Forests*, Melbourne: CSIRO Publishing

Forestry Tasmania (2005) 'Alternatives to clearfell silviculture in Tasmania's public old growth forests', www.forestrytas.com.au/forest-management/old-growth-forests, accessed 16 February, 2009

Franklin, Jerry F. and Richard T. T. Forman (1987) 'Creating landscape patterns by forest cutting: Ecological consequences and principles', *Landscape Ecology*, 1, 1, pp5–18

Franklin, J. F., D. Lindenmayer, J. A. MacMahon, A. McKee, J. Magnuson, D. A. Perry, R. Waide and D. Foster (1999) 'Threads of continuity', *Conservation Biology in Practice*, 1, pp9–16

Gaveau, David L. A., Hagnyo Wandono and Firman Seiabudi (2007) 'Three decades of deforestation in southwest Sumatra: Have protected areas halted forest loss and logging, and promoted re-growth?', *Biological Conservation*, 134, pp495–504

Gunningham, Neil and Peter Grabosky (eds) (1998) *Smart Regulation: Designing Environmental Policy* Edited by Oxford Socio-Legal Studies, Oxford: Oxford University Press

Gunningham, Neil and Darren Sinclair (2002) *Leaders and Laggards: Next-Generation Environmental Regulation*, Sheffield, UK: Greenleaf Publishing

Hall, Peter (1993). 'Policy paradigms, social learning, and the state: The case of economic policymaking in Britain', *Comparative Politics*, 25, 3, p275

Hansen, A. J., T. A. Spies, F. J. Swanson and J. L. Ohmann (1991) 'Conserving biodiversity in managed forests', *BioScience*, 41, 6, pp382–392

Hansen, Andrew and Ruth DeFries (2007) 'Ecological mechanisms linking protected areas to surrounding lands', *Ecological Applications*, 17, 4, pp974–988

Hay, Roger M. (1994) 'The development of a code of forest practice for forest roading', in D. P. Dykstra (ed) *Forest Codes of Practice*, Rome: International Union for the Conservation of Nature (IUFRO) and the Food and Agriculture Organization (FAO) of the United Nations

Hoberg, George (1993a) 'From logroll to logjam: Structure, strategy, and influence in the old growth forest conflict', paper read at Annual Meeting of the American Political Science Association, in Washington, DC

Hoberg, George (1993b) 'Regulating forestry: A comparison of institutions and policies in British Columbia and the US Pacific Northwest', Working Paper, Vancouver: Forest Economics and Policy Analysis Research Unit, University of British Columbia

Hoberg, George (1997) 'Governing the environment: Comparing policy in Canada and the United States', in K. Banting, G. Hoberg and R. Simeon (eds) *Degrees of Freedom: Canada and the United States in a Changing Global Context*, Montreal, Kingston: McGill-Queens

Hoberg, George (2001) 'The British Columbia forest practices code: Formalization and its effects', in M. Howlett (ed) *Canadian Forest Policy: Regimes, Policy Dynamics, and Institutional Adaptations*, Toronto: University of Toronto Press

Hoberg, George (2003) 'Alternative approaches to regulating forest practices: Lessons from British Columbia', paper read at XII World Forestry Congress, Quebec City

Hoekstra, Jonathan M., Timothy M. Boucher, Taylor M. Ricketts and Carter Roberts (2005) 'Confronting a biome crisis: Global disparities of habitat loss and protection', *Ecology Letters*, 8, pp23–29

Howlett, Michael, and Ben Cashore (2007) 'Re-visiting the new orthodoxy of policy dynamics: The dependent variable and re-aggregation problems in the study of policy change', *Canadian Political Science Review*, 1, 2, pp1–14

Humphreys, David (2006) *Logjam: Deforestation and the Crisis of Global Governance*, London: Earthscan

IPCC (2007) 'Climate change 2007, synthesis report, summary for policymakers', Geneva: Intergovernmental Panel on Climate Change

Iwata, Tomoya, Shigeru Nakano and Mikio Inoue (2003) 'Impacts of past riparian deforestation on stream communities in a tropical rainforest in Borneo', *Ecological Applications*, 13, 2, pp461–473

JCP Consortium (2000) 'Joint Certification Protocol (JCP) between LEI-accredited certification bodies and FSC-accredited certification bodies', Signatories: PT. T V International; Mutuagung Lestari, SmartWood, Forest Stewardship Council, SFMP-GTZ, Sucofindo, SGS ICS Indonesia, Lembaga Ekolabel Indonesia, GTZ

Johnson, Nels (1993) 'Introduction to Part I, Sustain what? Exploring the objectives of sustainable forestry', in G. Aplet, N. Johnson, J. T. Olson and V. A. Sample (eds) *Defining Sustainable Forestry*, Washington, DC: Island Press

Kanninen, M., D. Murdiyaso, F. Seymour, A. Angelsen, S. Wunder and L. German (2007) 'Do trees grow on money? The implications of deforestation research for policies to promote REDD', Bogor, Indonesia: Center for International Forestry Research

Kimmins, Hamish (1992) *Balancing Act: Environmental Issues in Forestry*, Vancouver: University of British Columbia Press

Kohm, Kathryn A. (1991) *Balancing on the Brink of Extinction: The Endangered Species Act and Lessons for the Future*, Washington, DC: Island Press

Landres, P. B., P. Morgan and F. J. Swanson (1999) 'Overview of the use of natural variability concepts in managing ecological systems', *Ecological Applications*, 9, 4, pp1179–1188

Lane, Marcus B. and Geoff McDonald (2002) 'Toward a general model of forest management through time: Evidence from Australia, USA and Canada', *Land Use Policy*, 19, 3, pp193–206

Laurance, W. F., A. K. M. Albernaz, G. Schroth, P. M. Fearnside, S. Bergen, E. M. Venticinque and C. Da Costa (2002) 'Predictors of deforestation in the Brazilian Amazon', *Journal of Biogeography*, 29, 5–6, pp737–748

LeMaster, D., D. Baumgartner and D. Adams (eds) (1982) *Sustained Yield*, Pullman, WA: Washington State University

Lindenmayer, David B. and Jerry F. Franklin (2003) *Towards Forest Sustainability*, Washington, DC: Island Press

Lippke, Bruce and Chadwick D. Oliver (1993) 'Managing for multiple values: A proposal for the Pacific Northwest', *Journal of Forestry*, 91, 12, December, pp14–18

Loftin, Cynthia, Michael Bank, John Hagan and Darlene Siegel (2001) 'Literature synthesis of the effects of forest management activities on riparian and in-stream biota of New England', Orono, ME: Cooperative Forestry Research Unit, Maine Agricultural and Forest Experiment Station, University of Maine

Lugo, A. E. and H. Gucinski (2000) 'Function, effects, and management of forest roads', *Forest Ecology and Management*, 133, 3, pp249–262

Mayers, James and Stephen Bass (1999) 'Policy that works for forests and people: Overview report', Policy that works for forests and people report series, Vol. 7, London: International Institute for Environment and Development (IISD)

McDermott, Constance L. (2003) 'Personal trust and trust in abstract systems: A study of Forest Stewardship Council-accredited certification in British Columbia', PhD, Vancouver: Department of Forest Resources Management, Faculty of Forestry, University of British Columbia

McDermott, Constance and Benjamin Cashore (2007) 'A global comparison of forest practice policies using Tasmania as a constant case', New Haven, CT: Yale Program on Forest Policy and Governance, Global Institute of Sustainable Forestry

McDermott, Constance L. and George Hoberg (2003) 'From state to market: Forestry certification in the U.S. and Canada', in B. Schindler, T. Beckley and C. Finley (eds) *Two Paths Toward Sustainable Forests: Public Values in Canada and the United States*, Corvallis: Oregon State University Press

McDermott, Constance, Aran O'Carroll and Peter Wood (2007) 'International forest policy – the instruments, agreements and processes that shape it', United Nations Forum on Forests, UN Department of Economic and Social Affairs

McDermott, Constance, Emily Noah and Benjamin Cashore (2008) 'Differences that "matter"? A framework for comparing environmental certification standards and government policies', *Journal of Environmental Policy and Planning*, 10, 1, pp47–70

Meidinger, Errol (1997) 'Look who's making the rules: International environmental standard setting by non-governmental organizations', *Human Ecology Review*, 4, 1, pp52–54

Millennium Ecosystem Assessment (2005) 'Ecosystems and human well-being: Biodiversity synthesis', Washington, DC: World Resources Institute

Naiman, R. J., R. E. Bilby and P. A. Bisson (2000) 'Riparian ecology and management in the Pacific Coastal Rain Forest', *Bioscience*, 50, 11, pp996–1011

Nepstad, Daniel, S. Schwartzman, B. Bamberger, M. Santilli, D. Ray, P. Schlesinger, P. Lefebvre, A. Alencar, E. Prinz, Greg Fiske and Alicia Rolla (2006) 'Inhibition of Amazon deforestation and fire by parks and indigenous lands', *Conservation Biology*, 20, 1, pp65–73

Newsom, D., B. Cashore, G. Auld and J. Granskog (2003) 'Certification in the heart of Dixie: A survey of Alabama landowners', in L. Teeter, B. Cashore and D. Zhang (eds) *Forest Policy for Private Forestry*, Wallingford, UK: CABI Publishing

Nilsson, C. and M. Svedmark (2002) 'Basic principles and ecological consequences of changing water regimes: Riparian plant communities', *Environmental Management*, 30, 4, pp468–480

Northwest Forest Resources Council, Association of O&C Counties (1991) 'A facade of science: An analysis of the Jack Ward Thomas report based on sworn testimony of members of the Thomas Committee', Preston Thorgrimson Shidler Gates & Ellis

Olson, Deanna H., Paul D. Anderson, Christopher A. Frissell, Hartwell H. Welsh Jr and David Bradford (2007) 'Biodiversity management approaches for stream-riparian areas: Perspectives for Pacific Northwest headwater forests, microclimates, and amphibians', *Forest Ecology and Management*, 246, pp81–107

Parkyn, S. M., R. J. Davies-Colley, N. J. Halliday, K. J. Costley and G. F. Croker (2003) 'Planted riparian buffer zones in New Zealand: Do they live up to expectations?', *Restoration Ecology*, 11, 4, pp436–447

Parry, B. T., H. J. Vaux and N. Dennis (1983) 'Changing conceptions of sustained-yield policy on the national forests', *Journal of Forestry*, 81, pp150–154

PEFC (2009) 'About PEFC', www.pefc.org, a web page of the Programme for the Endorsement of Forest Certification Schemes, accessed February 2009

Polasky, Stephen (1998) 'When the truth hurts: Endangered species policy on private land with imperfect information', *Journal of Environmental Economics and Management*, 35, pp22–47

Porter, Michael E. and Class van der Linde (1995) 'Green and competitive: Ending the stalemate', *Harvard Business Review*, September–October, pp20–38

Rayner, Jeremy and Michael Howlett (2003) National Forest Programmes as vehicles for next-generation regulation: Lessons from Canadian and European experiences, paper read at the International Seminar of COST Action E19, entitled 'Making NFPs work: Procedural aspects and supporting factors', 15–16 September, Vienna

Robbins, L. (2002) 'Managing riparian land for multiple purposes', Australian Capital Territory: Rural Industries Research and Development Corporation (RIRDC) Publication

Robertson, F. Dale (1992) 'Policy directive on ecosystem management of the national forests and grasslands', US Department of Agriculture, Forest Service

Rodrigues, Ana S. L., Sandy J. Andelman, Mohamed K. Bakarr, Luigi Boitani, Thomas M. Brooks, Richard M. Cowling, Lincoln D. C. Fishpool, Gustavo A. B. da Fonseca, Kevin J. Gaston, Michael Hoffman, Janice S. Long, Pablo A. Marquet, John D. Pilgrim, Robert L. Pressey, Jan Schipper, Wes Sechrest, Simon N. Stuart, Les G. Underhill, Robert W. Waller, Matthew E. J. Watts and Xie Yan (2004) 'Effectiveness of the global protected area network representing species diversity', *Nature*, 428, 8, pp640–643

Romm, Jeff (1993) 'Sustainable forestry, and adaptive social process', in G. Aplet, N. Johnson, J. T. Olson and V. A. Sample (eds) *Defining Sustainable Forestry*, Washington, DC: Island Press

Sample, Alaric and Steven Anderson (eds) (2008) *Common Goals for Sustainable Forest Management; Divergence and Reconvergence of American and European Forestry*, Durham: Forest History Society

Schmitt, C. B., A. Belokurov, C. Besançon, L. Boisrobert, N. D. Burgess, A. Campbell, L. Coad, L. Fish, D. Gliddon, K. Humphries, V. Kapos, C. Loucks, I. Lysenko, L. Miles, C. Mills, S. Minnemeyer, T. Pistorius, C. Ravilious, M. Steininger and G. Winkel (2008) 'Global ecological forest classification and forest protected area gap analysis: Analyses and recommendations in view of the 10% target for forest protection under the Convention on Biological Diversity (CBD)', Freiburg: Freiburg University

Scott, James C. (1998). *Seeing Like a State : How Certain Schemes to Improve the Human Condition Have Failed*, New Haven, CT and London: Yale University Press.

Semlitsch, R. D. and J. R. Bodie (2003) 'Biological criteria for buffer zones around wetlands and riparian habitats for amphibians and reptiles', *Conservation Biology*, 17, 5, pp1219–1228

Sharma, Sanjay (1996) 'A theory of corporate environmental responsiveness', research talk at the Research Seminar Series, Halifax: Faculty of Commerce, Saint Mary's University

Sist, Plinio, Robert Fimbel, Douglas Sheil, Robert Nasi and Marie-Hélène Chevallier (2003) 'Towards sustainable management of mixed dipterocarp forests of South-east Asia: Moving beyond minimum diameter cutting limits', *Environmental Conservation*, 30, 4, pp364–374

Spence, J. R. (2001) 'The new boreal forestry: Adjusting timber management to accommodate biodiversity', *Trends in Ecology and Evolution*, 16, 11, pp591–592

Spinelli, Raffaele and Enrico Marchi (1996) 'A literature review of the environmental impacts of forest road construction', paper read at Proceedings of the Seminar on Environmentally Sound Forest Roads and Wood Transport, Sinaia, Romania

Stanbury, William T., Ilan B. Vertinsky and Bill Wilson (1995) 'The challenge to Canadian forest products in Europe: Managing a complex environmental issue', Vancouver: Forest Economics and Policy Analysis Research Unit, University of British Columbia

Tacconi, Luca (2007) *Illegal Logging: Law Enforcement, Livelihoods and the Timber Trade*, London: Earthscan

Tarasofsky, Richard G. (1999) 'Assessing the international forest regime: Gaps, overlaps, uncertainties and opportunities', in R. G. Tarasofsky (ed) *Assessing the International Forest Regime*, Gland, Switzerland: International Union for the Conservation of Nature and Natural Resources

Tollefson, Chris (ed) (1998) *The Wealth of Forests: Markets, Regulation, and Sustainable Forestry* (2 vols), Vancouver: UBC Press

UNCTAD (2009) 'Tropical timber – quality', www.unctad.org/infocomm/anglais/timbertrop/quality.htm, United Nations Conference on Trade and Development, last accessed May 2009

UNFF (2004) 'Recent developments in existing forest-related instruments, agreements, and processes' (working draft, in *Background Document No. 2*, New York, NY: United Nations Forum on Forests (ad hoc expert group on Consideration with a View to Recommending the Parameters of a Mandate for Developing a Legal Framework on All Types of Forests)

United Nations (2009) '*The Millennium Development Goals Report 2009*', New York: United Nations

UNFCCC (2003) Methodological issues: Land use, land-use change and forestry: Definitions and modalities for including afforestation and reforestation activities under Article 12 of the Kyoto Protocol, United Nations Framework Convention on Climate Change, Subsidiary Body for Scientific and Technological Advice, Bonn, FCCC/SBSTA/2003/8

United States Congress Office of Technology Assessment (1992) 'Forest service planning: Accommodating uses, producing outputs, and sustaining ecosystems', Office of Technology Assessment, Report #OTA-F-505, Washington, DC: Government Printing Office

US CBP (2009) 'Guidance on the Lacey Act Declaration', www.cbp.gov/xp/cgov/trade/trade_programs/entry_summary/laws/food_energy/amended_lacey_act/guidance_lacey_act.xml, US Department of Homeland Security, Customs and Border Patrol, website, accessed June 2009

Vlosky, R. P. and J. E. Granskog (2003) 'Certification: A comparison of perceptions of corporate and non-industrial private forest owners in Louisiana', in L. Teeter, B. Cashore and D. Zhang (eds) *Forest Policy for Private Forestry: Global and Regional Challenges*, Wallingford, UK: CABI Publishing

Walker, J. L. (1990) 'Traditional sustained yield management: Problems and alternatives', *Forestry Chronicle*, 66, pp20–24

Westoby, J. C. (1983) 'Forest policy: Whose concern is it?', *Commonwealth Forestry Review*, 62, pp140–146

Wilkinson, Graham R. (1999) 'Codes of forest practice as regulatory tools for sustainable forest management', paper read at 18th Biennial Conference of the Institute of Foresters of Australia, 3–8 October, Hobart, Tasmania

Wilson, Jeremy (1998) *Talk and Log: Wilderness Politics in British Columbia*, Vancouver: University of British Columbia Press

Wood, Nancy (1971) *Clearcut: The Deforestation of America*, San Francisco: Sierra Club

Woodwell, G. M. and Ola Ullsten (2001) *Forests in a Full World*, New Haven, CT: Yale University Press

World Bank (2004) '*Sustaining Forests: A Development Strategy*,' Washington, DC: World Bank

World Bank (2008) 'Forest source book', http://web.worldbank.org/WBSITE/EXTERNAL/TOPICS/EXTARD/EXTFORESTS/EXTFORSOUBOOK/0,,menuPK:3745501~pagePK:64168427~piPK:64168435~theSitePK:3745443,00.html, accessed April 2009

WWF (2004) '"How effective are protected areas?" A preliminary analysis of forest protected areas by WWF – the largest ever global assessment of protected area management effectiveness', a report prepared for the Seventh Conference of Parties of the Convention on Biological Diversity, February 2004, Gland, Switzerland: Worldwide Fund for Nature (WWF) International.

Zhang, Daowei (2001) 'Endangered species and timber harvesting: The case of red-cockaded woodpeckers', working paper of the Forest Policy Center, Auburn University

Chapter 2

Selection and Global Context of the Case Study Countries[1]

CASE STUDY SELECTION

A central aim of this book is to facilitate inclusive, global-scale forest dialogue. However, in order to balance the breadth and depth of the analysis, it was necessary to select a subset of the world's 192 countries. The sampling methodology chosen was to focus on those countries of greatest global and regional significance in terms of forest extent and forest trade. Thus, we primarily sampled countries with:

1 the greatest forest area and/or
2 the greatest import/export value of forest trade in each of the world's major forested regions.

A total of 15 countries were selected on the basis of these criteria. A further five countries – Chile, Finland, Latvia, New Zealand and Portugal – were included to enhance regional representation and/or contrast.

Thus our study examines a total of 20 case study countries. The forest area and value of forest products trade for each of these countries, and their global ranking in these parameters, are presented in Table 2.1. These data are also shown graphically, and in rank order, in Figures 2.1 and 2.2. Together, the case study countries accounted for 70 per cent of the world's forest cover in 2005 (FAO, 2006a) and 61 per cent of the global forest products trade in 2006 (FAOSTAT, 2008). The area of plantation forest within the total forest area for each country is also presented in Table 2.1.

Forest area

Figure 2.1 shows the total forest and non-forest land area in all 20 case study countries as reported by the FAO in 2005. For this purpose, the FAO uses the following definition of 'forest': 'Land spanning more than 0.5 hectares with trees higher than 5 metres

Table 2.1 *Overview of case study countries*

FAO Regional Classification	2005 Forest area[2] (M ha)	2005 Global forest area ranking	2005 Plantation area[3] (M ha)	2006 Forest products trade value[3] ($M US)	2006 Global forest products trade ranking
Africa					
DRC	133.6	7	-	80.5	104
South Africa	9.2	57	1.4	2,159.0	33
America – Central & South					
Brazil	477.7	2	5.4	6,791.5	17
Chile	16.1	35	2.7	3,618.9	23
Mexico[*]	64.2	12	1.1	4,695.3	21
America – North					
Canada	310.1	3	-	33,355.4	3
USA	303.1	4	17.1	50,170.7	1
Asia					
China	197.3	5	31.4	30,572.8	4
India	67.7	10	3.2	2,689.4	31
Indonesia	88.5	8	3.4	7,590.5	16
Japan	24.9	23	10.3	15,359.4	8
Europe – Central & Eastern					
Latvia	2.9	90	0.0	1,453.0	39
Poland	9.2	58	0.0	5,742.7	19
Russia	808.8	1	17.0	10,437.1	12
Europe – Western					
Finland	22.5	25	0.0	16,146.2	7
Germany	11.1	47	0.0	34,499.3	2
Portugal	3.8	79	1.2	3,259.2	27
Sweden	27.5	22	0.7	17,073.5	6
Oceania					
Australia	163.7	6	1.8	3,442.1	25
New Zealand	8.3	60	1.9	2,457.9	32

Note: * formally classified as 'North America', but in this analysis considered with other Latin American countries, for reasons discussed in text.
Sources: FAO, 2006a; FAOSTAT, 2008; UNDP, 2008

and a canopy cover of more than 10 per cent, or trees able to reach these thresholds *in situ*. It does not include land that is predominantly under agricultural or urban land use' (FAO, 2004). This definition is notably broad and provides very little indication of the relative value and condition of each country's forested land.

As is clear from Figure 2.1, Russia has by far the largest forested area of any country worldwide – nearly twice as much as second-ranked Brazil. The next six in rank are Canada, the United States, China, Australia, the Democratic Republic of Congo (DRC) and Indonesia. These eight countries alone represent 63 per cent of the world's total forest cover. The other 12 in our sample account for an additional 7 per cent.

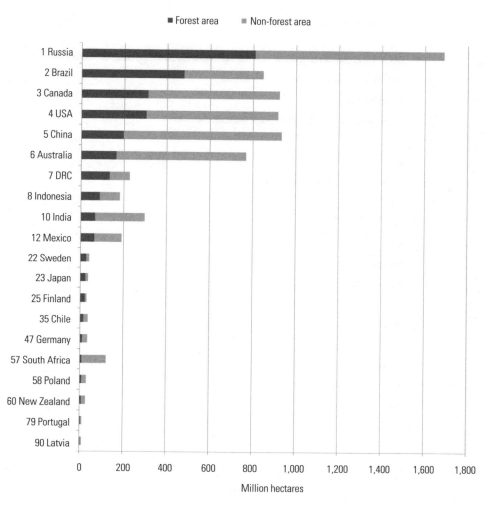

Source: FAO, 2006a

Figure 2.1 *Total forest and land area, and world ranking of forest area*

Forest products trade

In terms of the total value of forest products trade, measured as the sum of imports and exports as reported by the FAO (Figure 2.2), the US is ranked first worldwide, followed by Germany, Canada and China. The relative importance of imports versus exports varies; Canada is the largest exporter and the US the largest importer.

The FAO trade data below include all unprocessed and primary processed wood products, but exclude the very significant exchange of secondary processed products such as furniture, millwork and other high-value added products. In

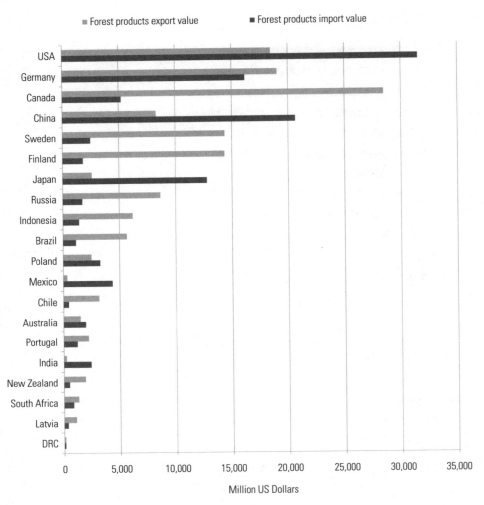

Source: FAOSTAT, 2008

Figure 2.2 *Forest product import/export value 2006*

regard to wooden furniture and parts, China is the world's top exporter, exporting US$ 7.1 billion in 2005. Germany follows in third place with exports of US$ 3.3 billion in the same year, Poland in fourth with US$ 3.1 billion, Canada in fifth with US$ 2.4 billion and Indonesia in ninth with US$ 1.2 billion. The US is the leading importer, importing US$ 15.7 billion in 2005, with Germany in third place at US$ 3.5 billion, Japan at fifth place with US$ 1.8 billion, and Canada in sixth place with US$ 1.4 billion (ITTO, 2006a).

Forest area, forest products trade and GDP

Table 2.2 compares forest area and the import/export value of forest trade, illustrating a wide range across both variables within our case study countries. Table 2.3 compares the import/export value of forest trade with per capita GDP as of 2004. It is notable that all countries with a per capita GDP of US$10,000 or greater are also leaders in global trade, with the exceptions of Australia and New Zealand.

Chapter order

Following a similar logic to our case study selection, the order of the empirical chapters begins with the regional case studies that together account for the largest combined import and export value of forest trade, i.e. Canada and the US, and proceeds to the regional case studies with the least, i.e. the DRC and South Africa. Within each chapter, the country case studies are arranged in alphabetical order.

Table 2.2 *Forest area and value of import/export forest trade*

Forest trade (cols)/ Forest area (rows)	< $5B	$5–10B	$10–20B	> $20B
< 10M ha	Latvia Portugal New Zealand South Africa	Poland		
10–20M ha	Chile India	Indonesia Mexico	Japan Finland Sweden	Germany
100–300M ha	Australia DRC			China
> 300M ha		Brazil	Russia	Canada USA

Sources: FAO, 2006a; FAOSTAT, 2008

Table 2.3 *Per capita GDP (2004) and value of import/export forest trade*

Forest trade (cols)/ per capita GDP (rows)	< $5B	$5–10B	$10–20B	>$ 20B
< $500	DRC			
$500–1000		Indonesia		
$1000–5000	India Latvia South Africa	Poland Brazil	Russia	China
$5000–10,000	Chile	Mexico		
$10,000–20,000	Portugal New Zealand			
> $20,000	Australia		Finland Japan Sweden	Canada Germany USA

Sources: FAO, 2006a; FAOSTAT, 2008

CHARACTERISTICS OF CASE STUDY COUNTRIES IN THE GLOBAL CONTEXT

Despite the shared characteristics of large forest areas and/or the importance of the forest products sector, the environmental, social and economic contexts for forest policy and management in the case study countries are very diverse. On a per capita basis, these countries range from forest-rich to forest-poor. In some, primary natural forests are the principal forest resource; others are entirely or mostly dependent on plantation forests. Socio-economically, our case studies include the world's wealthiest countries as well as some of the poorest. In terms of governance, they range from countries assessed as the most democratic to those that are lowest ranked in terms of various governance indicators.

The next sections of this chapter present and summarize case study data for four categories of indicator: key socio-economic indicators; key indicators of government capacity; indicators relating specifically to forests and their management; and indicators relating to harvesting and use of forest products.[4] In addition to providing a general background sketch of our diverse case study countries, we will also draw again on some of these criteria in Chapters 10 and 11 to reflect on possible correlation between such contextual variables and a given country's policy approach.

Key socio-economic indicators

Population

The 20 case study countries comprise 60 per cent of the world's population. China and India are by far the most populous, with 1.3 billion and 1.1 billion people respectively as of 2004. The third- and fourth-ranked countries, the United States and Indonesia, have populations of only 291 million and 128 million, respectively. The least populated country of our sample is Latvia, with only 2 million people. The median population of the case study countries is 50 million people.

Gross domestic product (GDP) per capita

The case study countries include the world's wealthiest and poorest, in terms of GDP per capita: in Japan, that figure is nearly US$40,000 and in the DRC it is less than US$100. Among the case study countries, only the US approaches Japan's wealth on a per capita basis, at US$37,000; five other case study countries – Australia, Canada, Finland, Germany and Sweden – have per capita GDPs of more than US$20,000; seven case study countries – DRC, India, Indonesia, Latvia, Poland, Russia and South Africa – have per capita GDPs of less than US$5000. The median per capita GDP of the case study countries is US$5700, compared to a global average of around US$6400.[5]

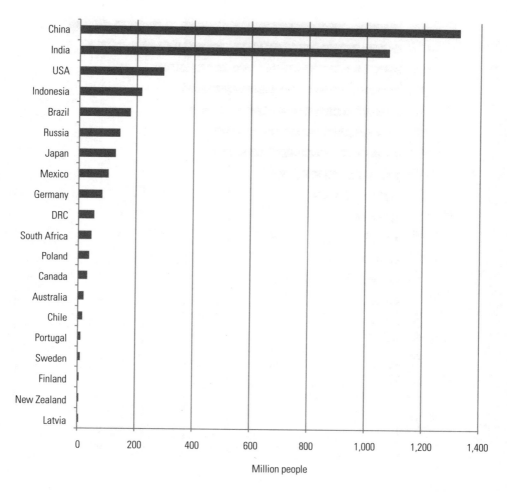

Source: FAO, 2006a

Figure 2.3 *Human population 2004*

Human Development Index (HDI)

The Human Development Index is a composite index that combines measures of life expectancy, educational attainment and income (UNDP, 2008). It provides a more holistic assessment of development than GDP alone. Our case study countries include four ranked in the top ten globally for HDI (Australia, Canada, Japan, Sweden) and one ranked in the lowest three (DRC). Fifteen of the case study countries are classified as having high human development (rankings 1–75), four as moderate (rankings 76–153) and one as low (rankings 154–179).

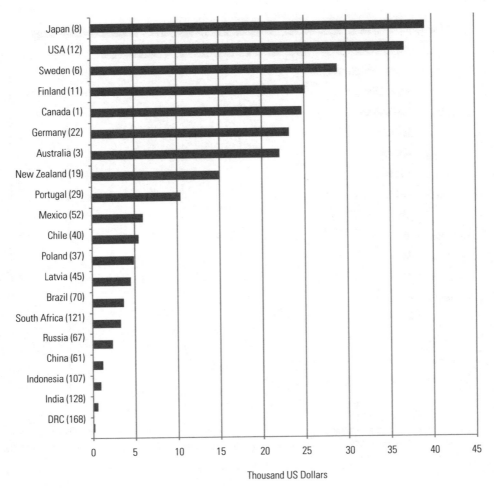

Source: FAO, 2006a

Figure 2.4 *Gross domestic product (GDP) per capita, 2004;*
Human Development Index (in parentheses) 2008

GDP + HDI as proxies for 'development'

Throughout this book, the terms 'developed' and 'developing' countries refer to a combined consideration of per capita GDP and HDI. 'Developed' refers to countries with a per capita GDP of US $10,000 or more and an HDI ranking in the top 33 countries worldwide. By this definition, the following countries are considered 'developed': Australia, Canada, Finland, Germany, Japan, New Zealand, Portugal, Sweden and the United States. While this simple binary distinction masks major socio-economic variation, it nevertheless appears to provide some important explanatory insights.

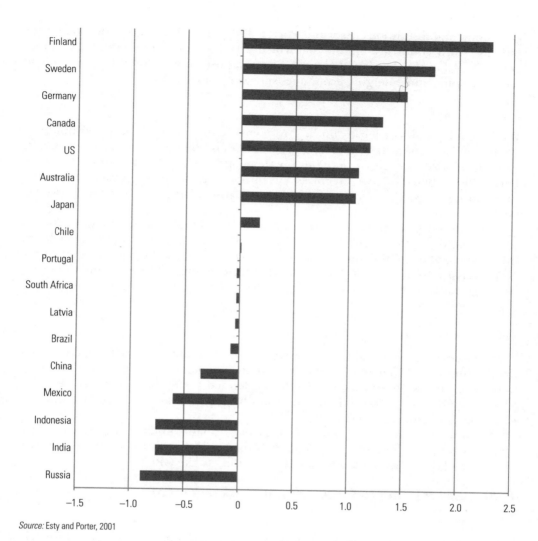

Source: Esty and Porter, 2001

Figure 2.5 *Environmental regime index*

Key indicators of government capacity

Environmental regulatory regimes

Yale and Columbia Universities have developed a suite of indices to assess countries' environmental performance (Esty et al, 2008). The most relevant of these to our work is the 2002 Environmental Regulatory Regimes Index, which integrates assessment of the stringency of environmental pollution standards, the sophistication of regulatory structure, the quality of available environmental information, the extent of subsidization of natural resources, the strictness of enforcement and the quality of environmental institutions (Esty and Porter, 2001). The pattern of index values for our case study countries mirrors that which Esty and Porter (2001) reported more generally. All of the developed countries show a positive score on this index and all the developing and emerging economies show a negative score.

Democracy index

Whilst 'there is no consensus on how to measure democracy, [and] definitions of democracy are contested' (Kekic, 2007), indices of democracy have nevertheless been developed. We use here the index developed by the Economist Intelligence Unit, which is based on five categories: electoral process and pluralism, civil liberties, functioning of government, political participation and political culture (Kekic, 2007).

Nine of our case study countries (Australia, Canada, Finland, Japan, Germany, New Zealand, Portugal, Sweden, USA) are ranked amongst the top 20 globally by this index, and one (DRC) in the lowest 25. According to the Economist Intelligence Unit's classification, nine of our case study countries are classified as full democracies, eight as flawed democracies, one as a hybrid regime and two as authoritarian regimes.

Corruption perceptions index

Transparency International, a non-governmental organization (NGO), has developed a corruptions perceptions index, a composite measure which ranks countries in terms of the extent to which corruption – defined as the abuse of public office for private gain – is perceived to exist amongst politicians and public officials (Transparency International, 2008).

Eight of our case study countries (Australia, Canada, Finland, Japan, Germany, New Zealand, Sweden, USA) are ranked in the top 20 (i.e. least corrupt) countries globally and one (DRC) in the lowest 20, according to the CPI.

Indicators relating to forest area and management

Forest area per capita

Forest area per capita provides one perspective on the human pressures on forests in different case study countries. There are some key caveats, however, to consider in interpreting the meaning of a country being 'forest-rich' or 'forest-poor' on a per capita basis. Firstly, as discussed above, the FAO's broad definition of 'forest' masks major variation in forest type and condition, ranging from dry, open forests with slow-growing, scattered trees and as little as 10 per cent crown cover, to dense, highly productive moist forests. Furthermore, countries vary considerably in their reliance on and use of forest products, their access to substitutes to these products and their production capacity. These caveats mean that forest area per capita describes only one element of the pressures on forests.

Figure 2.7 illustrates a range in forest area per capita from nearly 10ha in Canada to only 0.06ha in India. Two case study countries have more than 8ha of

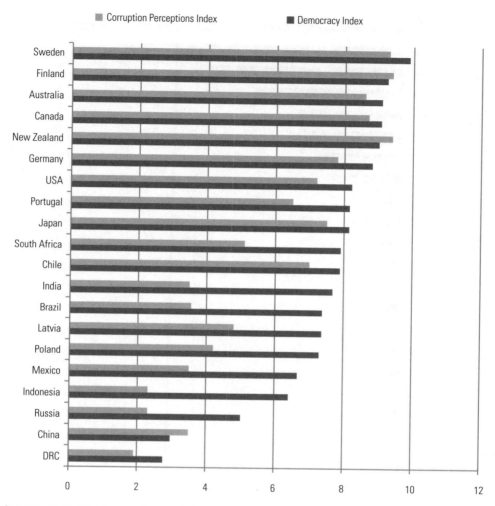

Note: Scale 0–10, 10 = 'best' (high democracy / low corruption).
Source: Transparency International, 2008

Figure 2.6 *Democracy and corruption perception indices 2007*

forest per person; two each fall within the ranges 6–8ha, 4–6ha and 2–4ha; 12 countries have less than 2ha of forest per person. The median forest area per capita for the case study countries is 1ha, compared to a global average of 0.62ha. The difference reflects our focus on the most-forested countries.

Extent and rate of forest cover change

Forest cover increased over the period 2000–2005 in ten of the case study countries, but the annual increases were small – less than 100,000ha per year – in all but the

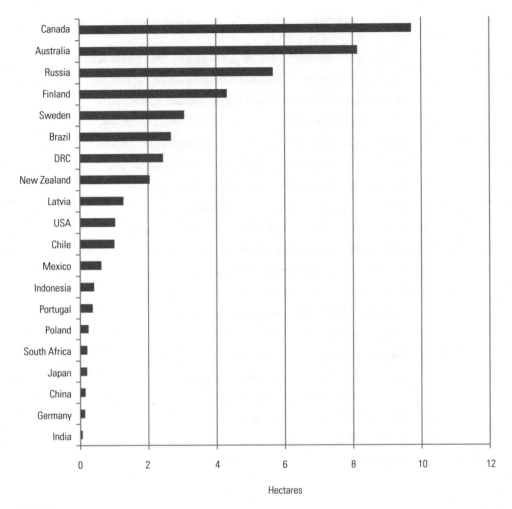

Source: FAO 2006a

Figure 2.7 *Forest area per capita*

USA (160,000ha per year) and China (4 million ha per year). Deforestation was greater than 100,000ha/year in five case study countries (Australia, Brazil, DRC, Indonesia, Mexico); it was nearly 2 million ha annually in Indonesia, and more than 3 million ha annually in Brazil. For those countries in which forest area was increasing, the median annual increase over the period 2000–2005 was 28,000ha; amongst those that were losing forest, the median annual loss over the same period was 260,000ha. Annual rates of forest cover change in case study countries ranged from +2 per cent for China to –2 per cent for Indonesia.

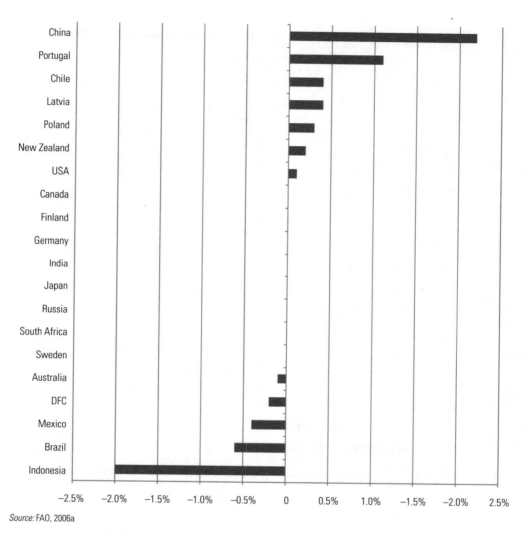

Source: FAO, 2006a

Figure 2.8 *Annual rate of forest cover change 2000–2005*

Forest ownership

Forest ownership is a complex issue, best understood as a 'bundle of rights' variously assigned to individuals, institutions, governments and/or the 'common good' (North, 1990; Ostrom, 1990; Luckert, 2005). The contents of this bundle may vary considerably, and are affected by the very forest practice policies covered in this book: i.e. lands classified as 'private' in one jurisdiction may conceivably be subject to more public constraint in the form of management restrictions than 'public' lands in another. Definitions of ownership are also dynamic, particularly in the case of developing countries, where tenure disputes may be widespread. 'Public' land in such

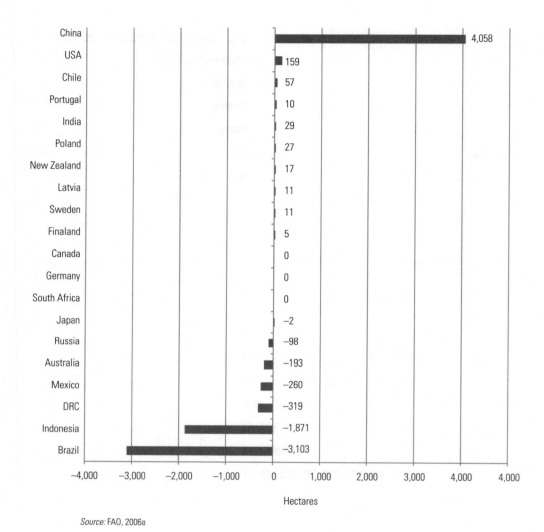

Source: FAO, 2006a

Figure 2.9 *Annual area of forest cover change 2000–2005*

cases may refer to land for which tenure is simply unresolved or unknown. These issues are variously addressed in each regional chapter and reflected upon as a whole in the concluding Chapter 11.

According to the FAO database, the reported distribution of forest ownership varies greatly between case study countries. Whilst public ownership dominates in the majority of cases and comprises a reported 100 per cent in four countries, private ownership exceeds 40 per cent of forest area in eight countries and is greater than 80 per cent in two. Mexico is uniquely distinguished in having a very large percentage of its forests under communal ownership, accounting for somewhere

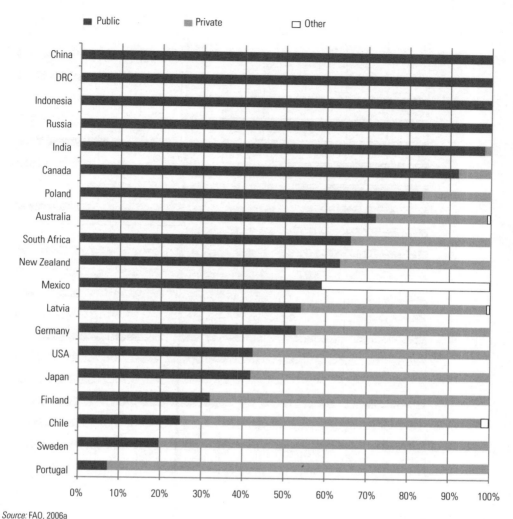

Figure 2.10 *Forest ownership*

between 41–80 per cent of total forestlands depending on definitions of both 'forest' and 'ownership' (FAO, 2006a; ITTO, 2006b) (see also Chapter 7). Most of the case study countries with high levels of private forest ownership are in Europe, but Japan and the USA each have nearly 60 per cent of their forests in private tenure.

Forest characteristics

The FAO's 2005 Global Forest Resource Assessment reported the proportion of countries' forest area in five categories: primary, modified natural, semi-natural, productive plantation and protective plantation. Four of the case study countries

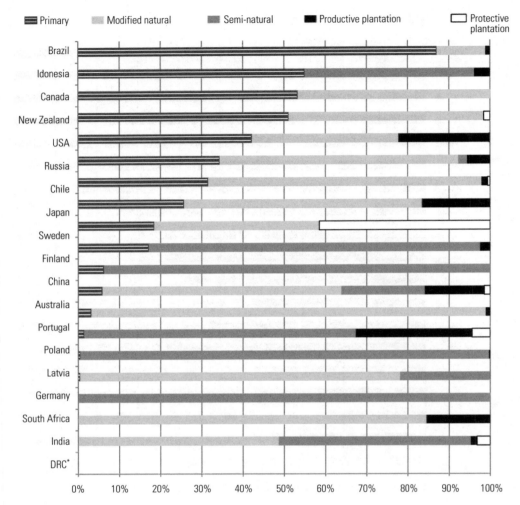

Note: * no available data for DRC.
Source: FAO, 2006a

Figure 2.11 *Forest characteristics*

(Brazil, Canada, Indonesia, Mexico) report more than 50 per cent of their forest as primary; that would also be the case for DRC, for which FAO-reported data were not available. New Zealand also reports a substantial proportion, around 40 per cent, of primary forest. Seven countries (Australia, China, Chile, Latvia, Russia, South Africa, USA) report a majority of modified natural forest. Six (Finland, Germany, India, Poland, Portugal, Sweden) report a majority of semi-natural forest. Japan reported the greatest proportion of protective plantations, at around 40 per cent of its forest area.

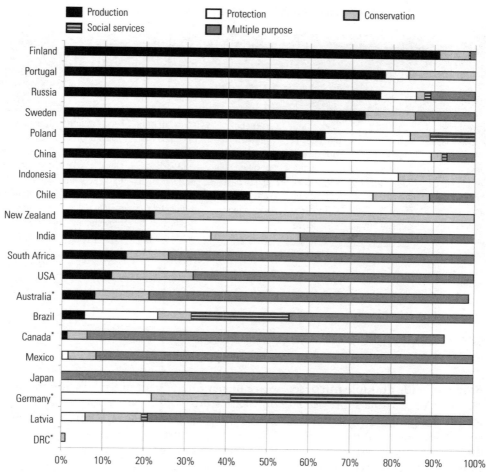

Figure 2.12 *Designated primary function of forests*

Designated primary function of forests

The FAO's 2005 Global Forest Resource Assessment also addresses the proportion of the countries' forest area designated for five categories of primary use: production, protection, conservation, social services and multiple purposes. Seven of the case study countries (China, Finland, Indonesia, Poland, Portugal, Russia, Sweden) report a majority of forest area designated primarily for production – around 90 per cent in the case of Finland. Forty-five per cent of Chile's forests are similarly designated. Seven case study countries (Australia, Canada, Japan, Latvia, Mexico, South Africa, USA) report that the majority of their forest area – or all, in Japan's case – is

Table 2.4 *Planted forests in the continuum of forest characteristics*

Naturally regenerated forests		Planted forests					Trees outside forests
Primary	Modified natural	Semi-natural		Plantations			
		Assisted natural regeneration	Planted component	Productive	Protective		
Forests of native species where there are no clearly visible indications of human activities and the ecological processes are not significantly disturbed	Forest of naturally regenerated native species where there are clearly visible indications of human activities	Silvicultural practices by intensive management: • Weeding • Fertilizing • Thinning • Selective logging	Forest of native species, established through planting or seeding, intensively managed	Forest of introduced and/or native species established through planting or seeding* mainly for production of wood or non-wood goods	Forest of introduced and/or native species, established through planting or seeding* mainly for provision of services		Stands smaller than 0.5ha; tree cover in agricultural land (agroforestry systems, home gardens, orchards); trees in urban environments; and scattered along roads and in landscapes

Note:* 'characterized by few species, straight tree lines and/or even-aged stands'.
Source: FAO, 2007

designated for multiple purposes, as are nearly 50 per cent of Brazil's and more than 40 per cent of India's forests. The only case study country to report a majority of its forests designated primarily for conservation was New Zealand, where nearly 80 per cent of forests are in this category. The largest proportion of German forests, around 40 per cent, is designated primarily for social services. No data were reported for most of DRC's forests.

Area of semi-natural planted forests and plantation

The distinction between semi-natural planted forests and plantation forests is often clear only at the extremes, and is not made consistently between countries. For example, the USA classifies 100 per cent of its planted forests, which consist mostly of native species, as 'plantation' and zero per cent as semi-natural planted forest; conversely, Finland classifies 94 per cent of its planted forests, which are also primarily native species, as 'semi-natural' and only 6 per cent as 'plantation'. Figure 2.13 includes all planted forests, both semi-natural and plantation, to provide a general picture of the area of forests under *relatively* intensive management.

The FAO's 2005 Global Forest Resource Assessment distinguishes between two broad objectives for planted forests: production and protection. The extent of each case study country's planted and plantation forests in each of these categories is presented in Figure 2.13. Most planted forests and plantations are designated

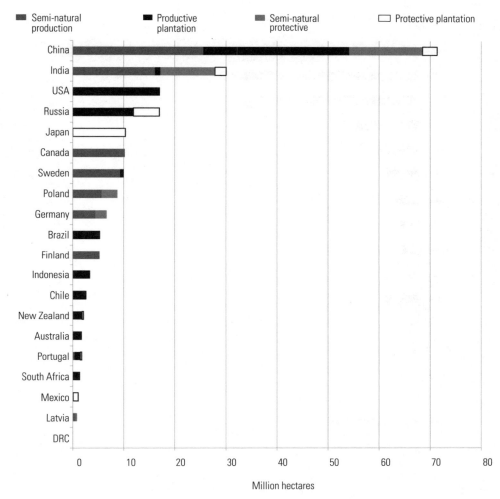

Source: FAO, 2006b

Figure 2.13 *Area of planted semi-natural forests and plantations intended for production and protection*

primarily for production in most case study countries. Exceptions are Japan and Mexico, where all plantations are classified as primarily for protection.

Among the case study countries, the area of plantation forest (Figure 2.14) ranged from nil in Finland and Germany and unrecorded in the DRC and Canada, to 31.4 million ha in China, and about 17 million ha in the US and Russia. The median reported plantation area was 1.9 million ha. The proportion of a country's forests that are in plantation ranged from nil, for those with no plantation area, to 42 per cent for Japan, 33 per cent for Portugal, and 22 per cent for New Zealand (see Figure 2.11).

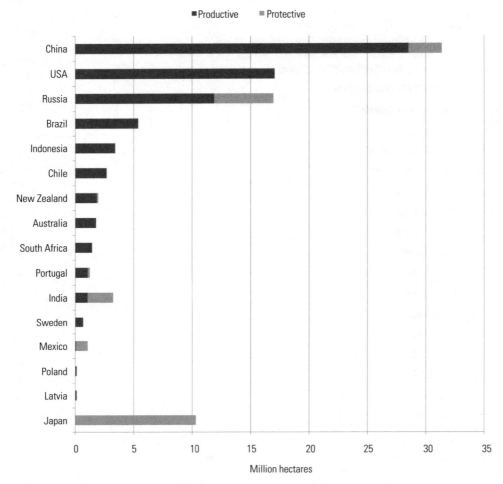

Source: FAO, 2006a

Figure 2.14 *Area of productive and protective plantations*

The greatest annual rates of plantation expansion in case study countries over the period 2000–2005 were in China (about 1.5 million ha), Russia (about 320,000ha) and the US (about 150,000ha); all others were less than 100,000ha annually. The median annual rate of plantation expansion in case study countries from 2000–2005 was 19,000ha.

As explained in Chapter 1, our study's forest policy analysis includes an examination of policies relating to plantation management in those case study countries where plantations provide a substantial portion of total wood production. For the purposes of that analysis, our focus is expressly on the top ten case study countries in terms of area of 'productive plantations', namely China, the US, Russia, Brazil, Indonesia, Chile, New Zealand, Australia, South Africa and Portugal.

Box 2.1 The IUCN Protected Areas Management Categories I–VI

CATEGORY Ia: Strict Nature Reserve: protected area managed mainly for science

Definition: Area of land and/or sea possessing some outstanding or representative ecosystems, geological or physiological features and/or species, available primarily for scientific research and/or environmental monitoring.

CATEGORY Ib: Wilderness Area: protected area managed mainly for wilderness protection

Definition: Large area of unmodified or slightly modified land, and/or sea, retaining its natural character and influence, without permanent or significant habitation, which is protected and managed so as to preserve its natural condition.

CATEGORY II: National Park: protected area managed mainly for ecosystem protection and recreation

Definition: Natural area of land and/or sea, designated to (a) protect the ecological integrity of one or more ecosystems for present and future generations, (b) exclude exploitation or occupation inimical to the purposes of designation of the area and (c) provide a foundation for spiritual, scientific, educational, recreational and visitor opportunities, all of which must be environmentally and culturally compatible.

CATEGORY III: Natural Monument: protected area managed mainly for conservation of specific natural features

Definition: Area containing one, or more, specific natural or natural/cultural feature which is of outstanding or unique value because of its inherent rarity, representative or aesthetic qualities or cultural significance.

CATEGORY IV: Habitat/Species Management Area: protected area managed mainly for conservation through management intervention

Definition: Area of land and/or sea subject to active intervention for management purposes so as to ensure the maintenance of habitats and/or to meet the requirements of specific species.

CATEGORY V: Protected Landscape/Seascape: protected area managed mainly for landscape/seascape conservation and recreation

Definition: Area of land, with coast and sea as appropriate, where the interaction of people and nature over time has produced an area of distinct character with significant aesthetic, ecological and/or cultural value, and often with high biological diversity. Safeguarding the integrity of this traditional interaction is vital to the protection, maintenance and evolution of such an area.

CATEGORY VI: Managed Resource Protected Area: protected area managed mainly for the sustainable use of natural ecosystems

Definition: Area containing predominantly unmodified natural systems, managed to ensure long-term protection and maintenance of biological diversity, while providing at the same time a sustainable flow of natural products and services to meet community needs.

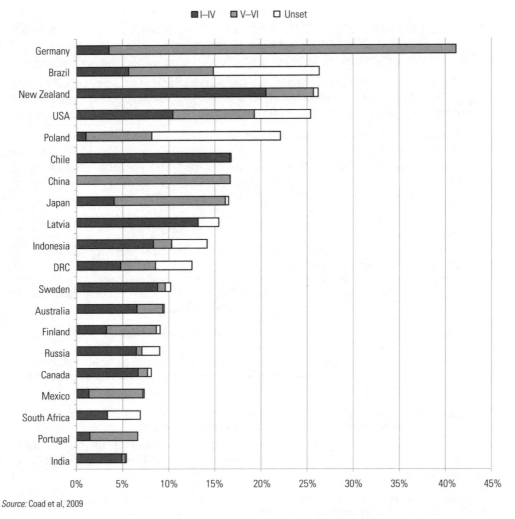

Source: Coad et al, 2009

Figure 2.15 *Protected areas as percentage of land base, IUCN Categories I–VI and IUCN Unclassified*

Protected areas

Although it is difficult to compile globally consistent and accurate data on the extent and management of protected areas, the United Nations Environment Programme – World Conservation Monitoring Centre (UNEP-WCMC) has been developing an increasingly comprehensive database of protected areas (WDPA Consortium, 2009).

Figure 2.15 illustrates the percentage of each case study country's land area categorized as protected areas in each of the International Union for the Conservation of Nature (IUCN) Management Categories I-VI and IUCN unclassi-

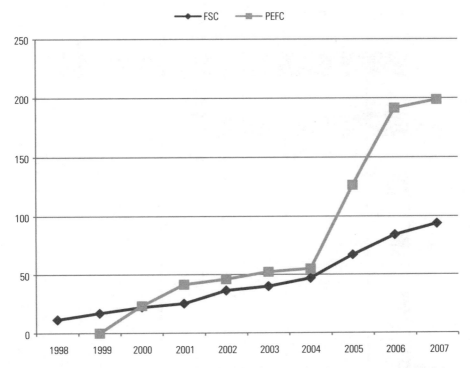

Note: * The figures for Russia are separated from those for the rest of Europe due to Russia's unusually large forest area and relatively low percentage of forest area certified.
Source: Auld, 2008

Figure 2.16 *Growth in area certified 1998–2007*

fied (Coad et al, 2009). There is wide variation among case studies both in terms of total percentage of land area protected (from 42 per cent for Germany to 5 per cent for India), as well as in the extent of protection. Considering only Categories I–IV, i.e. those categories of greatest protection that generally do not allow extraction of natural resources, the largest percentages of land area are found in New Zealand (at 21 per cent), followed by Chile (17 per cent), Latvia (13 per cent) and the US (10 per cent). China reports no protected areas for Categories I–IV, and Mexico, Poland and Portugal only 1 per cent.

Forest certification

Figures 2.16 and 2.17 present global data for forest certification, including the growth in area certified from 1998–2007, and area certified by region in 2008. PEFC-endorsed certificates grew particularly rapidly between 2004 and 2006, surpassing the growth in and extent of the FSC.

Figures 2.18 and 2.19 compare the case study countries in terms of total area certified and the percentage of forest area certified, and by certification scheme.

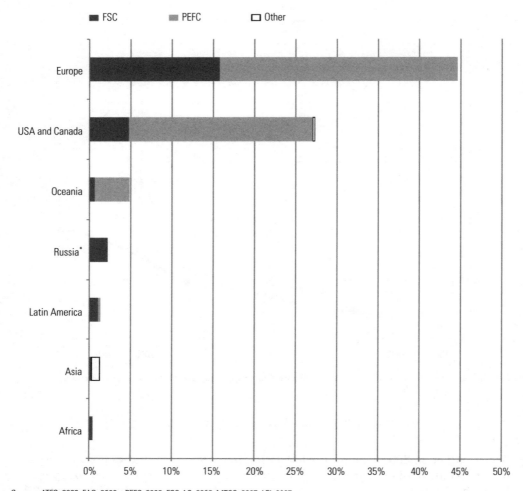

Sources: ATFS, 2006; FAO, 2006a; PEFC, 2008; FSC-AC, 2008; MTCC, 2007; LEI, 2007

Figure 2.17 *Percentage of forest area certified by world region 2008*

Both globally and among our case study countries, by far the largest certified area was found in Canada, totalling about 87 million ha, as of January 2008. At the other end of the scale among our case study countries were India (644ha) and the DRC (no certified forest area). The median certified area among all 20 case studies was 1.7 million hectares.

Placing certification in proportion to total forest area, more than 50 per cent of forests are certified in five case study countries – Finland, Germany, Sweden, Latvia and Poland; in Finland's case, more than 90 per cent are certified. In only four other case study countries – Canada, Chile, South Africa and the US – is more than 10 per cent of forest certified.

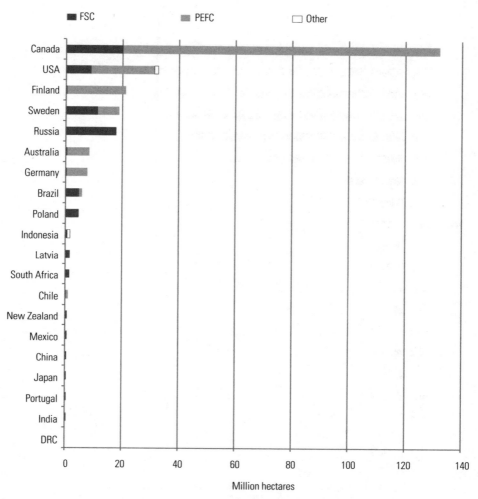

Sources: ATFS, 2006; FAO, 2006a; PEFC, 2008; FSC-AC, 2008; MTCC, 2007; LEI, 2007; SGEC, 2009

Figure 2.18 *Forest area certified by certification scheme*

With regard to certification schemes, PEFC-endorsed schemes cover the major-ity of certified forest area in Australia, Canada, Chile, Finland, Germany and the US; the Japanese SEGC covers the majority of certified forest area in Japan; the Indonesian LEI (which has a joint certification protocol with the FSC) leads in Indonesia; while the FSC covers the largest area in all other case study countries with certified forest.

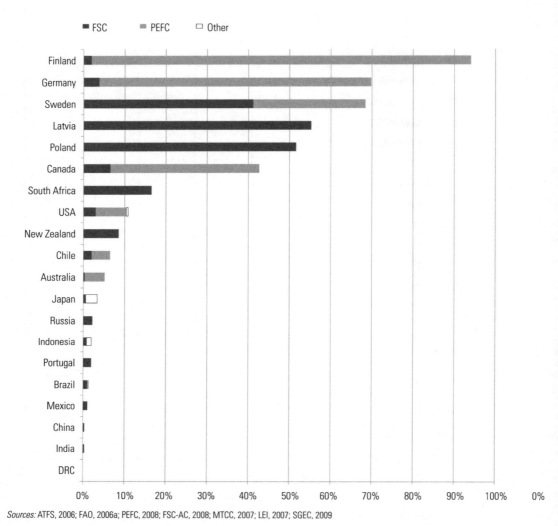

Sources: ATFS, 2006; FAO, 2006a; PEFC, 2008; FSC-AC, 2008; MTCC, 2007; LEI, 2007; SGEC, 2009

Figure 2.19 *Percentage of forest area certified by certification scheme*

Indicators relating to harvesting and use of forest products

Forest product production and consumption

As of 2005, the US, Canada, Russia and Brazil were the largest producers of industrial roundwood, while the US, Canada, China and Brazil were the largest consumers (see Figure 2.20).

With regard to fuel wood, India is the largest consumer, followed by Brazil, China and Indonesia.

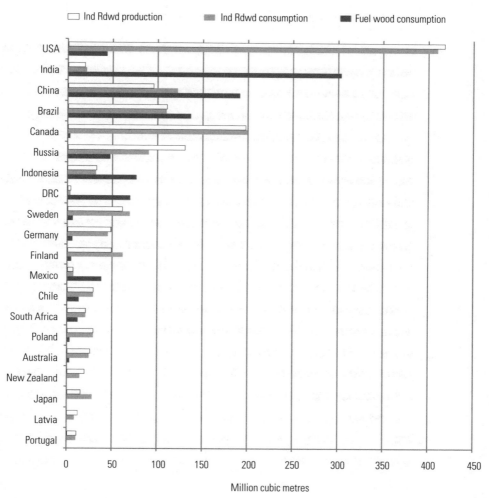

Note: Fuel wood production and consumption figures are identical for most countries.
Source: FAOSTAT, 2008

Figure 2.20 *Industrial roundwood and fuel wood production and consumption 2005*

Industrial roundwood versus fuel wood production

Figure 2.21 presents the proportion of each case study country's wood production that is industrial roundwood and fuel wood, as a crude indicator of the importance of commercial versus subsistence use. The DRC, Mexico, China and Brazil are all recorded as producing more than twice the volume of fuel wood as industrial round-wood, with the DRC producing 18 times as much. In general, the eight developing country case studies show the highest ratio of fuel wood to industrial roundwood and emerging and developed cases show lower ratios. However, there is considerable

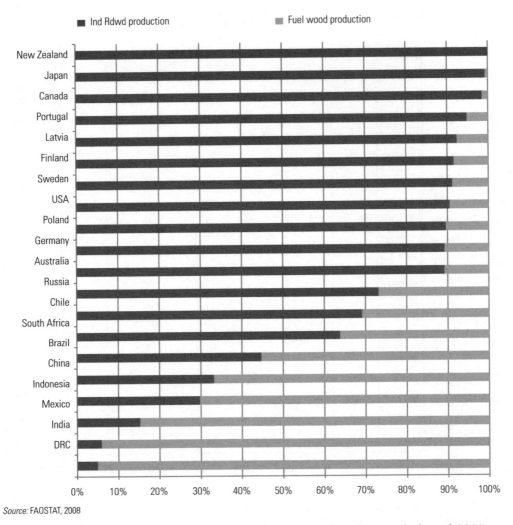

■ Ind Rdwd production   ▨ Fuel wood production

Source: FAOSTAT, 2008

Figure 2.21 *Percentage of production for industrial roundwood versus fuel wood 2005*

inter-country variation that is clearly not based on GDP alone. Some of this reflects the industrial use of fuel wood, such as in Brazil's steel industry (SBS, 2008).

Proportion of growing stock harvested

The proportion of growing stock harvested annually in case study countries ranges from more than 3 per cent, in Portugal, to 0.1 per cent, in India (see Figure 2.22). Five countries – Chile, Finland, Portugal, South Africa and Sweden – are harvesting

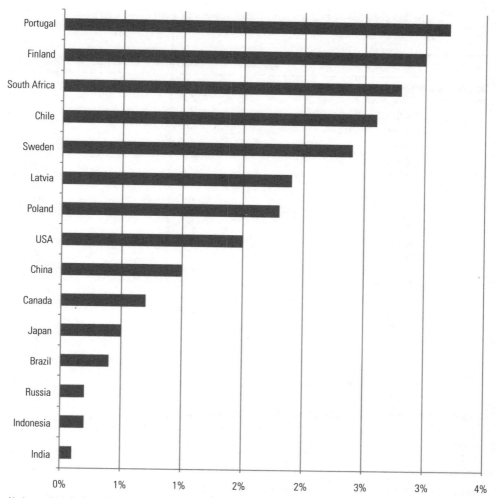

Note: No data available for Germany, Australia, New Zealand, Mexico and the DRC.
Source: FAO, 2006a

Figure 2.22 *Percentage of growing stock harvested annually*

more than 2 per cent of their growing stock annually, and five – Brazil, India, Indonesia, Japan and Russia – are harvesting 0.5 per cent or less. Data were not reported for Australia, DRC, Germany or New Zealand.

Table 2.5 *Case study jurisdictions, land ownerships, and forest types included in our standardized policy comparison*

Country	Jurisdiction	Public NF	Private NF	Public Plant	Private Plant
Chapter 3 US & Canada (20 case studies, 23 NF ownerships addressed)					
Canada	Alberta		✓✓		
	British Columbia		✓✓		
	New Brunswick		✓	✓	
	Ontario		✓✓		
	Quebec		✓✓		
United States	Alabama		✓✓		✓
	Alaska		✓✓		
	Arkansas		✓✓		✓
	California		✓✓		
	Georgia		✓✓		✓
	Idaho		✓✓		
	Louisiana		✓✓		✓
	Mississippi		✓✓		✓
	Montana		✓✓		
	North Carolina		✓✓		✓
	Oregon		✓✓		
	South Carolina		✓✓		✓
	Texas		✓✓		✓
	Virginia		✓✓		✓
	Washington		✓✓		
	US Forest Service, Pacific Coast	✓	✓✓		
Chapter 4 Western Europe (4 case studies, 6 NF ownerships addressed)					
Germany	Bavaria	✓	✓✓		
Finland		✓	✓✓		
Sweden			✓✓		
Portugal			✓✓		✓
Chapter 5 Asia (4 case studies, 5 NF ownerships addressed)					
China		✓✓		✓	
India	Madhya Pradesh	✓✓		✓	
Indonesia		✓✓		✓	
Japan		✓	✓✓		
Chapter 6 Central & Eastern Europe (3 case studies, 4 NF ownerships addressed)					
Latvia		✓✓	✓		
Poland		✓✓			
Russia		✓✓			
Chapter 7 Latin America (3 case studies, 4 NF ownerships addressed)					
Mexico		✓✓			
Brazil	Amazon Basin	✓✓	✓		✓
Chile		(no harvest)	✓✓		✓

Table 2.5 *continued*

Country	Jurisdiction	Public NF	Private NF	Public Plant	Private Plant
Chapter 8 Oceania (9 case studies, 15 NF ownerships addressed)					
Australia	ACT	✓✓(no harvest)		✓	
	New South Wales	✓✓	✓	✓	✓
	Northern Territory	✓	✓✓	✓	✓
	Queensland	✓✓	✓	✓	
	South Australia	✓✓(no harvest)	✓(no harvest)	✓	✓
	Tasmania	✓✓	✓	✓	✓
	Victoria	✓✓			✓
	Western Australia	✓✓		✓	✓
New Zealand		✓✓(no harvest)	✓	✓	✓
Chapter 9 Africa (2 case studies, 2 NF ownerships addressed)					
DRC		✓✓			
South Africa		✓✓		✓	✓

Note: NF = natural forest, Plant = plantation, ✓✓ = largest natural forest land ownership category. Total number of case study jurisdictions: 45. Total number of NF ownerships that cover at least 20 per cent of forest area (and thus addressed in this book): 59. Ratio of developed/developing country jurisdictions: 4:1.

OVERVIEW OF COUNTRY AND SUB-NATIONAL CASE STUDIES

In 14 of our 20 case study countries (Chile, China, DRC, Finland, Indonesia, Japan, Latvia, Mexico, New Zealand, Poland, Portugal, Russia, South Africa, Sweden), we were able to assess forest practice policies according to a single set of national-level policies. In the remaining six countries, the relevant forest practice policies were established at the subnational level. In three of these latter countries, we sampled one sub-national jurisdiction or region as an example of that country's forest practice policies: for Brazil, we assessed the Amazon Basin; for Germany, the state of Bavaria, and for India, the state of Madhya Pradesh. In the final three countries – Australia, Canada and the USA – we sampled more intensively across states or provinces,[6] assessing all eight Australian states and territories, five Canadian provinces and 15 US states and federal forestlands. In those cases where sub-national sampling was employed, we chose jurisdictions based on relative forest area and roundwood harvest. Thus, in total, we address forest policies in 45 jurisdictions.

Our analysis of forest practice policies in each case study country considered both public and private land as appropriate. All land ownerships were assessed that comprised at least 20 per cent of the forest area or contributed at least 20 per cent of the wood harvest. As mentioned above, plantation policy was assessed for those case study countries with the top ten largest areas of production-oriented plantations.

The resultant matrix of jurisdictions and forest types we assessed is shown in Table 2.5 (case study jurisdictions table), organized according to their presentation in the chapters of Part II of the book. In a few cases, i.e. Australian Capital Territory,

South Australia and New Zealand public forests, no commercial harvest is allowed in natural forests. We make note of these policies in country-level contextual discussions but exclude these ownerships from the standardized comparison of forest practice policies.

NOTES

1   This chapter draws on a range of global datasets, each of which has limitations in terms of reliability and consistency. The majority of forest and trade data is sourced from FAO's *Global Forest Resources Assessment 2005* and related studies (FAO, 2006a; 2006b; 2007; FAOSTAT 2008), which represents the most comprehensive and globally consistent dataset available for the majority of the forest and trade parameters used in this study. FAO notes various caveats to the quality and interpretation of these data (see, for example, FAO 2006a: 150-4). The World Database on Protected Areas (WDPA Consortium 2009) provides the most comprehensive and globally consistent source on protected areas; data on global corruption, democracy and environmental regulatory indices are drawn from established sources, as acknowledged; and certification data principally from the FSC and PEFC databases (FSC-AC 2008 and PEFC 2008, respectively). We acknowledge the limitations of these datasets, and also that the quantification of complex social phenomena such as 'democracy' is contested. Nevertheless we consider the data are adequate for the purposes of coarse filter comparison and analysis.

2   'Forest' is defined as: 'Land spanning more than 0.5 hectares with trees higher than 5 metres and a canopy cover more than 10 per cent, or trees able to reach these thresholds *in situ*. It does not include land that is predominantly agricultural or urban land use' (FAO, 2004).

3   Includes productive and protective plantations. See Table 2.4 for definitions of 'plantation'.

4   We have used the most recent consistent data sets available. For population and GDP, we use the data sets reported in FAO's Forest Resources Assessment 2005 (FAO 2006a), rather than more recent estimates, to maintain consistency with other data reported by that assessment.

5   holos.wgov.org/holosbank.com/unigov/World%202005%20total.pdf – 2005, accessed May 2009.

6   We were able to do this because of funding support from sponsors in these countries.

REFERENCES

ATFS (2006) Telephone conversation with Elizabeth Sandler, American Tree Farm System (ATFS) Certification Manager, Arvind Nagarajan, 25 October

Auld, Graeme (2008) 'Reversal of fortune: How early choices can alter the logic of market-based authority', PhD Thesis, New Haven, CT: School of Forestry and Environmental Studies, Yale University

Coad, Lauren, Neil Burgess, Lucy Fish, Corinna Ravillious, Colleen Corrigan, Helena Pavese, Arianna Granziera and Charles Besançon (2009) 'Progress towards the

Convention on Biological Diversity terrestrial 2010 and marine 2012 targets for protected area coverage', *Parks*, 17, 2, pp35–42

Esty, Daniel C. and Michael E. Porter (2001) 'Ranking national environmental regulation and performance: A leading indicator of future competitiveness?', in M. E. Porter, J. Sachs and A. M. Warner (eds) *The Global Competitiveness Report 2001*, New York: Oxford University Press

Esty, Daniel C., Christine Kim, Tanja Srebotnjak, Marc A. Levy, Alex de Sherbinin and Valentina Mara (2008) '2008 Environmental Performance Index', Yale Center for Environmental Law and Policy, Center for International Earth Science Information Network (CIESIN)

FAO (2004) 'Global forest resources assessment update 2005: Terms and Definitions (Final Version)', Rome: Food and Agriculture Organization of the United Nations

FAO (2006a) 'Global forest resources assessment 2005: Progress towards sustainable forest management', in *FAO Forestry Paper 147*, Rome: Food and Agricultural Organization of the United Nations

FAO (2006b) 'Global planted forests thematic study: Results and analysis', in A. Del Lungo, J. Ball and J. Carle (eds) *Responsible Management of Planted Forests: Vountary Guidelines*, Rome: United Nations Food and Agriculture Organization

FAO (2007) *State of the World's Forests – 2007*, Rome: United Nations, Food and Agriculture Organization

FAOSTAT (2008) 'ForesSTAT', http://faostat.fao.org/default.aspx, United Nations Food and Agriculture Organization, online database, accessed 8 August 2008

FSC-AC (2008) 'FSC certified forests', 10 January 2008, www.fsc.org, Forest Stewardship Council fact sheet, accessed January 2008

ITTO (2006a) *Annual Review and Assessment of the World Timber Situation*, Yokohama: International Tropical Timber Organization

ITTO (2006b) *Status of Tropical Forest Management 2005*, Yokohama: International Tropical Timber Organization

Kekic, L. (2007) 'The Economist Intelligence Unit's index of democracy', www.economist.com/media/pdf/Democracy_Index_2007_v3.pdf, accessed May 2009

LEI (2008) Lembaga Ekolabel Indonesia 2007, www.lei.or.id/english/akreditasi.php?cat=19, accessed January 2008

Luckert, Marty K. (2005) 'In search of optimal institutions for sustainable forest management: Lessons from developed and developing countries', in S. Kant and R. A. Berry (eds) *Institutions, Sustainability, and Natural Resources: Institutions for Sustainable Forest Management*, New York: Springer-Verlag

MTCC (2008) 'Frequently asked questions (FAQs)', www.mtcc.com.my/faqs.asp#FAQ2, Malaysian Timber Certification Council (MTCC), 1 April 2007

North, Douglass C. (1990) *Institutions, Institutional Change and Economic Performance*, New York: Cambridge University Press

Ostrom, E. (1990) *Governing the Commons: The Evolution of Institutions for Collective Action*, Cambridge: Cambridge University Press

PEFC (2008) PEFC Council Information Register, 31 January 2008, http://register.pefc.cz/

SBS May (2008) Fatos e Números do Brasil Florestal, Dezembro 2007, www.sbs.org.br/, Sociedade Brasileira de Silvicultura (SBS) 2008

SGEC (2009) Sustainable Green Ecosystem Council, www.sgec-eco.org/certforest/itiranhyou-synrin.pdf, accessed August 2009.

Transparency International (2008) 'Corruption perceptions index', www.transparency.org/policy_research/surveys_indices/cpi/2008/faq#general1

UNDP (2008) '2007/2008 human development index ratings', Washington, DC: United Nations Development Programme
WDPA Consortium (2009) 'Protected areas and world heritage', IUCN, UNEP

PART II

Regional Analyses

Chapter 3

Canada and the United States

INTRODUCTION

The forests of Canada and the United States cover 613 million hectares, or nearly 16 per cent of the world's total forest area. They are divided almost evenly between the two countries, making Canada and the US the third and fourth most forest-endowed countries worldwide (FAO, 2007). In Canada, boreal forests account for 76 per cent of forest area and the remaining 24 per cent are temperate. In the US about 48 per cent of forests are classified as temperate, 37 per cent as subtropical and 15 per cent as boreal (FAO, 2003).

Most forest loss in the region occurred in the 19th century during the peak of European immigration and the resulting clearance of land for agriculture. In both countries forests now cover about one-third of the land area, which in the US represents a net loss of about one-third of the original forest area and in Canada less than 10 per cent (Bryant et al, 1997b; Smith and Darr, 2004). While forest cover has remained fairly stable over the last 100 years, within the US there is considerable regional variation, with increases in forest area in the US north central and northeastern regions and continued loss in the US west (Smith and Darr, 2004; FAO, 2006a). The percentage of primary forests in Canada and the US, meanwhile, has continued to decline (FAO 2006a).[1]

Both countries are major forest products producers and the US is the world's largest consumer of wood products. In 2005, the US produced 541 million cubic metres of roundwood, with the majority consumed domestically (FAO, 2007). Canada produced 224 million cubic metres (FAO, 2007), the majority of which was exported, primarily to the US (NRC, 2007b).

The high environmental and economic significance of these countries' forests and forest trade is matched by considerable domestic and international controversy. To date, the most widely publicized conflicts over domestic timber harvest have been focused on the Pacific Coast temperate rainforest of both countries and, to a lesser extent, the Canadian boreal forests. While there are controversies elsewhere, including criticism of intensive plantation management and harvesting for woodchips in

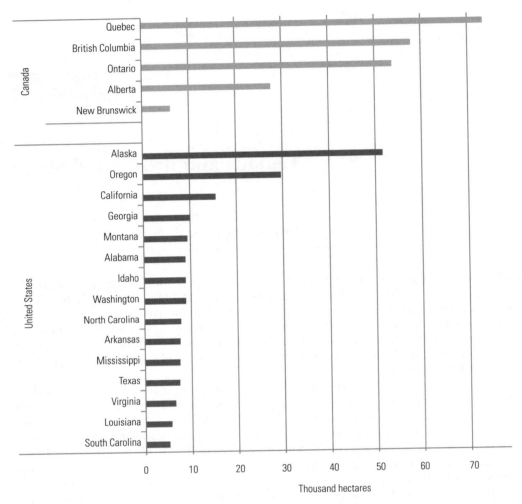

Note: As of March 2009, the 2001 forest inventory was the most recent forest inventory data available.
Sources: Canadian Council of Forest Ministers (CCFM), 2009; US Forest Service Forest Inventory and Analysis (USFS FIA), 2009

Figure 3.1 *Forest cover of select Canadian provinces and US states 2001*

the southeast, these issues have not received the same degree of national and international attention and scrutiny (Cashore and McDermott, 2004). One of the initial
catalysts for this book was to place the debate over North American Pacific Coast
forest management within the broader context of forest policies across the rest of the
continent and beyond.

Since forest governance is largely handled at a sub-national level in both Canada
and the US, five Canadian provinces and 15 US states were selected for detailed
policy analysis. Consistent with our overall approach, the province and state case
studies selected were those that contain the largest forest area and/or greatest

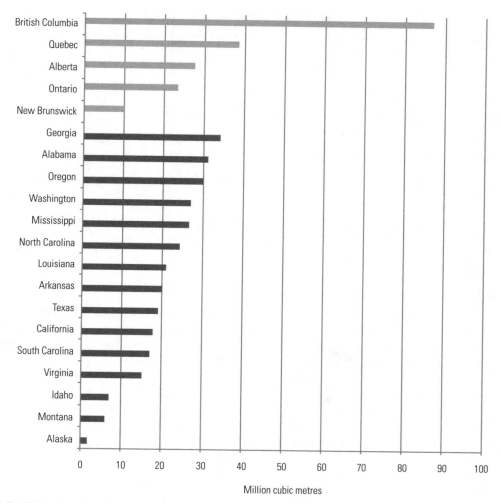

Note: The 2005 date for roundwood harvest was chosen to coincide with the date of the most recent FAO data available, in order to facilitate standardized comparison across countries.
Sources: CCFM, 2009; USFS FIA, 2009

Figure 3.2 *Total roundwood harvest of select Canadian provinces and US states 2005*

volume of timber harvest. The US sample also includes federal forestlands managed by the US Forest Service, as these cover a large percentage of the forest area within the US western case study jurisdictions. All of the selected US states are located either in the west or the southeast, reflecting the relative distribution of forest area and timber production in the country.

The total forest cover of the sub-national case studies represents about 57 per cent and 71 per cent of US and Canada respectively. These same states and provinces account for roughly three-quarters of forest production in both countries.

Table 3.1 *Forest ownership, Canada*

Case study	Private Total	Non-industrial	Industrial	Public Total	Provincial	Other public
BC	3%	3%	0%	97%	96%	1%
AB	4%	4%	0%	96%	87%	9%
ON	10%	9%	1%	90%	89%	1%
QC	11%	9%	1%	89%	89%	0%
NB	50%	29%	21%	50%	48%	2%
Canada	7%	6%	1%	92%	78%	14%

Source: CCFM, 2008

Figures 3.1 and 3.2 illustrate forest area and harvest volumes for all 20 selected provinces and states.

Canada

An overview of forests and forest ownership

Forests cover about 310 million ha of Canada. The boreal forests lie between about 50 and 70 degrees latitude and span from the western Yukon to the easternmost province of Newfoundland. Among the major temperate forest types are, from west to east, coast, subalpine, montane, Great Lakes and Acadian forests (Smith et al, 2001).

Over 90 per cent of Canada's forests are publicly owned and most of the authority for managing these forests rests with provincial governments. These publicly owned lands are commonly referred to as 'Crown lands' in reference to Canada's status as a member of the Commonwealth of Nations, and hence, at least figuratively, a subject of the British monarch. In regards to private ownership, there are significant areas of private forest in the Maritime Provinces, including the case study province of New Brunswick. In the Maritimes, the ratio of public to private forestland is more comparable to that of the US northwest than to the other Canadian provinces.

Some of the private forests in Canada are highly productive. In New Brunswick, private forests account for 50 per cent of forest ownership and produce 55 per cent of total timber volumes. In Quebec the contrast is greater. Private forestlands cover only about 11 per cent of Quebec's forests, but generate 21 per cent of total harvest volumes, thereby qualifying for inclusion in our policy comparison.

Native forests

Canada's boreal forests have been identified as the world's second largest 'frontier'[2] forests (Bryant et al, 1997). They are mostly coniferous and predominant species include black spruce (*Picea mariana*), tamarack (*Larix laricina*), balsam fir (*Abies balsamia*) and jack pine (*Pinus banksiana*). Aspen and poplar (*Populus* spp.) are the most common broad-leaved taxa (Gray, 1995; Smith et al, 2001).

British Columbia's coast forest region is the country's most productive forest type, extending from highly populated southwestern BC and southern Vancouver Island to the more rugged and remote central coast and Queen Charlotte Islands. These forests are predominantly coniferous, with common species including western hemlock (*Tsuga heterophylla*), western red cedar (*Thuja plicata*), Douglas fir (*Pseudotsuga menziesii*) and Sitka spruce (*Picea sitchensis*).

Other key temperate forests include the western interior lodgepole pine (*Pinus contorta*) and spruce (*Picea* spp.) forests of BC and Alberta. In southern Ontario and Quebec are found the mixed forests of the Great Lakes–St Lawrence region, with common conifers including eastern white and red pines (*Pinus strobus* and *resinosa*) and eastern hemlock (*Tsuga canadensis*) and broadleaf species such as yellow birch (*Betula alleghaniensis*) maple (*Acer* spp.), oak (*Quercus* spp.), aspen (*Populus* spp.), ash (*Fraxinus* spp.) and elm (*Ulmus* spp.). Common species in the Maritime or 'Acadian' mixed forests of New Brunswick include red spruce (*Picea rubens*), eastern pines (*Pinus strobus* and *resinosa*) and eastern hemlock (*Tsuga canadensis*), as well as diverse broadleaves such as yellow birch (*Betula alleghaniensis*), sugar maple (*Acer saccharum*) and ash (*Fraxinus* spp.).

Climate change and the legacy of a history of fire suppression are major threats to the Canadian forestry sector. The most spectacular manifestation of these has been a massive mountain pine beetle epidemic in the BC interior forests over the past decade. The beetle outbreak has been prolonged and compounded by mild winter weather during key phases in the beetle's life cycle, coupled with a legacy of fire suppression that has increased the average forest age within an otherwise fire-prone ecosystem. By 2009 the mountain pine beetle had impacted an estimated 15 million hectares of BC's interior lodgepole pine forestland and 620 million cubic metres of wood volume. It is estimated that by 2013 the mountain pine beetle will have killed 80 per cent of the merchantable timber in BC's central and southern interior (British Columbia, 2007). The beetle has also recently moved east into Alberta and south into parts of the US west (Logan and Powell, 2001; NRC, 2007b; Kurz et al, 2008).

Planted forests and plantations

According to an FAO working paper on planted forests and trees, Canada had established 10.2 million hectares of semi-natural planted forests as of 2005, reflecting the dominant method of regeneration following harvesting. No forest plantations were reported (FAO, 2006b).

Other data sources, besides the FAO, indicate modest plantation development in southern Canada. This includes small areas of poplar plantations located mostly on private lands in southern Quebec, eastern Ontario, Saskatchewan, southern British Columbia and Vancouver Island. Other species used in plantations include Norway spruce (*Picea abies*), red pine (*Pinus rositosa*), Larch (*Larix* spp.), White spruce (*Picea glauca*) and Trembling aspen (*Populus arunculus*) (NRC, 2008).

Canada's relatively cold climate and short growing seasons limit the country's global competitiveness in intensive plantation production. Nevertheless, the

Canadian Council of Forest Ministers has explored the potential for greater planta-tion development through the Forest 2020 Plantation Demonstration and Assessment Initiative. This initiative, completed in March 2006, revealed consider-able interest among Canadian investors in plantation development, but the findings suggested that 'while fast growing forest plantations can provide a range of timber supply and carbon benefits, market mechanisms alone are generally not high enough to drive significant amounts of private investment across the country' (NRC, 2008).

Forest governance and policy

All Canadian case study jurisdictions have established extensive legal frameworks for forest management of Crown lands. Forest Acts have been enacted in all case study provinces, with many dating back to the 19th century. The legal responsibilities contained in these Acts and associated regulations are handled by Provincial Forest Ministries.

Private forestlands in New Brunswick and Quebec are generally not subject to the Forest Acts and are governed by separate provincial and municipal-level legisla-tion. However, New Brunswick's Clean Water Act includes legally binding requirements for forestry activities on both Crown and private lands. As with all case studies in this book, our analysis does not include sub-provincial regulations (see Chapter 1), although local government policies may sometimes entail substantial additional requirements.

Wood Product Marketing Boards play an important role in marketing and forest stewardship for private woodlots in some provinces. Marketing Boards in both Quebec and New Brunswick have produced voluntary best management practice (BMP) guidelines.

Forest practice systems

The primary method for allocating harvest rights on Crown forestlands in Canada is through large-scale, long-term licences. This may involve the allocation of a defined forest area, or, as is common in British Columbia, the allocation of a set volume of timber. In addition, provinces have established mechanisms to promote small-scale licensing including both short-term area or volume-based tenures and longer-term woodlot licences.

All licensees are required to prepare management plans that align with strategic regional land use plans overseen by the province. Depending on the province and type of licence, these required plans may include longer-term (e.g. 20-year) plans as well as shorter-term planning documents, such as silvicultural prescriptions and road plans. Public participation is required for strategic level land use planning and mid-term management planning, as well as operational planning.

On private forests in New Brunswick and Quebec there are no universal require-ments for management plans or timber harvest permits. However, management plans are prerequisite for various tax credits and management subsidies and permits may be required at sub-provincial levels. Wood Product Marketing Boards provide voluntary technical guidance and marketing assistance.

Forest production and trade

British Columbia is Canada's leading forest products producer by volume, followed by Quebec, Ontario and Alberta (see Figure 3.2 above). In BC, recent harvest patterns have been impacted by the mountain pine beetle epidemic. The salvage of beetle infested trees and efforts to contain beetle spread have supported accelerated annual cut rates in interior forests. However, in 2006–2008 unfavourable market conditions contributed to an overall decline in harvest volumes. Such intersecting factors as the beetle epidemic, market fluctuations and changes in social priorities make it difficult to predict future volume yields. However, the combination of declining old growth stocks and large beetle losses have generated predictions of a medium term 'fall down' of BC's production levels pending forest recovery (Marchak et al, 1999; Pierce, 2001; British Columbia, 2007).

Harvests in boreal forests have been generally on the increase, including extensive harvest for pulp as well as solid wood (Smith et al, 2001). For Canada's eastern forests, access to US markets has been declining due to increasing international competition (Roberts, 2006; APEC, 2007).

The majority of Canada's wood production is exported. In 2006 (the year used for our global-scale statistics), exports comprised about 69 per cent and 66 per cent respectively of its hardwood and softwood lumber production, 94 per cent of its newsprint and 85 per cent of its printing and writing paper production. In 2008, these figures had dropped significantly, amounting to 44 per cent, 59 per cent, 85 per cent and 84 per cent respectively. The US was the leading importer, in particular for softwood lumber (82 per cent of Canadian production in 2006) and printing and writing paper (92 per cent of Canadian production in 2006) (NRC, 2007b).

This extensive trade with the US has been periodically disrupted by a 'softwood lumber dispute'. US timber companies, concerned with the major influx of relatively low cost sawn wood from Canada, have periodically and successfully lobbied the US government for the imposition of import quotas and duties on Canadian timber. This ongoing conflict has resulted in extensive bilateral negotiations as well as multilateral appeals to the North American Free Trade Agreement (NAFTA) and the World Trade Organization (WTO). These negotiations and appeals have led to fluctuating agreements between Canada and the US (including the recent Canada–US Softwood Lumber Agreement of 2006), involving variable levies and

Table 3.2 *Roundwood volume by ownership type, Canada*

Case study	Private	Provincial
Ontario	9%	91%
Alberta	11%	89%
British Columbia	11%	89%
Quebec	21%	79%
New Brunswick	55%	45%
Canada	19%	81%

Source: CCFM, 2008

even adjustments to Canadian forest policy (Cashore, 1997b; Yin and Baek, 2004; Mach and Shaw, 2007).

Indigenous and community forestry

The indigenous peoples of Canada are commonly referred to as 'First Nations' reflecting the sovereign authority of aboriginal governance. Between 1871 and 1921 11 numbered treaties were signed between the Canadian government and various First Nations. These treaty lands currently account for 0.39 per cent of all Canadian forestlands and are held in trust by the federal government (NRC, 2007a).

The first of the 'modern' First Nations agreements was the 1975 James Bay Agreement between the Cree and Quebec provincial government. The process of operationalizing the rights embedded in this agreement took years to unfold; in 2002, in response to a dispute over a hydroelectric project, *La Paix de Braves* (The Peace of the Braves) was signed with provisions for revenue-sharing and joint management of resource extraction activities such as mining, forestry and hydro-electric power generation.

There are still large areas of land in the west and the north of Canada that are not covered by numbered treaties or First Nations agreements, including most of British Columbia's public forestlands. As a result, the majority of forestlands in BC are currently under aboriginal tenure dispute. Based on pivotal federal court decisions, in 1993 the federal and BC provincial governments and the First Nations[3] Summit established a BC Treaty Commission to facilitate the development of treaties within the province. As of 2008, there were 58 First Nations participating in the BC treaty process, 41 of whom are in the stage of negotiating agreements-in-principle and eight who have signed agreements and are in the final stage of negotiations (British Columbia, 2008a).

During the 15 years since the establishment of the Commission, only one treaty, the Nisga'a Treaty, has been ratified. This treaty was negotiated outside the BC treaty process. Among the many components of the Nisga'a Treaty are the right to self-government and the authority to manage the land and resources across a traditional territory of 2019 square kilometres in northwestern BC (INAC, 2008). Should other treaties or similar agreements follow suit, as seems likely from recent political and legal developments, these are likely to have increasingly major impacts on the nature and structure of forest governance in the province.

With regard to aboriginal shares in access to Crown timber, a study completed in 2003 found that aboriginal groups and individuals held licences for about 7 million m^3 per year of Crown timber, or 4.1 per cent of the total national industrial allocation. The largest total volumes were allocated in British Columbia, followed by Quebec, Ontario and Alberta. British Columbia also led our case studies in terms of percentages allocated to aboriginal entities (6.1 per cent). British Columbia's new Community Forest tenure system (discussed below) has been identified as opening new avenues for aboriginal communities to 'operate in the forest sector as collective entities' (NAFA, 2003).

Community forestry initiatives have been adopted to varying extents, and on varying scales, among all of the Canadian case study provinces (see for example

Allan and Frank, 1994; Duinker et al, 1994; Haley and Luckert, 1998; Betts and Coon, 1999; Clark et al, 2003; Teitelbaum et al, 2006). In a recent national survey, Teitelbaum et al found 116 community forests of various kinds spread across four provinces. This figure was based on a carefully bounded definition of 'community forest' that excluded privately owned forests and aboriginal tenures and emphasized locally based control over decision-making, 'working forests' (i.e. forests where commercial timber harvest was one of the primary activities) and local benefit-sharing. About 60 per cent of the community forests thus defined occurred on Crown lands, and the remaining 40 per cent operated on land owned freehold by local governments.

The vast majority of community forests identified by Teitelbaum et al are located in Quebec, Ontario and British Columbia. Only one was noted in New Brunswick and none in Alberta. Quebec had the largest number, totalling 52 community initiatives with tenure or primary jurisdiction over a forest area. These took the form of either Territorial Management Agreements among local governments or Forest Management Contracts on Crown land. In Ontario, 51 initiatives were identified, mostly involving local governments and/or watershed-based municipal conservation authorities (Teitelbaum et al, 2006). In British Columbia, there are 33 active community forests, covering both Crown Forest tenures, and municipal lands, and totalling 900,000ha; it is envisaged that more than 1 million cubic metres of wood could be available from these forests (British Columbia, 2009). BC's community forestry arrangements variously allow management by community groups, First Nations and local governments.

The United States

An overview of forests and forest ownership

Forests cover about 303 million ha, or 33 per cent of the US land base. The majority of this forested area is located along the west and east coasts, the southeast, the northern lake states, the western Rocky Mountains and south and central Alaska.

The current distribution of land ownership in the US is a legacy of European settlement patterns. The eastern seaboard was the first to be colonized, resulting in large-scale deforestation and the establishment of private farms. Settlers then moved westwards to farm the fertile soils of the interior Great Plains. Settlers began to arrive on the west coast in substantial numbers by the 19th century, clearing forests to establish farms, as well as to finance the development of railways and generally supply the growing economy. By this time, concern was growing about uncontrolled resource exploitation and 'waste' of the nation's remaining forest resource, prompting the establishment of a number of federal land designations, including the US Forest Service (USFS) in 1905.

As a result, the vast majority of forestlands in our US southeastern case study states are privately owned and our policy analysis is restricted to private lands in this region. In our western case study states, USFS and private ownership are both dominant forest ownership types and thus both are addressed in our policy analysis. With the exception of protected areas in Alaska, no other forest ownership type

Table 3.3 *Forest ownership, United States*

Case Study	Private Total	Non-industrial	Industrial	Public Total	USFS	State	Other
Idaho	15%	10%	6%	85%	75%	5%	4%
Montana	27%	20%	7%	73%	63%	3%	7%
Alaska*	28%	28%	0%	72%	9%	19%	44%
Oregon	36%	18%	18%	64%	48%	3%	12%
California	43%	35%	8%	57%	43%	2%	11%
Washington	45%	25%	20%	55%	37%	10%	8%
Arkansas	81%	57%	24%	19%	13%	2%	3%
Virginia	84%	74%	10%	16%	10%	2%	4%
North Carolina	87%	75%	12%	13%	6%	2%	4%
Mississippi	90%	72%	17%	10%	6%	2%	3%
South Carolina	90%	71%	18%	10%	5%	2%	3%
Georgia	90%	72%	18%	10%	4%	1%	5%
Louisiana	91%	62%	28%	9%	4%	2%	3%
Alabama	95%	57%	24%	5%	3%	1%	2%
Texas	95%	75%	20%	5%	3%	0%	1%
US	58%	49%	9%	42%	20%	8%	14%

Note: 'Forest industry' (or 'industrial' forest) is defined as: 'An ownership class of private lands owned by companies or individuals operating wood-using plants' (Smith et al, 2004, p12). In addition, some owners of native trust lands may own wood-using plants but these are generally not included in the classification of 'industrial' forests.
Source: Smith et al, 2002

covers 20 per cent or more of the forest area and/or forest harvest volume of our case study states. Thus only USFS and private land policies are included in our case study comparisons.

Private forest ownership has been shifting significantly in recent years, resulting in increasing intensity of production in some areas, coupled with greater emphasis on non-timber objectives in others. Recent trends include large-scale divestiture of forestlands by the forest industry, involving company conversion or land sale to investment organizations focused on real estate and/or timber. Another ongoing trend is the subdivision of family forests into increasingly smaller properties, sometimes followed by conversion to non-forest residential and/or commercial developments. The traditional forestland ownership categories of 'industrial' and 'non-industrial', which refer only to the presence of a vertically integrated company mill, do little to capture these trends. Many investment organizations engage in intensive commercial forestry but do not own mills and hence are classified as 'non-industrial'.

Native forests

Forests of the western US are predominantly coniferous, with key timber species including – at the northern ends – lodgepole pine (*Pinus contorta*) and spruce (*Picea* spp.) and – further south, from southeastern Alaska to northern California – Douglas fir (*Pseudotsuga menziesii*), Ponderosa pine (*Pinus ponderosa*), hemlock (*Tsuga heterophylla*), true firs (*Abies* spp.) and western red cedar (*Thuja plicata*). The coastal redwoods *(Sequoia sempervirens)* of southern Oregon and northern California are also of major commercial value and include the tallest trees ever recorded (Koch et al, 2004) and among the oldest (Namkoong and Roberds, 1974; Waring and Franklin, 1979). The eastern forests support a larger proportion of hardwoods, including oak (*Quercus* spp.), maple (*Acer* spp.), ash (*Fraxinus* spp.) and poplar (*Populus* spp.), among many others, as well softwoods such as white and red pine to the north (*Pinus strobus, resinosa*) and shortleaf, loblolly, slash and longleaf pine to the south (*Pinus echinata, taeda, ellottii, palustris*).

As in Canada, climate change and the historical legacy of fire suppression are the leading forest threats in the US. These factors have contributed to severe forest fires and beetle outbreaks in the west. Warmer than usual winter weather has also contributed to pest problems elsewhere in the country (Joyce et al, 2001; Spencer et al, 2002).

Planted forests and plantations

The figures for 'planted' forests provided by the Forest Inventory and Analysis Program of the USFS appear to encompass the FAO categories for both planted semi-natural and plantation forests. According to this data, planted forests accounted for 18 per cent of forests in the south, 4 per cent in the west and about 3 per cent in the north (Smith et al, 2004).

In the US southeast, the most common plantation species is loblolly pine (*Pinus taeda*), although slash pine (*Pinus ellottii*) and other species are also used. The increase in the area of planted pine – and the intensive management practices associated with it – comprises perhaps the most controversial forestry issue in the US south (Wear and Greis 2002). The US Forest Service has conducted extensive research on the topic, including the production of a comprehensive Southern Forest Resource Assessment (SFRA).

According to the USFS SFRA, between 1953 and 1999, natural (i.e. non-planted) pine forests decreased from about 29 million ha to 13 million ha, while the area of planted pine increased from less than 1 million ha to over 12 million ha. By 1999, the proportion of all southern pine accounted for by planted stands increased from 11 to 47 per cent. Partly as a consequence, longleaf pine (*Pinus palustris*) forests, once a dominant forest type, have been reduced to one-third of their original range. This trend is expected to continue in our case study states, with the greatest losses occurring in North and South Carolina, each projected to lose 30 per cent and 35 per cent respectively of their remaining non-planted forests by 2040 (Wear and Greis, 2002).

In our western case study states there are two common types of planted tree farms, with only the latter type consistently classified as 'plantation'. The first type consists of relatively intensely managed native tree species planted mostly on industrial forestlands. In comparison with the 'plantations' of the US southeast, the rotation age of these tree farms tends to be longer, with less heavy use of chemicals and machinery, due in part to the cooler climate and more rugged topography. The second type of planted tree farm consists of plantations of hybrid poplar *Populus* spp. and other species managed as agricultural crops.

There are over 40,000 hectares of poplar plantations in Oregon and Washington, with the area split about evenly between the two states. About two-thirds of the poplar plantations are located to the east of the Cascades and one-third in the lower Columbia River basin (Moser, 2002; Shock et al, 2002; Stanton et al, 2002). Most of the poplar plantations are being established on land previously used for pasture or agricultural crops (Stanton et al, 2002; Biggs, 2003). The land is often owned or leased by paper companies and managed for the production of fibre for paper manufacturing. The market has been diversifying, however, to include other products such as engineered lumber (Stanton et al, 2002).

Forest governance and policy

Forest practices on private lands are governed by a wide variety of state laws and administrative rules. All of the case study states have established separate forest governance entities, whether they are agencies, divisions, departments and/or commissions. In addition, some states have established elected or government-appointed forestry boards. Some forestry issues may also involve interactions with a host of other environmental and administrative agencies.

Both county-level and federal policies frequently overlap with state authority. County-level forestry ordinances are commonplace in many states; however, as consistent with all of our case studies, we do not include such local-scale policies in our standardized comparison of forest practice rules. At the federal level, the US government holds authority over 'navigable waterways' and has taken legal measures to address air and water pollution, such as the reduction of non-point source pollution through the Clean Water Act of 1977. Many state laws were established to accommodate, or prevent, local and/or national law-making (Cashore, 1997a).

Eight out of fifteen case study states have enacted some form of general Forest Act or law (including Alaska, California, Idaho, Mississippi, Montana, Oregon, Virginia and Washington). In most cases these laws require, at a minimum, notification of timber harvest, submission of management plans, and/or harvest permits. In some cases state Forest Acts have set the stage for detailed forest practice regulations.

In addition to, or in place of, detailed forestry regulations, many case study states have developed voluntary 'best management practices' guidelines. While compliance is often voluntary, observance of best management practices (BMPs) may provide important legal protection for complying with federal laws. Many of the states' BMPs form an official part of non-point source management programmes developed under the 1972 US Clean Water Act section 319 (1987) (Ellefson et al, 2004). The implementation of BMPs, however, generally involves agencies not directly responsible for environmental regulation. For example, the Alabama Forestry Commission (1993) states explicitly that as the 'lead agency for forestry in Alabama' it is 'not an environmental regulatory or enforcement agency' (Ellefson et al, 2004, p1), but rather '[avoids] environmental problems through voluntary application of preventative techniques' (Ellefson et al, 2004).

The US Department of Agriculture (USDA) Forest Service is responsible for governing the majority of US federal forestlands. Because US Forest Service (USFS) forests (referred to in this book as 'National Forests') are largely independent of state control, we treat the USFS as a separate jurisdiction.

US National Forests are governed by a succession of key forestry Acts and regulations, as well as federal environmental legislation, such as the Endangered Species Act of 1973. As will be discussed later in this analysis, the sum total of this legislation has led to dramatic harvest reductions from US National Forests. Extensive public consultation and planning requirements must be met prior to logging in National Forests, although the exact nature and scope of these requirements has been a matter of great controversy and accompanying policy fluctuation.

Amongst the forest practice policies relevant to this analysis of USFS forestlands are those established by the northwest Forest Plan for Region 6. Region 6 covers Washington, Oregon and northern California from the west coast to the eastern side of the Cascade mountain range.

Forest practice systems

Each US National Forest must prepare a forest-wide management plan involving extensive public input. The plan includes timber inventory and sustained yield levels, strategies for the provision of multiple goods and services, strategies for the protection or rehabilitation of endangered species and habitats, as well as compliance with other relevant federal regulations.

The USFS has experienced continued staff declines since the early 1990s. An increasing percentage of responsibilities are now completed through contracts with private parties, including much of the silvicultural work. Timber is sold on the stump to private forestry firms and/or logging companies through competitive auction.

The forest practices systems for state-owned public forests and for private forests vary by state. Some states, and in particular those with Forest Acts, laws, or similar non-discretionary forest legislation, require management plans or cutting permits prior to harvest, while others do not. In addition, management plans may be a prerequisite for tax credits or management subsidies and cutting permits may be required at the county or municipal level.

Forest production and trade

The northwest and southeast are the top two timber harvesting regions, with the southeast accounting for an increasing share. The last 25 years have seen a shift, nationally, away from reliance on old growth rainforest in the Pacific northwest to increased use of intensively managed plantations.

About three-fifths of US production consists of softwoods, which dominate western production and account for a little more than three-fifths of southern production. The leading hardwood producing regions are the south, the lake states and the northeast (USDA Forest Service, 2009). Southern pine plantations, while occupying only about 16 per cent of the south's timberland area in 1999, accounted for 43 per cent of all softwood growth and 35 per cent of all softwood removals (Wear and Greis, 2002).

Table 3.4 *Percentage of roundwood production by forest ownership, United States*

Case study	Private Total	Other private	Forest industry	Undifferentiated private	Public Total	USFS	Other public
Idaho	75%	44%	31%		25%	9%	16%
Washington	80%	26%	54%		20%	2%	18%
Montana	82%	n/a	n/a	82%	18%	14%	4%
Oregon	85%	43%	42%		15%	5%	10%
Alaska	86%	86%	0%		14%	9%	5%
California	89%	34%	54%		11%	8%	3%
South Carolina	94%	68%	26%		6%	1%	5%
Arkansas	95%	51%	44%		5%	3%	2%
Virginia	96%	75%	22%		4%	1%	3%
Georgia	97%	68%	29%		3%	0%	3%
Mississippi	97%	76%	22%		3%	1%	2%
North Carolina	97%	79%	18%		3%	0%	3%
Louisiana	98%	56%	42%		2%	0%	2%
Texas	98%	60%	38%		2%	0%	2%
Alabama	98%	72%	26%		2%	0%	1%

Source: John S. Vissage, Research Forester, Forest Inventory and Analysis, North Central Research Station, US Forest Service, 2004. Data compiled for authors upon request.

Public forestlands, and in particular federal forests, currently play only a marginal role in forest production. As illustrated in Table 3.4, private forests account for 80 per cent or more of total roundwood production in almost all case study states. Non-industrial ownerships are the single largest producing ownership category, although the relative importance of industrial ownerships (i.e. properties with wood processing plants) varies considerably by state. The ongoing sale of forest industry lands to investment firms has generally served to reduce the vertical integration of production chains.

Indigenous and community forestry[4]

Treaties have been signed with over 140 tribal nations in the US. However, there are still many tribes that have not received federal recognition and disputes continue over the legality and interpretation of many of the existing agreements.

The total area of US Native American treaty lands is approximately 21.3 million ha. Lands in the mainland 48 states[5] are held in trust by the US federal government and may not be sold without approval of the Secretary of the Interior. The Bureau of Indian Affairs (BIA) of the US Department of the Interior is the federal agency responsible for implementing fiduciary duties regarding the administration and management of Indian trust lands. There are two different types of trust ownerships:

1   communal reservations; and
2   individually owned Indian allotments which may be held by numerous heirs of
    the original owner.

Forests cover about 7.3 million ha of Indian trust lands, with about 2.4 million ha classified as commercial timberlands. This amounts to about 2.4 per cent of total forestland ownership. Yet over the last five years, timber harvest from Indian lands has equalled about 30 per cent of harvest from national forestlands, while occurring in an area only one-twelfth the size of the USFS's operable land base (Vittello, 2008).

Tribal forests are distributed within 286 reservations across the US, with the heaviest concentrations in the lake states, intermountain west, southwest and north-west (Milakovsky, 2009). They are often intermixed with federal lands, including National Forests, National Parks and Bureau of Land Management parcels (Gordon et al, 2003). The 1975 Indian Self-Determination Act greatly expanded opportunities for tribal autonomy in forest management on tribal lands. This has led to a wide variety of governance arrangements, ranging from high capacity tribal forestry departments, to operations managed almost exclusively by the BIA (Milakovsky, 2009).

The state of Alaska contains the largest areas of Native tenure, granted in 1971 through the Alaska Native Claims Settlement Act (ANCSA). Currently 86 per cent of private forestlands in Alaska are held by Native Corporations (Alaska, 2007). Oil generally provides the largest revenue for these Corporations, although many are also engaged in timber harvest.

Indian trust lands are subject to federal environmental laws, such as the Endangered Species Act, and an approved forest management plan is required prior to timber harvest. Alaskan ANCSA lands, in contrast, are subject to the same require-ments as other private forest holdings.

In addition to indigenous tenures, the US supports an active community forest movement (Baker and Kusel, 2003; McCarthy, 2006; Charnley and Poe, 2007; Danks, 2009; McDermott, 2009). Community forestry in the US follows diverse forms and antecedents, ranging from centuries-old community-owned forests in the northeast to initiatives surfacing in the 1990s as an alternative to industrial forest management and a means of resolving environmental conflict on public forestlands in the west. In Appalachia and the southeast, community forestry efforts are geared at supporting the economic viability and collective social capacity of small private landowners and non-timber forest product producers and processors. Across the country, the types of implementing institutions include small, locally focused rural and urban-based non-profit groups, regional advocacy and economic development organizations, private–public collaboratives, forestry cooperatives and forest trusts and easements.

Without a national legal framework, these diverse forms of community forestry have in common the aim to expand the local share in decision-making about forest management and to increase its contribution to social well-being and ecosystem health. In some cases, they have contributed to broader institutional reforms both through the examples they set locally and through organized advocacy. Community forestry groups have pioneered innovations in the governance of public forests through informal collaborative groups, formal advisory committees and enhanced public participation processes. Some of the federal policies and legislation that community forestry groups have influenced include the Secure Rural Schools Act, which allocates Forest Service funds to projects prioritized by communities, the National Fire Plan, which mandates community wildfire protection plans, and a new federal authority under which the Forest Service can engage local communities in 'stewardship contracting' (Danks, 2009).

Stewardship contracting provides a particularly illustrative example of some core concepts common to community forestry in the US and elsewhere (see for example Teitelbaum et al, 2006; McDermott and Schreckenberg, 2009), including ecosystem management (Grumbine, 1994) and locally-based governance (Ostrom, 1990; Folke, 2006). Focused on forest restoration activities, stewardship contracts are typically multi-year and thereby encourage longer-term commitment to environmental outcomes. Contracts are awarded on the basis of the anticipated quality of performance as well as price, are subject to multi-party monitoring, and are evaluated based on results, thereby incorporating local knowledge. By packaging together multiple activities, the aim is to enable more holistic, systems-based stewardship. By permitting the investment of product sales into further stewardship activities, coupled with these other factors, the intent is to enable local parties to win contracts and contribute to community health (Davies et al, 2008).

BIODIVERSITY CONSERVATION

Protected areas

Canada

According to 2008 UNEP-WCMC data, protected areas cover 8.1 per cent of Canada's land area, with a large percentage (6.7 per cent) in IUCN Categories I–IV, which generally do not allow extraction of natural resources (Chapter 2, Figure 2.15). The total figure is slightly less than that reflected in an official 2005 national status report, which reported 8.6 per cent protected, with a further 1.3 per cent in 'interim' status (Government of Canada, 2006).

As in many federated countries worldwide, protected areas in Canada are administered at multiple scales and involve a variety of different agencies. Federally administered protected areas, including those administered by Environment Canada and those by Parks Canada, together constitute 49.2 per cent of all protected areas in the country. In addition, provinces and territories administer about 49.3 per cent of the protected areas and the rest are distributed among aboriginal and private tenures (Government of Canada, 2006).

The distribution of protected areas across Canada is very variable, covering as much as 22.6 per cent of the Arctic Cordillera, compared, for example, with 7.4 per cent of the Boreal Shield and 0.4 per cent of the Mixed Wood Plain ecozone (Great

Lakes–St Lawrence Valley). Eleven out of fifteen jurisdictions, including all of our case study provinces, have developed protected areas strategies to increase the size and effectiveness of their protected areas network, with British Columbia the first to have 'substantially implemented' its strategy (Government of Canada, 2006).

According to the 2005 status report, provincial protected area strategies have focused greatest attention on protecting representative habitats. Other core issues, more variably addressed, include conservation of biodiversity hotspots, protection of habitats for species at risk, maintenance of ecosystem structure and function (including connectivity and adaptation, for example to climate change) and protection of wide-ranging migratory species. Thirteen jurisdictions, including all of our case study provinces, have conducted some form of gap analysis to assess priorities for future action. There have also been varying levels of coordination across jurisdictions, with BC and Alberta among three provinces to develop inter-provincial protected areas (Government of Canada, 2006).

At the national level, the Canadian Council on Ecological Areas (CCEA) works 'to facilitate and assist Canadians with the establishment and management of a comprehensive network of protected areas representative of Canada's terrestrial and aquatic ecological natural diversity' (CCEA, 2009). To this end, it has recently developed the Conservation Areas Reporting and Tracking System (CARTS) to provide a framework for nationally consistent, and publicly available, protected areas information (Government of Canada, 2006).

Canadian civil society activists have been instrumental in driving the creation of new protected areas. One illustrative case is the popularly dubbed 'Great Bear Rainforest', a mountainous area of coastal BC totalling about 6.4 million ha. A consortium of environmental groups, First Nations, local communities, industry and government participants was engaged in formulating proposals for this area. Among the outcomes of these proposals was a 2006 agreement by the BC government to permanently protect 1.8 million ha from logging, and to engage in reduced intensity harvest in the remainder of the area (Clapp, 2004; Shaw, 2004; British Columbia, 2008b).

The United States

According to 2008 UNEP-WCMC data, protected areas cover 25.4 per cent of the US land base, of which 10.4 per cent is in IUCN Categories I–IV (Chapter 2, Figure 2.15). However, UNEP-WCMC and its US partners consider the figures for IUCN Categories V and VI and unclassified as somewhat inflated and efforts are being made to address the issue before the release of the World Database on Protected Areas (WDPA) 2010 dataset (Coad, 2009).

Nevertheless, even accounting for data discrepancies, the US ranks fairly highly among our case studies for IUCN Categories I–IV owing to a number of factors, including the country's relatively long history of protected areas development. The US was a pioneer in the creation of national parks, with dual objectives of protection and recreation (HFS NPS, 2005). As is true in Canada and elsewhere, many other types of protected areas have also been established, including wilderness areas, national monuments, national recreation areas, wildlife refuges and the national rivers systems. The area of state and local parks is also substantial in some regions of the country. For example, Adirondack State Park in New York is the largest protected area outside Alaska (Adirondack Park Agency, 2009).

The US Geological Survey (USGS) Gap Analysis Program (GAP), launched in 1989, has increasingly served to coordinate federal, state and private efforts at biodiversity conservation (USGS GAP, 2009c). There are currently seven regional projects within GAP, including four that overlap our study areas: the southeast and northwest region GAPs and the Alaska and California Mapping Projects (USGS GAP, 2009b). All of these projects include among their objectives the assessment of gaps in the adequacy of current land management to ensure the protection of all common species and the development of strategies necessary to prevent such species from becoming threatened or endangered. To aid in these efforts, USGS GAP, which operates in partnership with numerous governmental and non-governmental organizations, has developed standardized systems for data collection and analysis of species and habitat distribution and is beginning to coordinate mapping for all regions of the US (USGS GAP, 2009c).

In April 2009, USGS GAP, on behalf of the Protected Area Database Partnership (PAD-US), published the first comprehensive national database of federal and state conservation lands (USGS GAP, 2009a). This includes protected areas by IUCN category, as well as GAP status codes 1–4. These latter codes classify lands according to the level of human disturbance allowed, from those with a mandate to maintain natural ecosystem processes and disturbance patterns, to those where there are no restrictions on conversion of forests to other uses.

In addition to formal government partnerships and consortia, civil society initiatives have long played a key role in driving forest conservation, including some recent US and Canadian partnerships. Among these are the ambitious 'Yellowstone-to-Yukon' (Y2Y) project in the west, involving a continuous network of protected areas from California to Alaska, and the 'Two Countries, One Forest' initiative in the northern Appalachians, aimed at protecting forest corridors from New England through the Canadian Maritimes (Bateson, 2005).

While there are many catalysts for protected areas establishment, it is arguably the retention of federal land ownership that is the largest predictor of protected areas location. Once private property rights were well established across the country, the largest additions to the protected areas network have involved changes in the designation of public, and in particular federal, lands. For example, the extent of protected areas on the Pacific Coast more than doubled between 1997 and 2007, largely due to the redesignation of large areas of Alaskan Bureau of Land Management Lands as National Parks and Wildlife Refuges.

With regard to protected areas on US private lands, there has been considerable growth in the use of land trusts and conservation easements in some parts of the country. The Land Trust Alliance has estimated the total area of US land trusts at about 15 million hectares (37 million acres) (Land Trust Alliance, 2009).[6] The conservation outcomes of these tenure arrangements are variable and depend on such factors as the inclusion of specific biodiversity protection measures and the quality of auditing (Hagan et al, 2005). In general, national and international

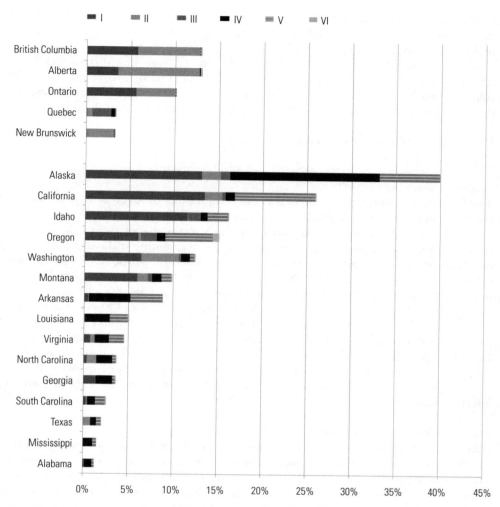

Note: Figure 3.3 reflects the most recent, nationally standardized data that were available for the case study states and provinces. The Canadian data reflects available figures as of 2005. Most of the US data has been updated to 2008, except for the northwest and California, which have not been updated since 2005 (Audin, 2009).
Sources: Government of Canada, 2006; PAD-US, 2009

Figure 3.3 *Protected areas by IUCN Categories I–VI as percentage of total land area for select US states and Canadian provinces*

reporting on protected areas by IUCN category excludes these private conservation arrangements.

Figure 3.3 illustrates the most recent data available on public protected areas by IUCN category in each of our Canadian and US case study jurisdictions. As is clear from the figure, there are significant differences in the extent and type of protected areas among the various province and state case studies. Alaska ranks clearly at the top in terms of total land area under protected status, followed by California and then Idaho. Three of the Canadian case study provinces and other US western states make up the remainder of the top half.

In contrast, New Brunswick and all of the US southeastern states except Arkansas have allocated 5 per cent or less of their land to protected status. The relatively small extent of protected areas in the US southeastern states contrasts with the high levels of biodiversity found in this region (Martin et al, 1993).

Protection of species at risk

Canada

Canadian federal legislation

The Canadian federal government has limited jurisdiction over provincial resource management. However, the federal government has enacted a central piece of legislation directed at species at risk, known as the Species at Risk Act (SARA). The Act signalled a new proactive and cooperative approach among industry and environmental stakeholders, who together developed policies and proposals for government consideration (Amos et al, 2001). SARA includes direct measures for federally owned lands and federal species (migratory birds and aquatic species) and provides 'fall back' policies for habitat protection in the absence of provincial government efforts. SARA also confers legal status on the previously advisory Committee on the Status of Endangered Wildlife in Canada (COSEWIC), which is responsible for preparing a list of threatened and endangered species. SARA leaves the federal government with more discretionary powers than does the US Endangered Species Act (ESA) (see below), placing greater emphasis on cooperation between interests in ensuring habitat recovery.

Provincial lands and regulations

Alberta handles the listing and protection of endangered species through its Endangered Species Conservation Committee (ESCC), established under the Wildlife Act (2000). The ESCC is strictly advisory in nature and was created to provide recommendations to the minister about which species should be established as endangered and to develop recovery plans for those species.

British Columbia has developed an 'Identified Wildlife Management Strategy' (IWMS). The strategy calls for identifying species at risk, and the extent of that risk, by using information from its conservation data centre. Wildlife Habitat Areas may then be designated for species whose protection requires special management measures. For species with very large ranges, the strategy calls for the establishment of Resource Management Zones, to be created through higher level planning processes. Conservation assessments and inventory and monitoring are required to assess the impact of conservation measures on species recovery. The implementation of BC's species protection strategy is limited, however, by the rule that the IWMS cannot reduce provincial harvest levels by more than 1 per cent. Exceptions to this rule may be made for species with particularly wide ranges, such as the mountain caribou, spotted owl, grizzly bear and marbelled murrelet (British Columbia, 2004a).

New Brunswick, Ontario and Quebec have all enacted provincial Endangered Species Acts. New Brunswick's first Endangered Species Act was enacted in 1976 and was later replaced by a new Act in 1996. These Acts list a growing number of animals and plants that are considered endangered, or regionally endangered, and require protection of these species and their critical habitats.

The Ontario Endangered Species Act calls for the protection of species of fauna or flora declared through regulations to be threatened with extinction. The Act states that the 'Lieutenant Governor in Council may make regulations declaring any species of fauna or flora to be threatened with extinction' and that 'no person shall willfully ... destroy or interfere with or attempt to destroy or interfere with the habitat of any species of fauna or flora, declared in the regulations to be threatened with extinction' (Ontario R.S.O., Chapter E.15, 1990a, Amended 1997).

Ontario also addresses habitat protection through the Crown Forest Sustainability Act, which calls for 'using forest practices that, within the limits of silvicultural requirements, emulate natural disturbances and landscape patterns' and conserve 'ecological processes and biodiversity'. The Act's approach to the maintenance of 'natural' habitats is essentially results-based, requiring that forest managers establish measures for monitoring and assessing the degree to which they have met the principles of sustainability set forth in the Act (Ontario, 1994).

Quebec enacted the 'Act respecting threatened or vulnerable species' in 1989. This Act established procedures for identifying species at risk and shoulders the provincial government with the responsibility to protect these species and their habitat. The Act requires both species protection and protection of habitat on all lands, public and private. According to the Act, 'No person may, in a wildlife habitat, carry on an activity that may alter any biological, physical or chemical component peculiar to the habitat of the animal or fish concerned' (Quebec, 2003a).

The United States

US federal lands

The US Endangered Species Act, enacted by Congress in 1973, has been considered a landmark piece of conservation legislation (see for example Kohm, 1991; Martin, 1994). It has also been a major source of public debate (Czech and Krausman, 2001). Administered by the US Fish and Wildlife Service (USFWS) and the National Marine Fisheries Service (NMFS), the ESA *requires* that these agencies list threatened and endangered species and their 'critical habitats'. The authority to list threatened or endangered species rests in the Secretary of the Interior or, as relates to the import or export of terrestrial plants, the Secretary of Agriculture. The Secretary's determination must be based 'solely on the best scientific and commercial data available' (section 4(b)(1)(A)), with explicit direction that the economic effects of such a decision not be given consideration. The strong, non-discretionary language of the ESA has proven a powerful tool for public interests to demand higher levels of environmental protection. Under the ESA, environmental groups have sought court injunctions against the US Forest Service, as well as sued private companies allegedly violating principles of the Act (Vogel, 1993, p256).

The ESA is not an entirely 'non-discretionary' piece of legislation, however, in that it creates avenues for legal exceptions. The Act empowers an 'Endangered Species Committee', which critics have labelled the 'God Squad' (Davis, 1992), with the authority to decide whether or not the 'economic and social benefits of the proposed action outweigh costs to the listed species'. In other words, the Committee holds the authority to create exemptions to the Act (Smith, 1993, p1039). This committee can only be established, however, when there are 'considerable' economic or social costs involved in implementing the Act and no 'feasible alternatives' exist (Smith et al, 1993, p1038).

Once a species has been listed, it is illegal under the ESA for a public or private landowner to 'take' that species. Section 3(18) of the Act defines the term 'take' to mean 'to harass, harm, pursue, hunt, shoot, wound, kill, trap, capture, or collect, or to attempt to engage in any such conduct'. In addition, the ESA specifically requires that the 'Secretary' create and implement recovery plans for the conservation and survival of endangered species.

The case of the northern spotted owl (*Strix occidentalis caurina*) listing provides a dramatic illustration of the Act's potential impact on federal lands. The combination of the Endangered Species Act and an old-growth forest-dependent owl led to a complete overhaul of the Forest Service approach to land management. As a result of the listing of the northern spotted owl as 'threatened', much of US federal land in the Pacific northwest is no longer considered accessible for harvest, and wood production from National Forests has dropped from nearly half to only 5 per cent of total US removals (Smith et al, 2004). The story of the northern spotted owl is well known and is often referred to as the model of environmental protection by environmental advocates. What have received less media attention, however, are the effects of the ESA on private land, from which 93 per cent of US softwood lumber is currently being harvested.

US private lands
While the enforcement of the US Endangered Species Act has resulted in a dramatic reduction in timber harvest across increasingly large areas of federal land, the Act's impact on private land management has been radically different. The federal-level ESA on private lands could be described as procedural rather than substantive or results-based. Section 10(a)(2) permits landowners to obtain an incidental take 'permit' that allows them to conduct management practices that harm threatened and endangered species and habitats, provided that the landowner also prepares a Habitat Conservation Plan that mitigates and minimizes the impacts of the taking. While public agencies may also apply for take permits, the USFS and other public agencies generally maintain a stricter interpretation of the federal ESA.

Figure 3.4 provides a graphic illustration of the differing impacts the 1992 spotted owl listing had on public versus private timber sales in Washington and Oregon.

In addition to the federal-level ESA, all of the case study states except Alabama have enacted state-level endangered species legislation. This legislation varies

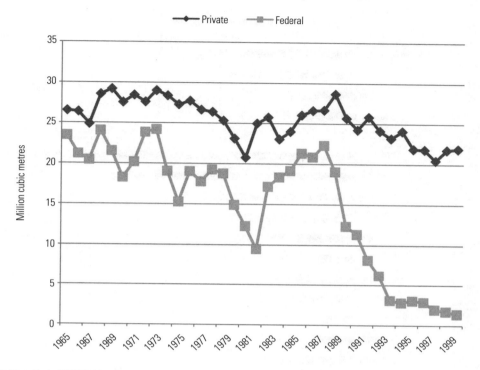

Note: 1 board foot = 0.00348 cubic metres.
Source: Warren, 2002

Figure 3.4 *Washington and Oregon timber harvest on private and federal lands 1965–1999*

significantly in the range of species covered and the extent of protection afforded, as well as the maximum civil or criminal penalties it entails (Pellerito and Wisch, 2008).

Thus the requirements of US endangered species legislation vary substantially both across land ownership types and between different sub-national regions. If we consider the known environmental distribution of endangered and threatened species (Figure 3.5), however, there is little evidence that these legislative differences correspond with environmental differences in the distribution of biodiversity. While major media, and judicial, attention has been focused on British Columbia and the Pacific northwest, other regions also contain many threatened and endangered species.

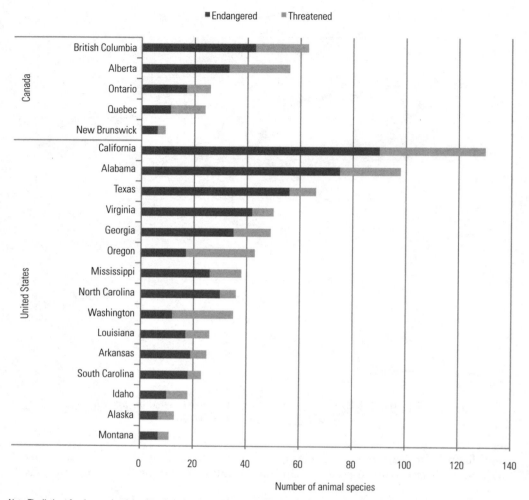

Figure 3.5 *Number of endangered and threatened animal species listed under the Canadian Species at Risk Act (SARA) and the US Endangered Species Act (ESA)*

Note: The listing of endangered and threatened species depends on a number of factors, including the level of research invested in identifying species and political negotiations. Thus the relative actual numbers of species at risk may vary significantly from those shown in this figure. On average, species richness increases closer to the equator, suggesting that natural habitats in the US southeast would be capable of supporting relatively high levels of biodiversity (Willig, 2000; Ihm et al, 2007).

Sources: US data: available at http://ecos.fws.gov/tess_public//pub/stateListing.jsp?state=AK&status=listed (last accessed 1May 2009)
Canadian data: available at www.sararegistry.gc.ca/sar/index/default_e.cfm (last accessed 1 May 2009)

FOREST PRACTICE REGULATIONS:
NATIVE FORESTS

Riparian zone management (Indicator: Riparian streamside buffer zone rules)

The US Clean Water Act has served as an important driver of riparian zone policies across the United States (Ellefson et al, 2004). The Act holds landowners liable for non-point source pollution of waterways that can be attributed to landowner negligence. As a result, even the 'voluntary' best management practices characteristic of some state riparian zone policies may hold legal implications. There is no federal legislation in Canada similar to the US Clean Water Act but, among our Canadian provincial case studies, a Clean Water Act has been enacted in New Brunswick.

On the Pacific Coast of Canada and the US, dramatic declines in wild salmon stock have driven some of the greatest controversy over riparian forest practices (Jankowski, 2000; Pacific Fishery Management Council, 2000; Washington Forest Protection Association, 2000; British Columbia, not dated). The late 1990s saw a flurry of new rulings listing numerous salmon stocks in Washington, Oregon and California under the Endangered Species Act. For example, in Washington alone, in 1999 the National Oceanographic and Atmospheric Administration (NOAA) Fisheries, National Marine Fisheries Service listed seven state salmon populations as 'endangered' or 'threatened'.

Several forest harvesting practices have been associated with reductions in Coho and other salmon populations (Ketcham, 1993; Northwest Renewable Resources Center, 1998). However, there has been little consensus over the relative impact of forestry versus other land management activities such as urban development, dams, pollution, commercial fishing and changes in ocean temperatures (Tschaplinski, 2000, 2004; Hinch, 2003). As we will see in the following analysis, the Pacific Coast jurisdictions have developed particularly complex riparian management policies under the influence of this ongoing controversy.

Figures 3.6–3.8 illustrate riparian zone rules for each of the Canadian and US case study jurisdictions. Table 3.5 compares the complexity of these requirements, including the total number of different buffer size classes per jurisdiction, as well as the number and range of bio-physical attributes considered in establishing the buffer zone requirements.

All case study jurisdictions in the US and Canada have developed substantive rules for buffer zone protection. However, these policies vary considerably in method (i.e. whether they are mandatory or voluntary) and in the extent of protection afforded, as well as in the type and complexity of bio-physical attributes used to determine buffer zone size. The most complex policies (see Table 3.5) are found on private lands in Washington state, and include a total of 20 different size specifications based on ten different categories of bio-physical attributes. Several of these attributes are unique to the Washington state classification system, including consideration of the type of management occurring outside the buffer area (for example,

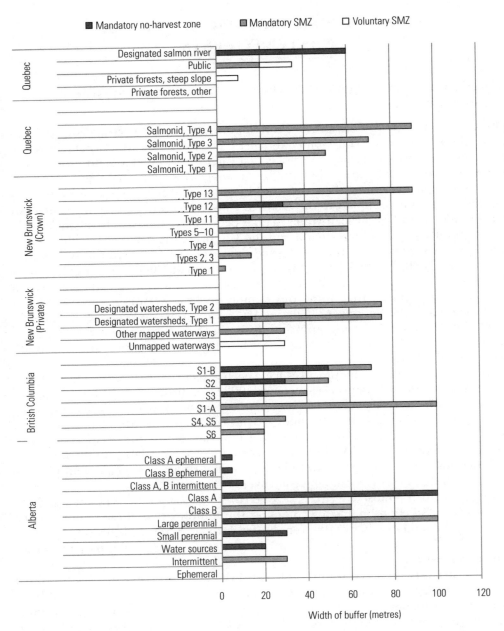

Figure 3.6 *Riparian buffer zone policies in Canadian case study jurisdictions*

Notes: Quebec private forestland BMPs recommend that culverts not 'reduce the width of the watercourse by more than 20 per cent, as measured by the natural high water mark' (Paquet and Groison, 2003).
Source: see Figure 3.8

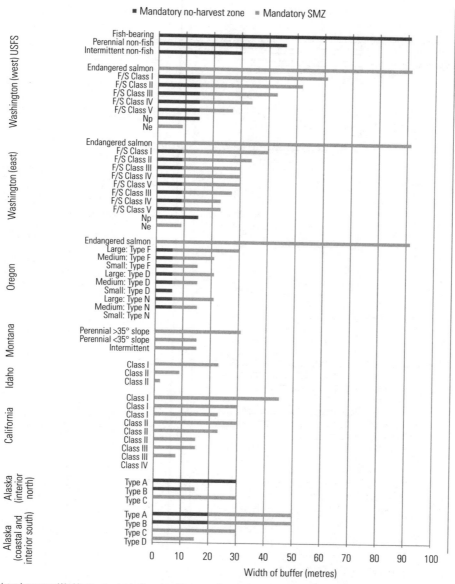

Notes: In western Washington, riparian buffers along Np streams (i.e. non-fish-bearing perennial streams whose rate of flow is less than or equal to 0.57 cubic metres per second (in other words streams that do not qualify as Washington 'shoreline')) must measure a minimum of 15 metres in width. The proportion of the stream for which the no-harvest rule applies depends on the distance of the Np stream from shoreline and/or fish-bearing streams.

In eastern Washington, 15-metre no-harvest zones are required along a portion of Np streams when clearcutting is used within the riparian special management zone.

Type N Waters are seasonal, non-fish habitat streams and are not located downstream from any stream reach that is a Type N_p Water. They must be physically connected by an above-ground channel system to Type S, F, or N_p Waters.

Source: see Figure 3.8

Figure 3.7 *Riparian buffer zone policies in US western case study jurisdictions*

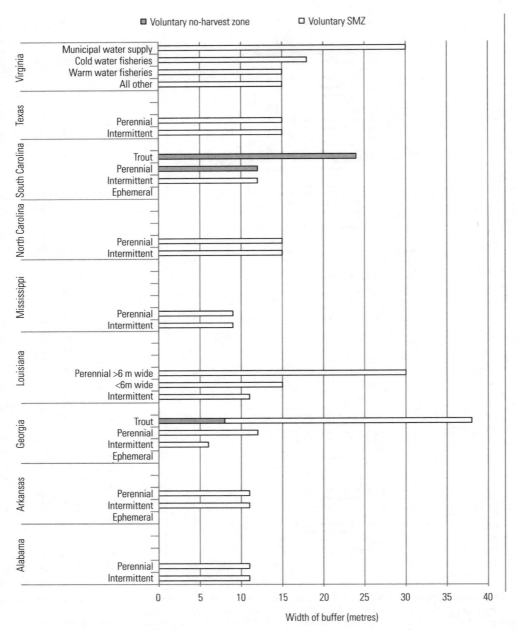

Figure 3.8 *Riparian buffer zone policies in US southeast case study jurisdictions*

Sources: for Figures 3.6–3.8 (for legislation in place as of 1 January 2007, the cut-off date for our standardized comparison): Alabama, 1993; Alaska, 2004; Alberta, 1994; Arkansas, 2002; British Columbia, 2004b; California, 2005; Georgia, 1999; Idaho, 1974; Idaho, 2000; Louisiana, 2000; Mississippi, 2000; Montana, 2001; Montana, 2006; New Brunswick 2004b; New Brunswick, not dated; North Carolina, 2003, 2006; Ontario, 1988; Ontario, 1991; Ontario Forest Industries Association 1998; Oregon, 1994; Paquet and Groison, 2003; Quebec, 1996; South Carolina, 1994; Texas, 2004; United States, 1994; Virginia, 2002; Washington, 2001

Table 3.5 *Number of different buffer size categories and number and type of attributes used to determine buffer zone sizes for each case study jurisdiction*

	No. of size distinctions	Fish	Season-ality	Width	Domestic water	Slope	Soil	Stream flow	Other	No. of attributes
Washington	20	1	1	1		1	1	1	4	10
New Brunswick (public)	13	1		1	1	1			4	8
Oregon	10	1			1			1		3
Alberta	10	1	1	1	1				2	6
California	9	1	1	1	1	1	1		2	8
Alaska	7	1				1				2
British Columbia	7	1	1	1	1	1			1	6
Georgia	4	1	1							2
South Carolina	4	1	1							2
Ontario	4	1				1				2
Virginia	4	1	1		1					3
New Brunswick (private)	4			1	1				1	3
Arkansas	3		1							1
Quebec (public)	3	1	1							2
Montana	3		1			1				2
USFS Region 6	3	1	1							2
Louisiana	3		1	1						2
Idaho	3	1		1	1					3
Alabama	2		1							1
Mississippi	2		1							1
North Carolina	2		1							1
Texas	2		1							1
Quebec (private)	2	1				1				2
Total # of cases		15	14	8	8	7	2	2	13	

clearcutting versus selective harvest). In general, the most common attribute shaping buffer size is the presence or absence of fish (considered in 65 per cent of the case studies), followed by seasonality (i.e. whether a stream is perennial or seasonal, considered in 61 per cent of the cases). Stream width, domestic water supplies, and/or slope are considered in roughly one-third of the cases.

The most restrictive requirements for riparian buffer areas are the very large 'no-harvest' zones found in all USFS watercourses, British Columbian provincial fish and domestic water channels equal to or greater than 1.5 metres wide and Quebec provincial salmon streams. Other jurisdictions with no-harvest buffers are New Brunswick (public and private), Alaska's Interior (public lands) and Coastal regions (public and private lands),[7] Oregon and Washington. Washington state, together with the USFS, are the only jurisdictions where no-harvest buffer zones are required on streams less than 1.5 metres, including some non-fish-bearing streams.[8]

Ontario, Montana, Idaho and California have established only mandatory Special Management Zones (SMZs), without accompanying no-harvest zones. SMZs as large as 90 to 100 metres are found in all Canadian public forests except Quebec, as well as on private lands in Washington and Oregon. Most mandatory SMZs involve some form of restriction on tree removal and machinery use, although the exact nature and extent of those restrictions varies by case study.

The southeastern US states and Quebec private forests, in contrast, have developed voluntary best management practices for buffer zone management. These BMPs include quantified buffer sizes, but the sizes are generally smaller with fewer management limitations than those found in other case study jurisdictions. As discussed above, adherence to these 'voluntary' BMPs holds some legal significance as proof of due diligence in meeting the requirements of the US Clean Water Act.

Roads (Indicators: Culvert size at stream crossings, road abandonment)

Road building and maintenance have been on the forest policy agenda in Canada and the United States for the past several decades. Debates have generally coalesced around two separate points of concern: the impacts of roads on water quality and the management of forest access.

With regard to water quality: road design, construction and maintenance began to gain increased attention in the 1980s, along with growing public awareness of 'non-point' sources of surface water contamination. As already mentioned, within the US this awareness has been heightened by the legal implications of the US Clean Water Act.

With regard to forest access, movements have arisen in Canada and the US to conserve 'intact' or 'roadless' forests. Various policy outcomes have emerged as a result, such as the Canadian Boreal Forest Initiative and the US 'Roadless Area Conservation Rule' for US National Forests. To a lesser degree, road decommissioning has been included in this policy arena as a means to further reduce human forest impacts.

Culverts at stream crossings

Two principal trends were found regarding culvert requirements in the US and Canada. Many states and provinces include nominal guidelines for culvert diameters based on peak stream flows that culverts must be built to accommodate. Mandatory requirements have been set by Alberta, BC, New Brunswick (public and private), Quebec (public), Alaska, California, Idaho, Oregon, USFS and Washington. Ontario does not have standardized, nominal rules governing culvert widths, but does nevertheless require that culvert size be 'adequate for fish passage' (Ontario 1990b).

In addition to, or in place of, peak flow requirements, many jurisdictions have established 'minimum' culvert sizes.[9] In the case of Quebec (private), Alaska, Idaho and Washington, minimum culvert diameter policies are mandatory, or what we classify as prescriptive (non-discretionary/substantive). In Alabama, Arkansas, Georgia, Louisiana, Mississippi, Montana, North Carolina, Texas and Virginia, diameter minimums are voluntary (discretionary-substantive).

The figures below illustrate the quantitative culvert sizing policies for all of the case study jurisdictions, distinguishing between mandatory or voluntary policies and peak flow and/or minimum diameter sizes.

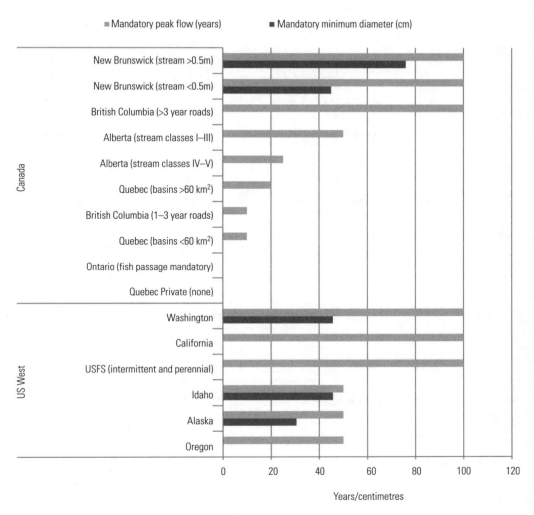

Sources: Alaska DNR, 2004; Alberta, 1994; British Columbia, 2004; California, 2005; Idaho, 1974; Idaho, 2000; Montana, 2001; New Brunswick, 2004; New Brunswick, not dated; Ontario, 1988; Ontario, 1990b; Ontario, 1991; Ontario Forest Industries Association 1998; Oregon, 1994; Paquet and Groison, 2003; Quebec, 2002; United States, 1994: Washington, 2001

Figure 3.9 *Culvert sizes at stream crossings: Canada and US west*

Road decommissioning

In terms of rules and guidelines for road closure, all of the Canadian provinces take a non-discretionary approach in requiring road decommissioning on provincial lands, as do Alaska, California, Idaho, Oregon, Washington and the USFS. Only Alberta, British Columbia, California and Washington, however, provide specific decommissioning prescriptions (for example, the requirement to remove all drainage structures and recontour the road). In contrast, policies in Alaska, Idaho,

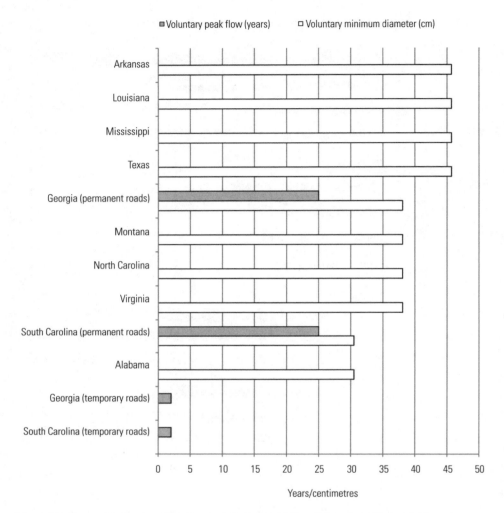

Sources: Alabama, 1993; Arkansas, 2002; Georgia, 1999; Louisiana, 2000; Mississippi, 2000; North Carolina, 2003, 2006; South Carolina, 1994; Texas, 2004; Virginia, 2002

Figure 3.10 *Culvert sizes at stream crossings: US southeast*

Oregon, USFS lands, Ontario and Quebec involve more generalized requirements (such as controlling for erosion) that allow room for discretion in determining the precise course of action. In order to distinguish the former, more prescriptive, non-discretionary policy from the latter, more flexible, policy we have classified the latter as a 'mixed' policy approach.

Road decommissioning standards that are voluntary in nature have been developed in all of the remaining case study jurisdictions, except for private lands in New Brunswick, Quebec and Alabama.

Clearcutting (Indicator: Clearcut size limits)

Clearcutting has been a predominant harvesting method in North America ever since the advent of enabling technologies in the early 20th century (Rajala, 1998). Clearcutting has fallen into increasing public disfavour, with the most heated controversies centred on large clearcuts in coastal old-growth forests of the Pacific northwest.

Figure 3.11 reveals that a number of jurisdictions along the west coast of North America have established limits for the size of the clearcut, including the province of British Columbia, the states of Washington, Oregon and California and the US Forest Service. Likewise, clearcut size limits have been established for provincial lands in Alberta, Ontario, Quebec and New Brunswick.

Figure 3.11 also reveals that the smallest (i.e. most restrictive) clearcut size limits are found on private lands in California, followed by US national forestlands, then British Columbia provincial lands and private lands in Washington and Oregon. The clearcut size limits on private lands in California range between 8.1 and 12.1 hectares, depending on harvest methods, with permission granted in some cases for 16ha. Limits on US national forests are 24.3 hectares in Douglas fir forest and 16.2 hectares in other forest types. BC has a maximum clearcut size policy of 40 hectares for the coastal and southern interior regions, and 60 hectares for northern interior regions. Washington and Oregon set limits of 48.5 hectares, with sizes up to 97 hectares possible with government approval.

The clearcut policy for Quebec provincial lands is particularly complex and difficult to capture in graphic form. The precise legal text from '*Standards of forest management for forests in the domain of the State, Regulation respecting, R.Q. c. F-4.1, r.1.001.1*. DIVISION VIII: SIZE AND LOCATION OF CUTTING AREAS AND APPLICATION OF SILVICULTURAL TREATMENTS' reads as follows:

> *74. In each of the 3 forest zones described in Schedule 1, the size of a single-block area of cutting with regeneration and soil protection or of the total area of the cut and residual strips of an area of strip cutting with regeneration and soil protection shall*
>
> *(1) in the hardwood forest zone a) be equal to or less than 25 hectares for at least 70% of the areas cut using those cutting methods; b) be equal to or less than 50 hectares for at least 90% of the areas cut using those cutting methods; and c) be equal to or less than 100 hectares for all areas cut using those cutting methods;*
>
> *(2) in the fir and mixed forest zone a) be equal to or less than 50 hectares for at least 70% of the areas cut using those cutting methods; b) be equal to or less than 100 hectares for at least 90% of the areas cut using those cutting methods; and c) be equal to or less than 150 hectares for all areas cut using those cutting methods;*
>
> *(3) in the spruce forest zone a) be equal to or less than 50 hectares for at least 20% of the areas cut using those cutting methods; b) be equal to or less than 100 hectares for at least70 % of the areas cut using those cutting methods; and c) be equal to or less than 150 hectares for all areas cut using those cutting methods.*

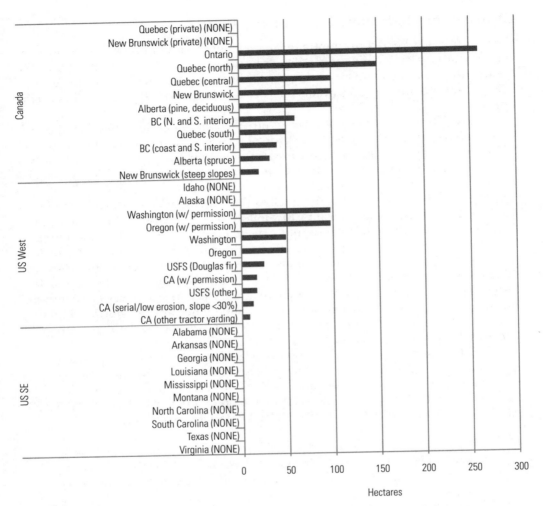

Notes: Alberta: Pine forest is defined as forest where pine makes up 40% or more of the stand, evenly distributed through Cutblock: from Harvest Plan Operating Rules.

BC (Coast and S. Interior): Coast Forest Region and Southern Interior Forest Region – Arrow Boundary Forest District; Cascades Forest District; Columbia Forest District; Headwaters Forest District, except the portion of the forest district that is in the Robson Valley Timber Supply Area; Kamloops Forest District; Kootenay Lake Forest District; Okanogan Shuswap Forest District; Rocky Mountain Forest District.

BC (N. and S. Interior): Northern Interior Forest Region and Southern Interior Forest Region – 100 Mile House Forest District; Central Cariboo Forest District; Chilcotin Forest District; the portion of the Headwaters Forest District that is in the Robson Valley Timber Supply Area; Quesnel Forest District.

California limits are smaller for Special Treatment Areas.

Sources: Alabama, 1993; Alaska DNR, 2004; Alberta, 1994; Arkansas, 2002; British Columbia, 2004; California, 2005; Georgia, 1999; Idaho, 1974; Idaho, 2000; Louisiana, 2000; Mississippi, 2000; Montana, 2001; New Brunswick, 2004; New Brunswick, not dated; North Carolina, 2003, 2006; Ontario, 1988; Ontario, 1991; Ontario Forest Industries Association 1998; Oregon, 1994; Paquet and Groison, 2003; Quebec, 1996; South Carolina, 1994; Texas, 2004; United States, 1994: Virginia, 2002; Washington, 2001

Figure 3.11 *Clearcut size limits in US and Canadian case study jurisdictions*

A single-block cutting area larger than 100 hectares shall be shaped so that its length is equal to or greater than 4 times its average width.

Thus, Quebec not only prescribes absolute size limits for individual clearcuts, but also requires that clearcuts be of a smaller size across a set percentage of the harvest area.

Alberta's limits compare in size with those of Oregon and Washington, while Ontario's requirements are the least restrictive of any of the size limits found in our case study jurisdictions.

There are no clearcut limits on private lands in the remaining US case study states: Alaska, Idaho, Montana and the southeastern states. Likewise, no provincial-level clearcut limits have been established on private lands in New Brunswick and Quebec.

Reforestation (Indicator: Requirements for reforestation, including specified time frames and stocking levels)

Most Canadian provincial lands, along with Alaska, California, Idaho, Oregon and Washington private lands and the USFS, are governed by the mandatory substantive policies on reforestation. Louisiana takes a procedural approach by requiring harvest plans for timber harvest on commercial forestlands, including prescriptions for reforestation. Reforestation policies in Virginia fall under our 'mixed' category, due to the fact that the state's Seed Tree Law applies only to forest areas 10 or more acres (4.05 hectares) in size on which loblolly or white pine constitutes 25 per cent or more of the live trees on each acre. This requirement comes with a procedural policy for exemption based on an approved reforestation plan. Voluntary guidance on reforestation is provided for the remainder of the private forestlands, although none of these voluntary BMPs provides, or refers to, quantitative guidelines on stocking levels and time frames.

Annual allowable cut (AAC) (Indicator: Cut limits based on maximum sustained yield or non-declining even flow)

Canada

Approaches to AAC regulations vary across the Canadian case study provinces. In Alberta, the Land and Forest Division of Alberta Sustainable Resource Development calculates cut quotas in individual Forest Management Units and Forest Management Areas. These calculations are to be based on 'perpetual sustained yield' timber management (Alberta Forests Act RSA, 2000), although no definition of 'perpetual' is provided. The sum of the individual cut quotas for local management units and areas are then compiled into a provincial AAC, subject to approval by the minister. Unit-level AAC levels are periodically recalculated at the time of timber licence renewals.

In BC, the Chief Forester of the Ministry of Forests and Range, a non-political appointee independent of the Forest Minister, establishes the AAC for provincially owned forest regions. The BC Ministry of Forests and Range annual allowable cut is not a 'calculation' as such but a mixture of qualitative and quantitative assessments of the composition and growth of forests, other forest uses, the long- and short-term implications of alternative harvest rates, mill productive capacities, economic and social objectives 'of the Crown' and abnormal disease or pest outbreaks and major salvage programmes.

The approach to annual cut calculations in BC has been the subject of intense scrutiny, as official policies and interpretation of the policy have evolved over time (Dellert, 1998; Cashore et al, 2001: Chapter Five). The BC Forest Act requires that the Chief Forester 'must consider' a range of factors including '… the rate of timber harvest that may be sustained on the area …' and 'the social and economic objectives of the government'. In other words, sustained yield does not by itself serve as the upper limit of annual allowable cut and hence BC rules governing AAC do not fit the most prescriptive category of our policy framework. Given the mixture of considerations required in setting sustained yield, together with the flexibility granted to the Chief Forester, BC policy is best described as a procedural requirement.

In Ontario, allowable cut levels are established at the management unit level. Under the Crown Forest Sustainability Act (Ontario, 1994) and accompanying manuals, sustainable forest licence (SFL) holders are required to inventory their licensed areas every 25 years. Allowable cut levels are calculated for each licence area, based on the required forest inventories. Twenty years after each licence area inventory, SFL holders must prepare re-inventory strategies and document these strategies in forest management plans. The management plans then establish allowable cut levels for the next 25 years. The implementation of re-inventory strategies and management plans is subject to approval by the Ministry of Natural Resources. The Ministry incorporates the information received from SFL management plans into its provincial planning processes.

Ontario introduced legislation in the late 1990s to expand its 'sustained yield' approach to 'sustainability', in an explicit effort to recognize non-timber aspects of sustainability. This means that timber yields can be set either higher or lower than they would have been under the old policy. Hence we classify Ontario's AAC policies, like those of BC, as essentially procedural.

The Quebec Minister of Natural Resources and Wildlife uses a 'conservation equation' as the basis for provincial level annual allowable cut determinations. A simulation model is used to consider a large array of factors, including the capping of AAC at sustained yield levels and the incorporation of non-timber objectives. Annual allowable cut levels on provincial lands are also calculated at the level of the forest management unit, based on forest protection and forest development objectives. Given the lack of an absolute limit on AAC based strictly on the non-declining even flow principle, however, we have categorized Quebec AAC policy as 'mixed', i.e. a mandatory substantive policy that still leaves considerable room for government discretion.

Annual allowable cut is calculated for New Brunswick Crown lands based on a range of issues. The stated goal is to maximize sustained yield rather than ensure an even flow of timber (Martin, 2003). There are no AAC requirements for New Brunswick private lands, although marketing boards provide estimates on potential wood supplies (New Brunswick, 2004a).

There are 17 regional agencies for private forest development in Quebec (Quebec, 2003a). Each of these agencies is responsible for preparing Private Forest Protection and Development Plans. The plans outline regional harvesting and management strategies for private lands, including establishment of annual cut levels. Observance of regional cut limits is not required at the provincial level (Paquet and Groison, 2003). However, some municipalities have exercised their rights to control overharvesting on private lands (Dansereau, 2004).

The United States

US Forest Service

The 1976 National Forest Management Act outlines a process for determining the 'Annual Sale Quantity' (ASQ) (the USFS term for AAC) at the national level, based on requirements for sustaining multiple forest uses and environmental protection over a 15-year planning period. The Act requires that ASQs adhere to the policy of 'non-declining even flow'. This even-flow policy applies at all administrative levels, from nationwide to the level of individual timber sales. In practice, however, the establishment of ASQs usually occurs through independent calculations made at the level of legally binding individual forest management plans, known as Land and Resource Management Plans (LRMPs) (Wilkinson and Anderson, 1985, p90).

The US Senate and House Appropriations Committee sets the overall funding for what is called 'annual programmed sale level', which must be equal to or less than the ASQs set by national forest plans. The US Forest Service is then empowered with the distribution of this funding among districts.

As discussed in the previous section on endangered species legislation, cut levels on national forestlands have declined dramatically since the 1992 listing of the northern spotted owl. Although procedures for calculating ASQ have remained essentially the same, actual annual timber harvest levels dropped from about 25 million cubic metres in 1992, to about 6.8 million cubic metres in 2003 (see Figure 3.4 above on declines in Washington and Oregon federal timber harvest).

US private lands

With two exceptions, the selected US states under review do not have policies – either mandatory or voluntary – regarding annual allowable cut or similar operational aspects on private lands. The first exception is California, whose Forest Practice Rules (Title 14 913.11 for the Coast District, 933.11 for the Northern District and 953.11 for the Southern Forest District) include provisions concerning Maximum Sustained Production of High Quality Timber Products (MSP). Forest operations in California must demonstrate achievement of MSP through one of several similar approaches (California, 2005). Californian landowners or managers are required to balance growth and harvest over time within an assessment area agreed to by the state forestry agency.[10]

US Indian trust lands are the second exception to which AAC policies apply. Forest management plans, including sustained yield commitments, are also required on lands held in federal trust.

FOREST PRACTICE REGULATIONS: PLANTATIONS

As discussed above, intensive plantation forestry is most prevalent in the US southeast and, to a lesser extent, in Washington and Oregon. Given that the 2005 FAO FRA reports negligible plantation area for Canada, we consider only the US cases in this section.

Forest practice policies and best management practices in the US southeast do not distinguish between native forests and plantations. In Washington and Oregon, planted forests would likewise be subject to the same regulations as non-planted forests. However, short rotation poplar plantations are considered agricultural crops and are subject to regulations and BMPs addressing agriculture (Copestake, 2003).

No state-level policies were identified that prohibit or restrict the conversion of natural forests to forest plantations.

ENFORCEMENT

Canada

Alberta's Land and Forest Division (LFD) of the Alberta Sustainable Resource Department (ASRD) conducts random audits of Crown licences. Between 1996 and 2002, audit activities and compliance results were posted on the government web page. Since 2002, systematic, routine audits have been replaced by more informal 'field checks' and no formal, systematic reporting system of audit activities is currently in place.

The auditing system in British Columbia is complex and multifaceted. The Compliance and Enforcement Branch (C&E) of the Ministry of Forests and Range is the government body responsible for ensuring compliance with forestry regulations. C&E conducts random and routine audits, the results of which are compiled in annual reports and posted on the government web page. MFR's 2008 Annual Report listed 15,688 inspections, 2,212 'compliance actions' (i.e. requests for corrective action), 403 'enforcement actions' and monetary penalties amounting to $112,385. The report also lists the names of licensees who have been subject to the enforcement actions, as well as the type of enforcement actions taken against them (British Columbia 2008c).

In addition to the C&E Branch of the Ministry of Forests and Range the 1996 Forest Practices Code of British Columbia Act established a Forest Practices Board to serve as a special independent monitoring organization. The Forest Practices Board (FPB) is charged with monitoring licensee compliance with forestry legislation, as well as evaluating the adequacy of government monitoring activities. The FPB is required by statute to conduct periodic random audits and to respond to public complaints regarding forest practices. FPB audits range from full assessments to issue-specific monitoring. The FPB prepares detailed reports of all its auditing activities and makes these reports available in hard copy and via the web. The creation of the BC Forest Practices Board appears to have had a positive impact on compliance as the Board has noted that 'Code compliance in all areas – including riparian management – has increased each year since the Board began auditing forest practices' (British Columbia Forest Practices Board, 2003).

The FPB does not hold the authority to prosecute auditees for non-compliance. Such authority is vested in the Ministry of Forests and Range, the Ministry of Energy and Mines, the Ministry of Sustainable Resource Management and the Ministry of Water, Land and Air Protection.

There have been some recent changes to the FPB designed to support the more 'results-based', rather than prescriptive, mandate of the 2004 Forest and Range Practices Act. According to the Board's 2006/7 Annual Report, the FPB is to 'shift … focus to assessing the effectiveness of forest practices in achieving government's legislated objectives for forest resource values'. In line with this requirement, as of March 2007 the FPB had conducted five Pilot Effectiveness Audits 'examining biodiversity and other forest values such as soil conservation, visual quality and stream riparian management' (British Columbia Forest Practices Board 2007).

In Ontario, the Ministry of Natural Resources (OMNR) is the agency in charge of forest practices compliance monitoring and auditing. 'Independent Forest Audits' are conducted every five years on all Crown tenures and sustainable forest licences to measure compliance with legislation, regulations, policies and forest management plans. OMNR auditing consists of three components:

1 compliance with legislation, regulations, policies and management plans;
2 an evaluation of the effectiveness of forest practices in meeting audit criteria and management objectives; and
3 compliance with the terms and conditions of a sustainable forest licence.

As part of their examination of the foregoing components, auditors are required to provide an assessment of forest sustainability in the audit report. Audit reports are prepared after each audit, and OMNR and the forest licensee are required to prepare an 'Action Plan' to address issues raised in the report.

In 2001, the first year of the current system of 'Independent Forest Audits', audits were carried out on 20 management units and documented in 19 audit reports. Eighteen of the audit reports concluded that forests were being managed in general compliance with legislation and policy and the principles of sustainable forest management, and one concluded that it was 'inadequately managed' (Ontario, 2002). Between 2002 and 2006 another 44 audits covered all of the remaining forest management units, while five units were audited for a second time (Ontario, 2009). The rate of compliance was 96 per cent in 2005/6 and similar in other years. Reportedly, most of the infractions have been minor (Ontario, 2008).

In Quebec, the Ministry of Natural Resources, Wildlife (MNRWP, English; MRNFP, French) is responsible for the enforcement of forestry regulations. The MNRWP conducts regular field inspections of licensee forest management operations. The Ministry also monitors the environmental impacts of forestry activities, as well as the impacts of external threats such as air pollution and climate change. As in British Columbia, the Quebec government monitors the effectiveness of its forest legislation, such as the 'Regulation respecting standards of forest management for forests in the public domain' in achieving environmental protection. For example, it studies the impacts of road building specifications on riparian areas and the effect of clearcut sizes on wildlife populations (Quebec, 2003b).

The United States

Across US private lands, three different monitoring methods are commonly used. The first is routine random and/or scheduled inspections of all or a portion of forest operations according to variable selection criteria (e.g. application for a harvesting permit, completion of harvest). The second type comes at the behest of landowners who seek advice from the forest landowner assistance division of their state forestry departments, who then perform 'courtesy' visits to their properties. The final monitoring method, which has become increasingly common in the last decade, is compliance surveys. Compliance surveys involve systematic, statistical audits of forest operations which are then compiled to produce average performance rates. Such surveys vary greatly in methodology (e.g. qualitative or strictly quantitative) and are often performed only sporadically due to the high cost of survey implementation (Ellefson et al, 1995).

The majority of states conduct their monitoring with forest agency staff and resources, though a few states employ an interdisciplinary team comprised of different agencies, academics and/or other specialists. In a few cases, a separate agency – such as a council on environmental quality or a water quality protection agency – is in charge of forest practice monitoring. For instance, in Oregon recent legislation directs the Oregon Watershed Enhancement Board (OWEB) to develop and implement a statewide monitoring programme in coordination with state natural resource agencies for activities conducted under the Oregon Plan for Salmon and Watersheds. California employs a similar approach. Many of the monitoring efforts of states are informal and so no official reports are released (USDA Forest Service Southern Research Station and USDA Forest Service Southern Region, 2002).

In some cases, citizen involvement may also trigger monitoring of forest practice standards. For example, in 1990 the Idaho Department of State Lands received 163 citizen complaints about forestry practices, 126 of which resulted in on-site inspections (Ellefson et al, 1995). Around the same time in Oregon, 89 per cent of the complaints that resulted in on-site inspections originated from private citizens (Ellefson et al, 1995).

It appears that for those states with voluntary BMP regimes, monitoring procedures for public lands may be more developed than those for private lands. In South Carolina, for instance, the state's Forestry Commission conducts BMP monitoring annually on state forestlands, covering at least a minimum of 10 per cent of forestry operations completed during the year (South Carolina, not dated). Thus, even in the US southeast, where our case study states have generally been subject to less stringent regulations, public lands (in this case owned by the state) are subject to greater scrutiny and perhaps higher performance expectations.

In a few states, including Alaska, California and Washington, statutes have been enacted (e.g. Alaska Forest Resources and Practices Act, the Washington Forest Practices Act, and the California Public Resource Code Section 4604), which provide specific authority for government agency foresters to access a landowner's property for inspections at agency discretion. In other states, government inspectors cannot force landowners to participate in all monitoring efforts. For example, in Arkansas, state forestry agency staff must gain specific permission from landowners to inspect their properties (Wear and Greis, 2002).

In terms of formal compliance survey efforts, the survey methodology employed and the sampling intensity vary considerably. Nevertheless all states show strikingly similar monitoring results, with compliance falling roughly within the range of 80 per cent and above and an average overall compliance or implementation rate of 90 per cent. Eight of the states fall in the 90s range, and the other seven fall in the 80s range. Public lands earn the highest compliance or implementation rates, followed by industrial timberland sites and then non-industrial private lands (USDA Forest Service Southern Research Station and USDA Forest Service Southern Region, 2002).

In general, state forestry agencies report increasing overall BMP or forest practice implementation or compliance rates, which they attribute to such factors as the value of technical assistance and increased partnerships with the forest industry (Ice et al, 2002; USDA Forest Service Southern Research Station and USDA Forest Service Southern Region, 2002).

Written recommendations to improve practices are the most common product of state monitoring of forest practices (Ellefson et al, 1995). These recommendations may or may not come in the form of official notice of violations. In some cases, state agencies will issue corrective action orders to address damage done by unsatisfactory forest practices and landowners will be obligated to cover the financial burden of those activities. In South Carolina, even though the routine monitoring takes place in the form of landowner-initiated 'courtesy exam' visits, the Forestry Commission provides a monthly summary of these visits to the state's Department of Health and Environmental Control (SCDHEC). If the exam finds that water quality has been affected by a forest operation, SCDHEC may institute an enforcement action under the South Carolina Pollution Control Act.

States vary in their application of penalties for violations of mandatory forest practices. Most states, however, appear to avoid legal redress. For example, in 1991, 184 civil penalty citations affecting 146 operations were imposed in Oregon, while only one legal action was initiated around the same time period in Idaho (Ellefson et al, 1995). Georgia, which has a non-regulatory policy regime, has referred cases of water quality impairment to its Environmental Protection Division for enforcement action only five times between 1998 and 2000 (USDA Forest Service Southern Research Station and USDA Forest Service Southern Region, 2002). Arkansas employs a similar process to that of Georgia and the state's forestry commission has pursued enforcement action about four times on forestry operations between 1998 and 2000 (USDA Forest Service Southern Research Station and USDA Forest Service Southern Region, 2002). Georgia also has adopted a system for imposing legal sanctions (e.g. penalties, licence suspensions) against registered professional foresters found to be contributing to BMP non-compliance (USDA Forest Service Southern Research Station and USDA Forest Service Southern Region, 2002). Texas has a 'bad actor' provision in its water quality law that enables the state to pursue legal action against repeat offenders, but this provision has rarely, if ever, been utilized with regard to a forestry operation (USDA Forest Service Southern Research Station and USDA Forest Service Southern Region, 2002). In Louisiana, no formal process exists for dealing with forestry operations suspected of impairing water quality.

While almost all of the southern states lack direct regulatory forest practices policies (e.g. a forest practices Act), two of the case study states – North Carolina and Virginia – have developed interesting policy variations that are revealed through an examination of their enforcement programmes. North Carolina has instituted voluntary BMPs to ensure the achievement of nine mandatory forest practice guidelines (FPGs) (White, 1992). These nine FPGs are mandatory, however, in the sense that they must be observed if an operation is to be exempt from permitting and other requirements under the Sediment Pollution Control Act passed by the state in the 1970s (White, 1992). Therefore, findings of non-compliance with the voluntary BMPS may trigger not legal action, but rather additional regulatory requirements. The state of Virginia has a similar situation in that its BMPs are not regulatory but must be followed in order for forestry operations to maintain exemptions from requirements of the Chesapeake Bay Preservation Act (where applicable) (USDA Forest Service Southern Research Station and USDA Forest Service Southern Region, 2002).

FOREST CERTIFICATION

Civil society actors in both the US and Canada have played instrumental roles in launching forest certification as both a domestic and global movement. Forest certification efforts in these countries date as far back as the 1980s, including several 'grass roots' locally based certification schemes along the Pacific Coast, as well as the 'SmartWood' programme of the New York-based Rainforest Alliance that completed its first forest certification in Indonesia in 1990 (McDermott and Hoberg, 2003). A few years later, the 1993 founding meeting of the Forest Stewardship Council (FSC) was held in Toronto, Canada, foreshadowing Canada's rapid rise to global first place in terms of total forest area certified.

Since the launch of the FSC, the growth of certification in both countries has also been fraught with conflict among several competing certification schemes. In 1994, Canadian representatives from government and major industrial firms put considerable support behind developing a Sustainable Forest Management (SFM) certification standard under the Canadian Standards Association (CSA) (Cashore et al, 2004). In the US, the American Forest and Paper Association (AF&PA) launched the Sustainable Forestry Initiative (SFI). The SFI soon expanded its operations into Canada as well. These two schemes have since received endorsement by the Europe-based Programme for the Endorsement of Certification Schemes (PEFC).

As illustrated in Figure 2.18, Chapter 2, the PEFC-endorsed schemes to date have grown substantially larger than the FSC in terms of total area certified in Canada and the US. Considering relative growth within each scheme, Canada leads the world in area certified under both the FSC and PEFC. CSA certification was designed with public forest management expressly in mind and accounts for the largest percentage of Canada's certified area. In addition, SFI certification has been applied to Crown land and to some Canadian private industrial ownerships.

The US has the second largest area certified worldwide under the PEFC and the fourth largest under the FSC. Most of the PEFC certified area is owned by industrial members of the SFI, while much of the FSC certified area consists of state lands and smaller-scale producers.

The importance of non-industrial forest ownership and production in the US presents a significant barrier to certification's expansion, due to economies of scale, limited capacities, etc. A fourth North American certification scheme, the American Tree Farm System (ATFS), has therefore created a certification system expressly for family forest ownerships. The ATFS, which has recently received endorsement by the PEFC, still covers only a very small percentage of forestlands.

SUMMARY

This analysis supports Cashore and his colleagues' earlier work (Cashore 1997c, 1999; Teeter et al, 2003; Cashore and McDermott, 2004) identifying both important similarities and differences between and within Canada and the United States. Though forest policy is both complex and varied, as illustrated by the wide variety of policy approaches, discernible trends have emerged.

The first is that it is a mistake to draw conclusions about these countries' regulatory approaches based on broad brush comparisons. In the past, the governance of US national forestlands has been broadly characterized as litigious, US private lands as more cooperative and Canadian provinces as emphasizing government discretion (Cashore 1997c; Hoberg, 1997). However, a closer analysis reveals different patterns. In many cases, from riparian rules to clearcutting, the Canadian provinces under review had some of the most prescriptive policies in place. Furthermore, regulation of forest practices on US private lands varied considerably depending on the state in question (see the Tables 3.6 to 3.8).

While US federal lands were under severe harvesting restrictions owing to court battles in the 1990s, the approach of US state governments, which regulate the bulk of US timber production, varies greatly. Key US western states compare most closely with the Canadian case study provinces, while those in the US south emphasize voluntary guidelines on some issues, such as water protection in response to the Clean Water Act, and do not promulgate guidelines for other issues.

Protected area efforts are also quite different. According to the latest data available (Figure 3.3), the western US states, western Canadian provinces, Ontario and Arkansas fit within the top half of case study jurisdictions in terms of the percentage of land area protected, and New Brunswick, Quebec and the rest of the US southern states occupy the bottom half.

Our discussion of enforcement mechanisms addresses the question of policy 'outputs', i.e. the degree of assurance provided that policies are actually implemented as written. This analysis indicates that those jurisdictions with more prescriptive policies have also established more formal, mandatory, mechanisms of enforcement, although there is substantial variation in the extent and frequency of on-the-ground auditing.

KEY TO TABLES 3.6–3.8

Mandatory substantive
Mandatory procedural
Mandatory mixed
Voluntary
No policy

Table 3.6 *Policy approach to five forest practice criteria in Canadian case studies*

Case Study	1) Riparian	2) Roads	3) Clearcuts	4) Reforestation	5) AAC
Alberta (Public)					Mandatory mixed
British Columbia (Public)					Mandatory procedural
New Brunswick (Private)		Mandatory mixed	No policy	Voluntary	No policy
New Brunswick (Public)					Mandatory mixed
Ontario (Public)		Mandatory mixed			Mandatory procedural
Quebec (Private)	No policy	No policy	No policy	Voluntary	No policy
Quebec (Public)		No policy			Mandatory mixed

Table 3.7 *Policy approach to five forest practice criteria in US western case studies*

Case Study	1) Riparian	2) Roads	3) Clearcuts	4) Reforestation	5) AAC
Alaska (Private)		Mandatory mixed		Mandatory mixed	No policy
California (Private)					Mandatory mixed
Idaho (Private)		Mandatory mixed	No policy	Voluntary	No policy
Montana (Private)		Voluntary		Voluntary	No policy
Oregon (Private)		Mandatory mixed			No policy
USFS Region 6* (Public)		Mandatory mixed			
Washington (Private)					No policy

Notes: * USFS lands account for the following percentages of forest cover in the US western case study states = 75% ID; 27% MT; 9% AK; 48% OR; 37% WA; 43% CA. Harvest = 9% ID; 4% MT; 9% AK; 5% OR; 1% WA; 8% CA

Table 3.8 *Policy approach to five forest practice criteria in US southeast case studies*

Case Study	1) Riparian	2) Roads	3) Clearcuts	4) Reforestation	5) AAC
Alabama (Private)					
Arkansas (Private)					
Georgia (Private)					
Louisiana (Private)					
Mississippi (Private)					
North Carolina (Private)					
South Carolina (Private)					
Texas (Private)					
Virginia (Private)					

Notes: As discussed above, complying with best management practices (BMPs) for riparian buffer zones and road building in the US may be used as proof of 'due diligence' of complying with the Clean Water Act and thus be considered to have some legal standing. As a result, in the summary in Chapter 10 where we rate jurisdictions by level of prescriptiveness we rate the US southeast riparian policies as moderately more prescriptive than BMPs for other policy criteria and/or jurisdictions without equivalent legal frameworks.

NOTES

1   Primary forests are defined as forests of native species, where there are no clearly visible indications of human activities and the ecological processes are not significantly disturbed. They include areas where collection of non-wood forest products occurs, provided the human impact is small. Some trees may have been removed (FAO, 2006a, p171).

2   A frontier forest must meet seven criteria:

   1   It is primarily forested.

   2   It is big enough to support viable populations of all indigenous species associated with that forest type (measured by the forest's ability to support wide-ranging animal species, such as elephants, harpy eagles or brown bears).

   3   It is large enough to keep these species' populations viable even in the face of the natural disasters – such as hurricanes, fires, and pest or disease outbreaks – that might occur there in a century.

   4   Its structure and composition are determined mainly by natural events, though limited human disturbance by traditional activities of the sort that have shaped forests for thousands of years – such as low-density shifting cultivation – is acceptable. As such, it remains relatively unmanaged by humans and natural disturbances (such as fire) are permitted to shape much of the forest.

   5   In forests where patches of trees of different ages would naturally occur, the landscape exhibits this type of heterogeneity.

   6   It is dominated by indigenous tree species.

   7   It is home to most, if not all, of the other plant and animal species that typically live in this type of forest.

3   Aboriginal peoples in Canada are commonly referred to as 'First Nations'.

4   With the contribution of Dr Melanie H. McDermott, Department of Human Ecology, Rutgers University, New Jersey.

5   The mainland 48 states exclude Alaska and Hawaii.
6   Land trusts and easements have also been emerging in Canada, although as a conservation tool for private lands they play a much smaller role due to the relative percentage of public land ownership in the country. For information on land trusts in Canada see, for example, the Nature Conservancy Canada's announcement of new land trusts, available at www.natureconservancy.ca/site/PageServer?pagename= ncc_work_impact_feature20 (last accessed July 2009).
7   Riparian zone rules for Alaska include those governing state lands, since state lands account for 19 per cent of forestlands. 'Public' land rules are also provided, since this category establishes a minimum standard for the 72 per cent of Alaska's forestlands that are publicly owned (including federal, state, municipal and other public lands) (Alaska, 2000).
8   Requirements for no-harvest buffer zones on small, non-fish-bearing streams in Washington are based on such factors as the distance of the stream from the confluence of fish-bearing streams (in western Washington) and harvesting patterns within special management zones (in eastern Washington).
9   In some cases minimum culvert diameters apply to all culverts and in some cases only to stream crossings. However, our indicator of policy approach is restricted to stream crossings.
10 The projected inventory resulting from harvesting over time shall be capable of sustaining the average annual yield achieved during the last decade of the planning horizon. The average annual projected yield over any rolling ten-year period, or over appropriately longer time periods, shall not exceed the projected long-term sustained yield. Additionally, the projected yield is required to meet minimal stocking and basal area standards; protect soil, air, fish and wildlife, water resources and any other public trust resources and give consideration to recreation, range and forage, regional economic vitality, employment and aesthetic enjoyment.

REFERENCES

Adirondack Park Agency (2009) 'The Adirondack Park', www.apa.state.ny.us/About_park/ index.html, New York State Adirondack Park Agency 2009, accessed August 2009
Alabama (1993) 'Alabama's best management practices for forestry', Montgomery, AL: Alabama Forestry Commission
Alaska (2000) Alaska Forest Resources and Practices Act
Alaska (2004) 'Alaska forest resources & practices regulations', Alaska Department of Natural Resources, Division of Forestry
Alaska (2007) 'Forest ownership in Alaska', Alaska Department of Natural Resources, Department of Forestry, Forest Stewardship Program
Alberta (1994) 'Timber harvest planning and operating ground rules', Edmonton, AB: Alberta Environment Protection, Land and Forest Service
Alberta (2000) Alberta Forests Act
Allan, K. and D. Frank (1994) 'Community forests in British Columbia: Models that work', *Forestry Chronicle*, 70, 6, pp721–724
Amos, William, Kathryn Harrison and George Hoberg (2001) 'In search of a minimum winning coalition: The politics of species at risk legislation in Canada', in K. Beazley and

R. Boardman (eds) *Politics of the Wild: Canada and Endangered Species*, Toronto: Oxford University Press

APEC (2007) 'Atlantic Canada's forest industry under pressure', Report Card (Quarterly), Atlantic Provinces Economic Council, available at http://nbwoodlotowners.ca/// uploads//Website_Assets/APEC's_Atlantic_Lumber.pdf, accessed November 2009

Arkansas (2002) 'Arkansas forestry best management practices for water quality protection', Arkansas Forestry Commission

Audin, Lisa (2009) Personal communication with Lisa Audin, Stewardship Coordinator, US Gap Analysis Program, 31 July

Baker, Mark and Jonathan Kusel (2003) *Community Forestry in the United States: Learning from the Past, Crafting the Future*, Washington, DC: Island Press

Bateson, Emily M. (2005) 'Two countries, one forest – deux pays, une forêt: Launching a landscape-scale conservation collaborative in the northern Appalachian region of the United States and Canada', *Conservation Practice at the Landscape Scale*, 22, 1, pp35–45

Betts, M. and D. Coon (1999) 'Working with the woods. Restoring forests and community in New Brunswick', in P. Wolvekamp (ed) *Forests for the Future: Local Strategies for Forest Protection, Economic Welfare and Social Justice*, London: Zed Books

Biggs, Jeff (2003) 'Short rotation hybrid poplar plantations in the Pacific Northwest', paper read at International Forest Conservation Field Course 2003 at Oregon and Washington

British Columbia (not dated) 'British Columbia salmon', Victoria, British Columbia: British Columbia Fisheries Secretariat

British Columbia (2004a) 'Identified wildlife management strategy: Procedures for managing identified wildlife', in *Version 2004*, British Columbia Ministry of Water, Land and Air Protection

British Columbia (2004b) Forest and Range Practices Act: Forest Planning and Practices Regulation, British Columbia Ministry of Forests and Range

British Columbia (2007) 'Mountain pine beetle action plan: Sustainable forests, sustainable communities: Annual progress report 2006/2007', Ministry of Forests and Range, British Columbia, Mountain pine beetle emergency response division

British Columbia (2008a) 'Negotiation update', British Columbia Treaty Commission, www.bctreaty.net/files/updates.php, accessed April 2008

British Columbia (2008b) 'Province announces a new vision for coastal B.C.', 7 February, British Columbia Ministry of Agriculture and Lands

British Columbia (2008c) 'Annual report 2008: Providing statistics of compliance enforcement activities recorded by the Ministry of Forests and Range from April 1, 2007 to March 31, 2008', Ministry of Forests and Range – Compliance & Enforcement Program

British Columbia (2009) 'Community forests', British Columbia Ministry of Forests and Range, www.for.gov.bc.ca/hth/community/, accessed May 2009

British Columbia Forest Practices Board (2003) 'Streamside protection', Victoria, BC: British Columbia Forest Practices Board

British Columbia Forest Practices Board (2007) 'Forest Practices Board 2006–2007 annual report', Victoria, BC: British Columbia Forest Practices Board

Bryant, Dirk, Daniel Nielsen and Laura Tangley (1997) 'The last frontier forests: Ecosytems and economies on the edge. What is the status of the world's remaining large, natural forest ecosystems?', report, Washington, DC: World Resources Institute, Forest Frontiers Initiative

California (2005) California Forest Practice Rules. The California Department of Forestry and Fire Protection, Resource Management, Forest Practice Program

Cashore, Benjamin (1997a) 'A tale of two journeys: Environmentalism and the politics of forest policy change in the US Pacific Northwest', paper read at the University of British

Columbia Forest Economics and Policy Analysis Research Unit (FEPA) Conference, An International Comparison of Forest Institutions, 23 January, Vancouver, BC

Cashore, Benjamin (1997b) 'Flights of the phoenix: Explaining the durability of the Canada–US softwood lumber dispute', *Canadian–American Public Policy*, 32, December, pp1–63

Cashore, Benjamin (1997c) 'Governing forestry: Environmental group influence in British Columbia and the US Pacific Northwest', PhD, Toronto: University of Toronto

Cashore, Benjamin (1999) 'US Pacific Northwest', in B. Wilson, K. V. Kooten, I. Vertinsky and L. Arthur (eds) *Forest Policy: International Case Studies*, Washington, DC: CABI Publications

Cashore, Benjamin and Constance L. McDermott (2004) 'Global environmental forest policy: Canada as a constant case comparison of select forest practice regulations', Victoria: International Forest Resources

Cashore, Benjamin, George Hoberg, Michael Howlett, Jeremy Rayner and Jeremy Wilson (2001) *In Search of Sustainability: British Columbia Forest Policy in the 1990s*, Vancouver: University of British Columbia Press

Cashore, Benjamin, Graeme Auld and Deanna Newsom (2004) *Governing Through Markets: Forest Certification and the Emergence of Non-State Authority*, New Haven, CT: Yale University Press

CCEA (2009) 'Mission statement', www.ccea.org/en_mission.html#, accessed August 2009

CCFM (2008) 'Compendium of Canadian forestry statistics', http://nfdp.ccfm.org/, accessed January 2008

CCFM (2009) 'National forestry database, forest products', http://nfdp.ccfm.org/index_e.php, Canadian Council of Forest Ministers, accessed February 2009

Charnley, Susan and Melissa R. Poe (2007) 'Community forestry in theory and practice: Where are we now?', *Annual Review of Anthropology*, 36, pp301–336

Clapp, R. A. (2004) 'Wilderness ethics and political ecology: Remapping the Great Bear Rainforest', *Political Geography*, 23, 7, pp839–862

Clark, T., S. Harvey, G. Bruemmer and J. Walker (2003) 'Large-scale community forest in Ontario, Canada – A sign of the times', paper read at XII World Forestry Congress, Quebec City

Coad, Lauren (2009) Personal communication with Lauren Coad, former research staff at UNEP-WCMC, 14 August

Copestake, Martha (2003) 'Short rotation hybrid poplar plantations in the Pacific Northwest', paper read at International Forest Conservation Field Course 2003, Oregon and Washington

Czech, Brian and Paul R. Krausman (2001) *The Endangered Species Act: History, Conservation Biology and Public Policy*, Baltimore, MD: Johns Hopkins University Press

Danks, C. M. (2009) 'Benefits of community-based forestry in the US: Lessons from a demonstration programme', *The International Forestry Review*, 11, 2, pp171–185

Dansereau, Jean-Pierre (2004) Directeur général of the Fédération des producteurs de bois du Québec, email communication

Davies, B., M. Geffen, M. Kauffman and H. Silverman (2008) *Redefining Stewardship: Public Lands and Rural Communities in the Pacific Northwest*, Portland, OR: Ecostrust and Resource Innovations

Davis, Phillip A. (1992) 'Critics say too few jobs, owls saved under "God Squad" plan', *Congressional Quarterly Weekly Report*, 50, 1 June, pp1438–1439

Dellert, Lois (1998) 'Sustained yield: Why has it failed to achieve sustainability?', in C. Tollefson (ed) *The Wealth of Forests: Markets, Regulation and Sustainable Forestry*, Vancouver: UBC Press

Duinker, P. N., P. W. Matakala, F. Chege and L. Bouthillier (1994) 'Community forests in Canada: An overview', *Forestry Chronicle*, 70, 6, pp711–720

Ellefson, Paul, Anthony Cheng and R. J. Moulton (1995) 'Regulation of private forestry practices by state governments', *Station Bulletin 605–1995*, Minnesota Agricultural Experiment Station, St Paul, University of Minnesota

Ellefson, Paul, Michael Kilgore, Calder M. Hibbard and James E. Granskog (2004) 'Regulation of forestry practices on private land in the United States: Assessment of state agency responsibilities and program effectiveness', Department of Forest Resources, College of Natural Resources and Agriculture Experiment Station, University of Minnesota

FAO (2003) *State of the World's Forests – 2003*, Rome, United Nations: Food and Agriculture Organization

FAO (2006a) 'Global forest resources assessment 2005: Progress towards sustainable forest management', in *FAO Forestry Paper 147*, Rome: Food and Agricultural Organization of the United Nations

FAO (2006b) 'Global planted forests thematic study: Results and analysis', A. Del Lungo, J. Ball and J. Carle (eds) Rome: United Nations Food and Agriculture Organization

FAO (2007) *State of the World's Forests – 2007*, Rome, United Nations: Food and Agriculture Organization

Folke, Carle (2006) 'Resilience: The emergence of a perspective for social-ecological systems analyses', *Global Environmental Change*, 16, pp253–267

Georgia (1999) 'Georgia's best management practices for forestry', Georgia Forestry Commission

Gordon, John, Joyce Berry, Mike Ferucci, Jerry Franklin, K. Norman Johnson, Calvin Mukumoto, David Patton and John Sessions (2003) *An Assessment of Indian Forests and Forest Management in the United States*, Portland, OR: Clear Water Printing

Government of Canada (2006) 'Canadian protected area status report: 2000–2005', Minister of Environment

Gray, S. L. (1995) 'A descriptive forest inventory of Canada's forest regions', Chalk River, ON: Natural Resources Canada, Canadian Forest Service

Grumbine, Edward R. (1994) 'What is ecosystem management?', *Conservation Biology*, 8, 1, p31

Hagan, John M., Lloyd C. Irland and Andrew A. Whitman (2005) 'Changing timberland ownership in the Northern Forest and implications for biodiversity', Brunswick: Manomet Center for Conservation Sciences

Haley, David and Martin K. Luckert (1998) 'Tenures as economic instruments for achieving objectives of public forest policy in British Columbia', in C. Tollefson (ed) *The Wealth of Forests: Markets, Regulation and Sustainable Forestry*, Vancouver: UBC Press

HFC NPS (2005) 'The national parks: Shaping the system', Washington, DC: Harpers Ferry Center National Park Service

Hinch, Scott (2003) 'Salmon at their southern edge: Current challenges to survival and prognosis for long-term sustainability', paper read at Jubilee Lecture Series, Vancouver

Hoberg, George (1997) 'Governing the environment: Comparing policy in Canada and the United States', in K. Banting, G. Hoberg and R. Simeon (eds) *Degrees of Freedom: Canada and the United States in a Changing Global Context*, Kingston McGill-Queens and Montreal: University Press

Ice, George, Liz Dent, Josh Robben, Pete Cafferata, Jeff Light, Brian Sugden and Terry Cundy (2002) 'Programs assessing implementation and effectiveness of state forest practice rules and BMPs in the West', paper read at Forest BMPs Research Symposium, Atlanta, Georgia

Idaho (1974) Idaho Forest Practices Act

Idaho (2000) 'Best management practices: Forestry for Idaho: Forest Stewardship guidelines for water quality', in cooperation with Idaho Department of Lands, Bureau of Forestry Assistance; University of Idaho, Cooperative Extension System; Idaho Forest Products Commission; Montana Department of Natural Resources and Conservation; USDA Forest Service, State and Private Forestry

Ihm, Byung-Sun, Jeom-Sook Lee, Jong-Wook Kim and Joon-Ho Kim (2007) 'Relationship between global warming and species richness of vascular plants', *Journal of Plant Biology*, 50, 3, pp321–324

INAC (2008) 'Fact sheet: The Nisga'a Treaty', Indian and Northern Affairs Canada', http://www.ainc-inac.gc.ca/pr/info/nit_e.html, Indian and Northern Affairs Canada, accessed April 2008

Jankowski, Pat (2000) 'Oregon ecologist tells what it will take to save NW salmon', *Seattle Daily Journal*, 11 April 2000

Joyce, Linda, John Aber, Steve McNulty, Virginia Dale, Andrew Hansen, Lloyd Irland, Ron Neilson and Kenneth Skog (2001) 'Potential consequences of variability and change for the forests of the United States', in National Assessment Synthesis Team, *Climate Change Impacts on the United States: The Potential Consequences of Climate Variability and Change*, Washington, DC: US Global Change Research Program, www.usgcrp.gov/usgcrp/nacc/default.htm, accessed November 2009

Ketcham, Paul (1993) 'Testimony before the Oregon Board of Forestry on the draft rules for stream classification and protection', Portland, OR: Audubon Society of Portland

Koch, George W., Stephen C. Sillett, Gregory M. Jennings and Stephen D. Davis (2004) 'The limits to tree height', *Nature*, 428, 22 April, pp851–854

Kohm, Kathryn A. (1991) *Balancing on the Brink of Extinction: The Endangered Species Act and Lessons for the Future*, Washington, DC: Island Press

Kurz, W. A., C. C. Dymond, G. Stinson, G. J. Rampley, E. T. Neilson, A. L. Caroll, T. Ebata and L. Safranyik (2008) 'Mountain pine beetle and forest carbon feedback to climate change', *Nature*, 452, 24, pp987–990

Land Trust Alliance (2009) 'About land trusts', www.landtrustalliance.org/conserve/about-land-trusts, accessed July 2009

Logan, Jesse A. and James A. Powell (2001) 'Ghost forests, global warming, and the mountain pine beetle (Coleoptera: Scolytidae)', *American Entomologist*, 47, 3, pp160–172

Louisiana (2000) 'Recommended forestry best management practices for Louisiana', Louisiana Forestry Association, Louisiana Department of Environment Quality, and the Louisiana Department of Agriculture and Forestry

Mach, Helmut and Williams J. Shaw (2007) 'Information bulletin: A CN Canada–US trade relations forum', paper read at the Canada–US softwood lumber agreement of 2006, 'The experience of the Western Canadian industry under the softwood lumber agreement 2006', June, University of Alberta, Edmonton

Marchak, M. Patricia, Scott L. Aycock and Deborah M. Herbert (1999) *Falldown: Forest Policy in British Columbia*, Vancouver, BC: David Suzuki Foundation and Ecotrust Canada

Martin, Catrina M. (1994) 'Recovering endangered species and restoring ecosystems: Conservation planning for the twenty-first century', *Ibis*, 137, ppS198–S203

Martin, G. (2003) 'Management of New Brunswick's Crown forests', Fredericton, New Brunswick: Department of Natural Resources

Martin, William Haywood, Stephen G. Boyce and A. C. Echternacht (eds) (1993) *Biodiversity of the Southeastern United States: Lowland Terrestrial Communities*, Vol. 2, New York: John Wiley & Sons

McCarthy, James (2006) 'Neoliberalism and the politics of alternatives: Community forestry in British Columbia and the United States', *Annals of the Association of American Geographers*, 96, 1, pp84–104

McDermott, Constance L. and George Hoberg (2003) 'From state to market: Forestry certification in the U.S. and Canada', in B. Schindler, T. Beckley and C. Finley (eds) *Two Paths Toward Sustainable Forests: Public Values in Canada and the United States*, Corvallis: Oregon State University Press

McDermott, Melanie H. (2009) 'Equity first or later? How community-based forestry distributes benefits', *International Forestry Review*, 11, 2, pp207–220

McDermott, M. H. and K. Schreckenberg (2009) 'Equity in community forestry: Insights from North and South', *International Forestry Review*, 11, 2, pp157–170

Milakovsky, Brian (2009) 'Marketing and utilization of timber from tribal forest lands: Survey results 2008', New Haven, CT: Yale School of Forestry and Environmental Studies

Mississippi (2000) 'The best management practices handbook for forestry in Mississippi', Mississipi Forestry Association

Montana (2001) 'Water quality best management practices (BMPs) Montana Forests', authored by Robert Logan, Forestry Specialist, Montana State University Extension Service in cooperation with Montana Department of Natural Resources and Conservation, Forestry Division and Montana Logging Association

Montana (2006) 'Montana guide to the streamside management zone law and rules', Montana Department of Natural Resources and Conservation, Forestry Assistance Bureau; Montana Department of Environmental Quality; Montana Logging Association; Montana Wood Products Association; Plum Creek Timber Company, LP; USDA Forest Service; USDA Bureau of Land Management. 2006 Revision by Dan Rogers, Forest Stewardship Specialist, Montana Department of Natural Resources and Conservation. Originally compiled and edited by Norman Fortunate, then Forest Practices Specialist, Montana Department of Natural Resources and Conservation

Moser, B. (2002) 'Reproductive success of northern saw-whet owls nesting in hybrid poplar plantations', *Northwest Science*, 76, 4, pp353–355

NAFA (2003) 'Aboriginal-held forest tenures in Canada', Ottawa: National Aboriginal Forestry Association

Namkoong, G. and J. H. Roberds (1974) 'Extinction probabilities and the changing age structure of redwood forests', *The American Naturalist*, 108, 961, pp355–368

New Brunswick (no date) 'Understanding the law: A guide to New Brunswick's watershed protected area designation order', Fredericton: Department of the Environment and Local Government

New Brunswick (2004a) 'Final report on wood supply in New Brunswick', New Brunswick, Legislative Assembly, Select Committee on Wood Supply

New Brunswick (2004b) 'Forest management manual for New Brunswick Crown land', Fredericton: New Brunswick Department of Natural Resources, Forest Management Branch

North Carolina (2003) 'Forestry activities under the river basin rules', Raleigh: North Carolina Department of Environment and Natural Resources; Division of Forest Resources

North Carolina (2006) 'North Carolina forestry best management practices manual to protect water quality', Raleigh: North Carolina Division of Forest Resources

Northwest Renewable Resources Center (1998) 'Report from the first annual review of timber/fish/wildlife', *Northwest Renewable Resources Center Newsletter*, Summer, 1

NRC (2007a) 'Canada's national forest inventory', Natural Resources Canada, http://cfs.nrcan.gc.ca/subsite/canfi, accessed February 2007

NRC (2007b) 'The state of Canada's forests: Annual report 2007', Ottawa: Natural Resources Canada

NRC (2008) 'Forest 2020 PDA', Natural Resources Canada, http://cfs.nrcan.gc.ca/subsite/afforestation/forest2020pda, accessed July 2009

Ontario (1988) 'Timber management guidelines for the protection of fish habitat', Ministry of Natural Resources, Fisheries Branch, Toronto: Queen's Printer for Ontario

Ontario (1990a) Endangered Species Act, Chapter E.15, amended by 1997, c. 41, s. 116. Government of Ontario

Ontario (1990b) 'Environmental guidelines for access road and water crossings', Ministry of Natural Resources, Toronto: Queen's Printer for Ontario

Ontario (1991) 'Code of practice for timber management operations in riparian areas', amended April 1998, Ministry of Natural Resources, Toronto: Queen's Printer for Ontario

Ontario (1994) Crown Forest Sustainability Act, Chapter 25, amended by 1996, c.14, s.1; 1998, c.18, Sched. I, ss. 15–18; 2000, c.18, s.64; 2000, c.26, Sched. L, s.3; 2001, c.9, Sched. K, s.2, Government of Canada

Ontario (2002) 'Annual report on forest management 2000/2001', Ministry of Natural Resources

Ontario (2008) 'Annual report on forest management: For the year April 1, 2005 to March 31, 2006', Ministry of Natural Resources

Ontario (2009) 'Audit reports tabled in legislature', Ministry of Natural Resources, www.mnr.gov.on.ca/en/Business/Forests/1ColumnSubPage/STEL02_167055.html#2006, accessed February 2009

Ontario Forest Industries Association (1998) 'Guiding principles and code of forest practices', Toronto, Ontario: Ontario Forest Industries Association

Oregon (1994) *Water Protection Rules: Purpose, Goals, Classification, and Riparian Management Areas*. Contains Oregon Administration Rules (OARs) filed through November 15, 2007. Oregon Administrative Rules, Department of Forestry, Division 635

Ostrom, E. (1990) *Governing the Commons: The Evolution of Institutions for Collective Action*, Cambridge: Cambridge University Press

Pacific Fishery Management Council (2000) 'Amended sections of the Pacific Coast salmon plan', Portland, OR: Pacific Fishery Management Council

PAD-US (2009) 'Protected areas database of the United States (PAD-US)' www.protectedlands.net/padus/, online database, accessed July 2009

Paquet, Jean, and Valérie Groison (2003) 'Sound forestry practices for private woodlots', Field Guide, Quebec City: Foundation de la Faune du Québec

Pellerito, Ryan and Rebecca Wisch (2008) 'State endangered species chart', Michigan State University College of Law, available at www.animallaw.info/articles/ddusstateesa.htm, accessed April 2009

Pierce, Peter (2001) 'Ready for change: Crisis and opportunity in the coast forest industry. A report to the MoF on BC's coastal forest industry sector', www.for.gov.bc.ca/hfd/library/documents/phpreport/, last accessed November 2009

Quebec (1989) 'An Act respecting the conservation and development of wildlife', Chapter C-61.1, Government of Quebec, available at www.canlii.org/en/qc/laws/stat/rsq-c-c-61.1/latest/rsq-c-c-61.1.html, accessed April 2009

Quebec (2003a) 'Nature in Québec: An experience to share', Government of Quebec, report for the XII World Forestry Congress in Quebec City.

Quebec (2003b) 'Protecting the forest environment', Quebec Ministry of Natural Resources, Wildlife and Parks, www.mrn.gouv.qc.ca/english/forest/quebec/quebec-system-management-protection.jsp#suivis, accessed January 2004

Rajala, Richard Allan (1998) *Clearcutting the Pacific Rainforest: Production, Science, and Regulation*, Vancouver: University of British Columbia Press

Roberts, Don (2006) 'Global trends in the forest products sector: China's boom', paper read at 'China's Boom: Implications for investment and trade in forest products and forestry', 18 January, Vancouver, BC

Shaw, Karena (2004) 'The global/local politics of the Great Bear Rainforest', *Environmental Politics*, 13, 2, pp373–392

Shock, C., E. Feibert, M. Seddigh and L. Saunders (2002) 'Water requirements and growth of irrigated hybrid poplar in semi-arid environment in Eastern Oregon', *Western Journal of Applied Forestry*, 17, 1, pp46–53

Smith, Andrew A., Margaret A. Moote, Cecil R. Schwalbe (1993) 'The Endangered Species Act at Twenty: An analytical survey of federal endangered species protection', *Natural Resources Journal* vol 33, no 4, pp1027–1076

Smith, W. Brad and David Darr (2004) 'US forest resource facts and historical trends', Washington, DC: USDA Forest Service

Smith, Wynet, Bryan Evans, Tim Wilson, Andrew Nikiforuk, Karen Baltgalis, Allison Brady, Laurel Brewster, Tim Gray, Will Horter, Peter Lee, Robert Livernash, Claire McGlynn, Susan Minnemeyer, Aran O'Carroll, Alan Penn, Geoff Quaile, Mike Sawyer, Elizabeth Selig and Dirk G. Bryant (2001) 'Canada's forests at a crossroads: An assessment in the year 2000', Washington, DC: World Resources Institute

Smith, Brad W., Patrick D. Miles, John S. Vissage and Scott A. Pugh (2004) 'Forest resources of the United States, 2002: A technical document supporting the USDA Forest Service 2005 update of the RPA Assessment', St Paul, MN: USDA, Forest Service, North Central Research Station, www.ncrs.fs.fed.us/pubs/gtr/gtr_nc241.pdf

South Carolina (no date) 'State forests: Long range plan', South Carolina Forestry Commission

South Carolina (1994) *South Carolina Forestry Commission Best Management Practices for Forestry*, South Carolina Forestry Commission

Spencer, Shannon, Gary Lauten, Barrett Rock, Lloyd Irland and Tim Perkins (2002) 'The impact of climate on regional forests', in *The New England Regional Assessment of the Potential Consequences of Climate Variability and Change: A Final Report*, NERA Foundation, New England Regional Assessment

Stanton, Brian, Jake Eaton, Jon Johnson, Don Rice, Bill Schuette and Brian Moser (2002) 'Hybrid poplar in the Pacific northwest: The effects of market-driven management', *Journal of Forestry*, 100, 4, pp28–33

Teeter, Lawrence, Benjamin Cashore and Daowei Zhang (eds) (2003) *Forest Policy for Private Forestry: Global and Regional Challenges*, Wallingford, UK: CABI Publishing

Teitelbaum, Sarah, Tom Beckley and Solange Nadeau (2006) 'A national portrait of community forestry on public land in Canada', *The Forestry Chronicle*, 82, 3, pp416–428

Texas (2004) 'Texas forestry best management practices', Texas Forestry Association; Texas Forest Service

Tripp, D., A. Nixon and R. Dunlop (1992) 'The application and effectiveness of the coastal fisheries forestry guidelines in selected cut blocks on Vancouver Island', prepared for the Ministry of Environment, Lands and Parks, Nanaimo, BC: D. Tripp Biological Consultants Ltd

Tschaplinski, P. J. (2000) 'The effects of harvesting, fishing, climate-variation, and ocean conditions on salmonid populations in Carnation Creek, Vancouver Island, British Columbia', in E. E. Knudsen, C. R. Steward, D. D. MacDonald, J. E. Williams and D. W. Reiser (eds) *Sustainable Fisheries Management: Pacific Salmon*, Boca Raton: CRC Press

Tschaplinski, P. J. (2004) 'Fish-forestry interaction research in coastal British Columbia – The Carnation Creek and Queen Charlotte Islands studies', in G. F. Hartman and T. G. Northcote (eds) *Fishes and Forestry – Worldwide Interactions and Management*, London: Blackwell Publications

United States (1994) Standards and guidelines for management of habitat for late-successional and old-growth forest related species within the range of the northern spotted owl. Attachment A to the record of decision for amendments to Forest Service and Bureau of Land Management Planning documents within the range of the northern spotted owl; Interagency Supplemental Environmental Impact Statement (SEIS) Team

USDA Forest Service (2009) 'Forest inventory and analysis national program, timber products output', http://fia.fs.fed.us/program-features/tpo/, online database, United States Department of Agriculture, Forest Service, accessed May 2009

USGS GAP (2009a) 'A protected areas database for the United States', United States Geological Survey Gap Analysis Program, http://gapanalysis.nbii.gov/portal/community/GAP_Analysis_Program/Communities/GAP_Projects/Protected_Areas_Database_of_the_United_States/, last accessed August 2009

USGS GAP (2009b) 'Regional projects', United States Geological Survey Gap Analysis Program, http://gapanalysis.nbii.gov/portal/server.pt?open=512&objID=1484&PageID=5128&cached=true&mode=2&userID=2, accessed August 2009

USGS GAP (2009c) 'USGS GAP analysis program history and overview', United States Geological Survey Gap Analysis Program, www.gap.uidaho.edu/Portal/gap_fs2004.pdf, accessed November 2009

Virginia (2002) 'Best management practices for water quality', Virginia Department of Forestry

Vittello, J. R. (2008) Email communication with Senior Forester, Bureau of Indian Affairs, Washington, DC, 6 May

Vogel, David (1993) 'Representing diffuse interest in environmental policy making', in R. K. Weaver and B. A. Rockman (eds) *Do Institutions Matter?* Washington, DC: The Brookings Institution

Waring, R. H. and J. F. Franklin (1979) 'Evergreen coniferous forests of the Pacific northwest', *Science*, 204, 4400, pp1380–1386

Warren, Debra D. (2002) Harvest, Employment, Exports, and Prices in Pacific Northwest Forests, 1965-2000, USDA Forest Service, Pacific Northwest Research Station, Portland, OR, Report No. PNW-GTR-547

Washington (2001) 'Forest practices rules', Chapter 222-30 WAC, July 2001, Government of Washington

Washington Forest Protection Association (2000) 'Forest Practices Board passes salmon protection', http://resources.ca.gov/00-03-16BOFRegsAdoption.html, last accessed December 2009

Wear, David N., and John G. Greis (2002) 'The southern forest resource assessment: Summary report', Asheville, NC: USDA Forest Service Southern Research Station and Southern Region

White, Fred (1992) 'History of forest practices guidelines in North Carolina', Raleigh, NC: North Carolina Division of Forest Resources

Wilkinson, C. F. and H. M. Anderson (1985) *Land and Resource Planning in the National Forests*, Vol. 64, Washington, DC: Island Press

Willig, M. R. (2000) 'Latitudinal gradients in diversity', in S. Levin (ed) *Encyclopedia of Biodiversity*, San Diego: Academic Press

Yin, R. and J. Baek (2004) 'The US–Canada softwood lumber trade dispute: What we know and what we need to know', *Forest Policy and Economics*, 6, pp129–143

Chapter 4

Western Europe: Finland, Germany, Portugal and Sweden[1]

INTRODUCTION

Western Europe encompasses only about 4 per cent of the worlds' forests, but the region is a major force in global forest trade. Germany, Finland and Sweden rank third, fifth and seventh worldwide in total value of wood product imports and exports. The economic importance of the forest sector in these countries is matched with strong political and social engagement in global forestry dialogue. Portugal, the fourth case study in this chapter, plays a much smaller role in global trade. However, it is included in the analysis for its significant role in the global market for plantation pulp and as an example of forest policy and practice in southern Europe.

Western European governments have leveraged their countries' key role as importers through international forest-related initiatives aimed at reforming forest practices in developing countries. They have done this through both European Union-wide actions, such as the EU's leadership in FLEG(T) initiatives and through individual national initiatives such as public procurement policies (see European Commission, 2009). Likewise, European civil society has launched numerous environmental campaigns protesting forest practices worldwide (Stanbury et al, 1991; Stanbury, 2000).

The role of European countries as forest exporters, however, has received relatively little attention in global environmental debates. Forest product export revenues from Germany, Finland and Sweden are the third, fourth and fifth largest worldwide, behind only Canada and the US. This high value trade reflects very well-developed processing and manufacturing capacities, which draw on global sources of raw materials.

Western Europe's forest history differs substantially from most other case studies covered in this book. The region's natural forests have been shaped by centuries of intensive human use, leaving little primary forest to spark the battles over wilderness preservation common in many other regions of the world. Likewise, land tenure issues are relatively settled, with a high percentage of private lands and a long tradi-

Table 4.1 *Forest ownership*

Case study	Total Private	Private Non-Corporate	Corporate	Total Public	State	National	Other	Total Communal*
Germany	46%			34%	31%	3.5%		20%
Bavaria	67%	57%	10%	33%	31%	2%		*
Finland	71%	62%	9%	24%		24%		
Portugal	84%	77%	8%	2%		2%		13%
Sweden	81%	57%	24%	19%		18%	1%	

Note: * The German national data distinguish between private and communal ownerships while the Bavarian state data do not. There are also communally held forestlands in Bavaria.

Sources: Direccao-Geral das Florestas, 1991, 2001; INE, 1997; Bundesministerium für Verbraucherschutz Ernährung und Landwirtschaft, 2001; Mendes and da Silva Dias, 2002; Finnish Forest Industries Federation, 2003a; FAO, 2006a; SFIF, 2007

tion of family ownership. These family-owned forests, furthermore, are relatively small. The average private forest landholding across Europe is 13ha (MCPFE, 2003), with many holdings less than 5ha (CPI, 2008). Publicly owned forest units are somewhat larger, with an average (outside Russia) of 1,300ha (MCPFE, 2003). Table 4.1 above summarizes forest ownership for each of the Western European case study countries.

European forest stakeholders face controversies of their own, however. According to Nabuurs et al, 'Forests are generally seen by the public as refuges in Europe, even though the European forests sometimes strongly deviate from the potential natural vegetation' (Nabuurs et al, 2003; also Uuttera et al, 1997). In response to public concerns for more environmentally oriented forest policies and practices, as well as to emerging international inter-governmental initiatives (Ottitsch and Palahi, 2001), a number of European countries passed new Forest Acts and policies in the 1990s (Thelander, 2000). These policies generally reframed the broad goals of forest management, from a primary emphasis on timber production to an equal balance of environmental and economic outcomes. In addition, many new policies called for the devolution of decision-making to more local levels, as well as for an increase in public participation and inter-agency cooperation (Thelander, 2000). In some countries, such as Sweden, forest rules were 'simplified' towards a more results-based approach, allowing managers more discretion in how they achieve particular forest management goals (Thelander, 2000).

Meanwhile, the European forest products industry is facing rapidly changing world markets for both its resources and its products. Among these are increasing wood demands from Eastern Europe and other rapidly growing economies, increasing global competition among wood producers and growth in markets for woody biomass as various forms of bio-fuel. Stakeholder responses to these changes are varied. For example, some stakeholders are concerned about the environmental and economic effects of bio-fuel production relative to 'traditional' forestry, while others see bio-fuels as an opportunity to address a range of issues including the decline in the competitiveness of European forest production, overall growth in Europe's native forest wood resources and the growing need for alternative energy (Nilsson, 2007).

The following overview reviews forests and forestry issues across the four European case study countries. In three cases, our analysis is focused on the national level. However, in the federal republic of Germany authority over forests largely rests within its 16 states. Thus, we have selected Bavaria, located in the southeast corner of Germany, as a sub-national case study. Bavaria contains the largest expanse of forest and harvests the largest volume of wood products of all the German states.

FORESTS AND FORESTRY IN FINLAND, GERMANY AND BAVARIA, PORTUGAL AND SWEDEN

Finland

An overview of forests and forest ownership

Forests cover about 74 per cent of Finland, the highest percentage of any of our case study countries and of all countries in Europe (FAO, 2006a). Sixty-two per cent of Finland's 22.5 million ha of forests are owned by non-industrial private landowners, 24 per cent by the state and 9 per cent by forest industries. The average area of forest ownership is 26ha (Finland MAF, 2001) and there are approximately 440,000 private forest holdings of at least 1ha. The number of individual private owners is estimated at nearly 900,000, 'which means that one out of every five Finns is a forest owner' (Mikkelä et al, 2001). Consequently, Finnish forestry is commonly termed 'family forestry', defined as small-scale non-industrial forest management undertaken by families in their own forests (Finland MAF, 2001; Mikkelä et al, 2001).

Finnish forest industries own 9 per cent of the country's forests. They include some of the world's largest forestry companies with holdings in numerous countries worldwide. The three largest domestically – Stora-Enso, UPM-Kymmene and Metsäliitto – rank among the world's top ten in sales volume for forest, paper and packaging companies (PwC, 2006).

State forestlands account for about 24 per cent of Finland's forest area and are managed by the state enterprise Metsähallitus. These forests are mostly located in the northern part of the country, with roughly one-quarter of their total area available for wood production (Metsähallitus, 2009a). Since public lands surpass the 20 per cent threshold for ownership types included in our analysis, we include both public and private lands in our assessment of forest practice policies in Finland.

Native forests

Most of Finland lies between 60 and 70 degrees north, with parts extending into the Arctic Circle. Gulfstream waters exert a warming influence, allowing forest ecosystems to extend into the far north of the country. Boreal forests are the dominant forest type and the most common tree species are Scotch pine (*Pinus silvestris*) (66 per cent), Norway spruce (*Picea abies*) (24 per cent) and pendulous and downy/white birch (*Betula pendula* and *pubescens*) (9 per cent) (FFRI, 2007). Small areas in the south and southwest support hardwood forests of oak, maple, ash and elm (Boreal Forest, 2008a).

Finland's history of forest use and management has resulted in few areas of old growth and relatively low tree species diversity in the southern half of the country, coupled with less intensively managed forests in the north. The percentage of forests classified as old growth (over 140 years) ranges from almost 25 per cent in the north-ernmost forest vegetation type to 0.3 per cent in the southernmost (Punttila and Ihalainen, 2006). Single species forests (defined as the volume of dominant species greater than 95 per cent) constitute about 38 per cent of all forests, 31 per cent are 'some mixed species' (volume of dominant species 75–95 per cent) and the rest are mixed stands (FFRI, 2006).

The loss of biodiversity in southern Finland, and climate change, are assessed as the greatest threats to Finland's forests (Finland MAF, 2008a).

Planted forests and plantations

None of Finland's forests is classified as a plantation (FAO, 2007). Among the 94 per cent of forests classified as 'semi-natural', about 5.3 million ha are defined as 'planted' and 15.8 million ha as 'assisted regeneration'. Almost all species planted are native. In 2007, Norway spruce was the most common species planted and less than 0.1 per cent of the planted species were non-native (FFRI, 2008). Scotch pine and pendulous birch are also popular planted species (FAO, 2006b).

Forest governance and policy

The principal laws governing forestry in Finland are the Forest Act 1996 and the associated Act on the Financing of Sustainable Forestry 1996. These Acts apply to land designated for forestry purposes. The Forest Act regulates forest practices and the Financing Act provides public grants or loans to private landowners to promote forest conservation and sustainable management (Finland MAF, 2009a, 2009b).

The Ministry of Agriculture and Forestry (MAF) and its subsidiary Department of Forestry oversee forest policy and planning. MAF policies are implemented by a network of 13 regional forestry centres (Grey, 1988). The centres also advise and train private landowners, forest workers and forestry entrepreneurs (Finnish Forest Certification Council, 1999; Hytönen, 2002) and cooperate with representatives of forest industries and with the Forest and Park Service (Hytönen 2002).

Metsähallitus is a state enterprise operating within the administrative sector of the Ministry of Agriculture and Forestry. It serves as the business wing of MAF responsible for 'sustainable and profitable' forestry activities on state lands. Metsähallitus also has responsibility for nature conservation and for this purpose is guided by the Ministry of the Environment (Act on Metsähallitus 1378/2004).

Finnish forest owners are represented in policy and governance processes by nested organizations at the national level (the Central Union of Agricultural Producers and Forest Owners: MTK), regional level (Regional Unions of Forest Management Associations) and local level (Forest Management Associations (FMAs)). FMAs are part of a long-standing tradition of cooperation among forest owners throughout Finland (MTK, 2009). The Act on Forest Management Associations, passed in 1950 and since revised, outlines the purposes of the associations: in essence, 'FMAs offer training and guidance and provide professional assistance in forestry issues, thus protecting forest owners' interests and helping to achieve the set objectives' (MTK, 2009). The FMAs are governed and financed completely by forest owners, who pay an obligatory fee depending on the size of the holding and current stumpage prices. Almost all private landowners in Finland with holdings greater than 5ha are members of one of over 200 local FMAs (Finnish Forest Certification Council, 1999; MTK, 2009).

Forest practices system

Finland's forest planning and practices system reflects the structure of its forest sector and governance arrangements. Private forest planning, at scales from the regional to the farm, is undertaken by the regional forestry centres. These centres are charged with preparing, implementing, monitoring and revising forestry programmes within their region, and maintaining regional environmental data and records.

The preparation of a management plan for private lands is voluntary. About one-third of private non-industrial forest owners have a farm level forestry plan and those plans cover nearly half the total area of NIPF ownerships (FDC Tapio, 2007). Plans for corporate forests are typically prepared by the company and Metsähallitus prepares management plans for state forests (Finland MAF, 2009c).

Forest production and trade

Although small and fragmented, private holdings are critical to the forest industry because they supply domestic processors with over 80 per cent of their total raw material. Each year, between 100,000 and 150,000 individual wood contracts are made between private forest owners and industrial companies (Finnish Forest Certification Council, 1999; Finland MAF, 2001; Finnish Forest Industries Federation, 2003b). Forest owners are able to grant Forest Management Associations (FMAs) the power of attorney for wood sales and deliveries. In 2001, approximately 40 per cent of timber sales from private forests were facilitated by attorney sales (MTK, 2009). In addition, more than 80 per cent of activities relating to timber production in private forests, and 70 per cent of preliminary timber sale planning, were carried out by FMAs (MTK, 2009).

Harvest from a wood producing family forest is usually carried out every three to four years and logs are sorted to their highest use. There are over 150 industrial sawmills and hundreds of small mills (Finnish Forest Industries Federation, 2009). Large mills dominate exports while small mills largely sell to the domestic market (Boreal Forest, 2008a).

Around three-quarters of the wood consumed by Finland's forest industries is harvested domestically; the balance is imported, mostly from the Baltic states and Russian Federation. The extent of illegally sourced wood amongst these imports is disputed (Chatham House, 2009; Finnish Forest Industries Federation, 2009).

Forest product exports generate one-quarter of Finland's total export revenues. The forest industry is highly international, with over half of Finnish forest company production located outside the country (FFA, 2004). Pulp and paper, and in particular high grade paper, are a primary emphasis and account for the majority of exports.

Indigenous peoples, reindeer herding and forestry

The Saami people of Sápmi – an indigenous domain which encompasses parts of present day northern Finland, Norway, Russia and Sweden – claim rights of full ownership to their traditional lands in northern Finland, but these are not recognized to their satisfaction by Finnish law (Baer, 1996; Saami Council, 2006).

Reindeer husbandry is commonly practised in the north of Finland by both indigenous and non-indigenous peoples and the differing priorities of herders and forest owners can be a source of conflict. Metsähallitus and the Finnish Reindeer Owners' Association have signed an agreement to improve cooperation between the forest sector and reindeer herders and annual meetings are held with each local herding cooperative to discuss forest operations. However, conflicts over tenure rights continue in some areas (Metsähallitus, 2009b).

Germany and Bavaria

Forests and forest ownership

About 32 per cent, or 11 million ha, of Germany is forested (FAO, 2006a). These forests cover a varied topography, progressing north to south from lowlands, through rolling hills, to the Bavarian Alps. The climate is oceanic in the north and northwest, transitional in the central and south and continental in the east (Hofmann et al, 2000).

The proportion of Bavaria that is forested is about the same as the national average. In the north of Bavaria are the Franconian forests; in the east, the Bohemian and Bavarian forests and to the northwest are wooded sandstone hills. In the southeast, forests grow up to the tree line of the Bavarian Alps, where mountains reach a height of almost 3000 metres.

Across Germany as a whole, private forest ownership predominates (46 per cent), followed by state (31 per cent), communal (20 per cent) and lastly federal (3.5 per cent) (Bundesministerium für Verbraucherschutz Ernährung und Landwirtschaft, 2001). Almost all forests in the former East Germany were state-owned pre-transition and are now being re-allocated as private property. The average size of private woodlots is 5ha (Volz and Bieling, 1998) and nearly 10 per cent of private forests are in blocks of less than 1ha on which no forestry practices occur (Hofmann et al, 2000).

In Bavaria, the proportion of forests that are privately owned is reported at 67 per cent; this figure includes forests in communal ownership, a distinction which is not reported at the state level (see Table 4.1). Ten per cent of these holdings are 'corporate' and over 600 properties exceed 200ha; however, the average forest ownership is only 3.7ha (Beck and Krafft, 2001). State (31 per cent) and federal (2 per cent) forests comprise the remainder of Bavaria's forest area and thus public forests are included in our analysis of forest practice policies.

Native forests

Hardwood forests account for about 13 per cent of Germany's forestlands, softwoods 44 per cent and mixed forests 43 per cent (Hofmann et al, 2000). The major tree species include spruce (primarily Norway spruce, *Picea abies*, 28.2 per cent), pine (including Scotch Pine, *Pinus sylvestris*, 23.3 per cent), European beech (*Fagus sylvatica*, 14.8 per cent) and oak (*Quercus* spp., 9.6 per cent) (Roering, 2004). As in many other European countries, Germany's forest area has been expanding due to increasing urbanization and the abandonment of small-scale agricultural farms. The vast majority of German forests have been shaped by at least a millennium of human management, leaving virtually no primary forests (FAO 2006a).

Global warming and air pollution pose substantial threats to German forests. Rising temperatures and drought have spurred the spread of spruce bark beetle, gypsy moth and oak procession moth. In Bavaria, bark beetle infestations have affected up to 50 per cent of the spruce harvest on state forests, with the heaviest losses occurring in Franconia. In addition, climate change is anticipated to shift the distribution of spruce northwards, to be replaced by beech and other hardwoods. Nitrogen accumulation from air pollutants continues to threaten the productivity of forest soils (BSLF, 2006).

Planted forests and plantations

None of Germany's forests is classified as a plantation and about 6.6 million ha are classified as 'semi-natural' planted forest. About one-third of the planted forest area is designated primarily for protective purposes and the remainder for production (FAO, 2007). In state-owned forests where timber production is allowed, there has been a strong movement away from intensive forest management (Tschacha, 2007).

Forest governance and policy

Legislation and regulations at each of European, federal and state levels are important for forest management in Bavaria. Regulations for the protection of species and ecosystems under the Birds and Habitats Directives of the European Union (EU, 2009) have a particularly strong influence on forest management, imposing specific restrictions that go substantially beyond those imposed by the National Forest Act and state level legislation.

The German Federal Forest Law (*Bundeswaldgesetz*) was enacted in 1975 and amended in 1997 (Cashore, 1997). This law sets broad goals for 'sustainable' (*nachhaltig*) forest management, including the maintenance of forest cover, and upholds public rights to forest access. The law requires government permission to convert forestlands to agriculture and other uses and prohibits the granting of deforestation permits if it is considered to be counter to the public interest (Roering, 2004).

Numerous state-level Forest Acts have been passed and there is considerable variation in their content. The Bavarian Forest Law was first established in 1852 and replaced by a new law in 1982 (Bayerisches Waldgesetz). The main stated goals of the law are to:

1   Maintain total forest area and increase it if necessary;
2   Maintain forests that are appropriate to local conditions or restore them;
3   Increase and strengthen the protective function of the forest;
4   Secure and increase the yield of forest products through sustainable forestry;
5   Provide and enhance opportunities for recreation in the forest;
6   Support and further forest owners in the pursuit of these goals.

While the National Forest Act focuses principally on maintaining the extent of forests, the Bavarian Forest Act contains more specific forest management provisions. For example, clear felling is generally prohibited in high-altitude forests and requires special permission in protection forests.

At the federal level, the Ministry of Consumer Protection, Food and Agriculture (BMVEL) provides national oversight of forest management and production. In Bavaria, the Ministry of Agriculture and Forestry, which has replaced the previous Forest Ministry, has oversight of forests and the forest sector. The 'business and economic' aspects of forest management have been assigned to a new institution, Bavarian State Forests (*Bayerische Staatsforsten*, BaySF). BaySF oversees 41 forest enterprises with mandates to operate as for-profit forestry firms. The state plays primarily an advisory role in relation to private forests.

Forest practices system
Historically, the forest practices system in Bavaria, as elsewhere in Germany, is based more on incentives – extension, advice and expert support – than on sanctions. Forest management plans are the principal means by which the Bavarian forest practices system is implemented and planning requirements differ between tenures. Management plans are mandatory for public forest owners, including cities and municipalities. They are not required of private forest owners. However, consistent with the incentive-oriented policy regime, tax concessions for forestry activities can only be obtained on the basis of forest management plans. Consequently, management plans are very common among both large private forest owners and small-scale, traditional, forest farmers. The main aim of these plans is to encourage good forest practices rather than control the behaviour of forest enterprises (Schraml, 2009).

When the responsibilities of the Bavarian State Forest Service were redefined in 2004 to focus on 'the common good', the state forestry administration largely withdrew from extension services and subsequently encouraged the development of private service providers, in particular the Forestry Associations, to take this role. This situation in Bavaria contrasts with that in many other German states, where extension and support services are still being offered for free or at reduced fee levels. The impact of these changes in Bavaria's forest practices system on forest management is not yet fully apparent.

Forest production and trade

Germany has the second lowest per capita forest area of all case study countries, after India. While its population is nearly 3.5 times that of Finland, Sweden and Portugal combined, the country contains only a fifth of their combined forest area. Consequently, the German forest products industry relies heavily on wood product imports and high value-added trade. Germany reports that 83 per cent of its forest-lands are designated primarily for purposes other than production: 42 per cent are designated primarily for social services, 22 per cent for protection and 19 per cent for conservation (FAO, 2006a).

In general, forest properties are highly fragmented, particularly on private lands (Schraml and Winkel, 1999; Schraml and Thode, 2000). Landowners have tried to achieve economies of scale through hundreds of voluntary 'forest enterprise cooperatives' (*Forstbetreibsgemeinschaft*), which coordinate forestry and marketing activities. Forest ownerships associated with such cooperatives cover about 70 per cent of Bavaria's private and communal forest area (BSLF, 2006). These coopera-tives are represented through state-level associations that are in turn affiliated with the national 'Alliance of German Forest Owner Associations' (Roering, 2004).

The species harvested across Germany, in roughly decreasing volume, are Norway spruce, fir, Douglas fir, pine, larch, European beech, other deciduous, and oak, red oak (Roering, 2004). Bavarian state forests report a harvest of 46 per cent spruce, 17 per cent beech, 13 per cent pine, 5 per cent oak and 19 per cent other species (Tschacha, 2007).

There are very few vertically integrated forestry operations in Germany (i.e. operations that own forestland, harvest trees and process them). Rather, harvesting is done by private contractors or (more frequently) by the state and processing is done by private sector sawmills or paper companies (Schraml and Thode, 2000). There are still large numbers of small sawmills. The largest companies are found in the more capital-intensive enterprises, such as the wood-based panel industry, cellu-lose and paper industries (Bundesministerium für Verbraucherschutz Ernährung und Landwirtschaft, 2001). However, there has been a trend towards increasing concentration across all product categories (Schraml and Thode, 2000).

Germany is Europe's largest importer and exporter of forest products by value. In 2006 it ranked second worldwide, after Canada, in the value of its wood product exports (FAOSTAT, 2008). Market growth in recent years has been strong, due in part to expanding exports to Eastern Europe. Germany's paper production, in particular, remains a growth sector (ACPWP, 2006).

Portugal

Forests and forest ownership

Portugal's forests, totalling some 3.8 million ha, comprise around 40 per cent of its land area. Semi-natural forests comprise two-thirds and plantation forests almost one-third of Portugal's forest area (FAO, 2007).[2]

Eighty-five per cent of Portugal's forests are privately owned, in 400,000 holdings; 93 per cent of these are of less than 10ha and held by non-industrial owners. Fifty-five per cent of private forestland is held by 1 per cent of owners. Pulp and paper companies own 6 per cent, local communities 7 per cent and the state 2 per cent of Portugal's forests (ISA, 2006).

Native forests

Historically maritime pine (pinheiro-bravo/*Pinus pinaster*) forests, most heavily concentrated in the northern half of the country, were the largest forest type in Portugal, followed by cork oak (*Quercus suber*) systems, largely concentrated in the southern half. However, between the 1995/98 and 2005/06 national inventories, about 245,000ha of maritime pine forest were lost, leaving the 745,000ha of cork oak forests as the largest forest type (CELPA, 2007). Holm oak (*Quercus ilex*) and other oak species comprise a further quarter. Total area classified as forest in Portugal continues to increase, consistent with a long-term trend that has seen a tripling of the country's forest cover since the mid-19th century (Mendes and da Silva Dias, 2002), although much of the expansion since the 1960s has been due to plantation forestation.

Forest fires are a recurrent and substantial threat to Portugal's forest. While frequent fires are a natural part of the maritime pine disturbance regime (Fernandes and Rigolot, 2007), the number and extent of fires has increased in recent years, resulting in greater loss of standing forest from fire over the past four decades than gains from extensive planting of deforested and/or non-forested areas (Mendes and da Silva Dias, 2002; JRC-IES, 2007).

Consequently, managing fire and reducing its adverse impacts are significant challenges to sustainable forest management in Portugal. Other challenges include the restoration of native species, reforesting burnt regions and old agricultural lands and addressing the declining competitiveness of forest industries resulting in part from losses due to fire (Mendes and da Silva Dias, 2002).

Planted forests and plantations

Plantations play a central role in Portuguese forestry. In 2005, there were 1.1 million ha of productive plantation and 0.2 million ha of protective plantation, corresponding to about one-third of the country's total forest area. Some 16 per cent of Portugal's semi-natural forests are planted: 0.3 million ha for production and 0.1 million ha for protection.

Eucalypt plantations, principally *E. globulus* grown for pulpwood, comprise nearly a quarter of all Portugal's forests. These plantations are most heavily concentrated in the centre and north of the country (CELPA, 2007). Industry acquisition and renting of properties for eucalypt plantations has become common with rising out-migration from rural areas (Alves, no date).

On average, industry plantations are somewhat larger than non-industrial plantations. However, individual plantation management units across both ownership types are commonly between 1 and 20ha and rarely exceed 100ha (Mendes and da Silva Dias, 2002). The Portucel Soporcel Group is the single largest industry firm, which manages 4.5 per cent of Portugal's natural forests and 20 per cent of its eucalypt plantations (Portucel Soporcel, 2004).

Forest governance and policy

The Portuguese Forest Service was established in 1886 and the first national laws related to forests and forestry were promulgated in 1901, 1903 and 1905. These early laws emphasized both forest protection and reforestation. The 1938 Forest Law (No. 1971) was significant in enabling subsequent large-scale forestation by the state.

Portugal's first comprehensive Forest Act was introduced in 1996 and a National Strategy for Forests (ENF) was approved in 2008. The 1996 Forest Policy Act (No. 33/96) outlines broad principles of sustainable forest management and places a strong emphasis on social rights and benefits (EFC, 2000). Decree-Laws No. 204/99 and 205/99 set out requirements for preparing Regional Forest Management Plans (Planos Regionais de Ordenamento Florestal, PROF) and operational level Forest Management Plans (Planos de Gestão Florestal, PGF) (AFN, 2009b, 2009c).

The government body responsible for forestry activities is the Autoridade Florestal Nacional (formerly the Direcção-Geral das Florestas, DGF), of the Ministério da Agricultura, Desenvolvimento Rural e Pescas (Ministry of Agriculture, Rural Development and Fisheries). Given the small size of most Portuguese forest holdings, forestry associations are also important as a means of representing small forest owners and assisting their engagement with the forestry sector; 179 such organizations are registered, the majority in northern and central Portugal (AFN, 2009a).

Forest practices system

Regional Forest Management Plans have a 20-year planning horizon and are developed consultatively for each of 28 regions. They identify, amongst others, preferred approaches to: the expansion and conversion of forests; species mixes; forest harvest; the identification of areas critical for the conservation of natural and cultural importance; and the management of fires and soil erosion (AFN, 2009b).

At the level of the forest management unit, proposed forest practices are specified in a Forest Management Plan (PGF). Best Practices for Sustainable Forestry and Criteria and Indicators were recently established under the direction of the Plan for the Sustainable Development of Portuguese Forests (Comissão Técnica de Normalização, 2004).

Forest production and trade

According to FAO, total industrial roundwood production in Portugal was 10.9 million m³ in 2004 (FAO, 2007). Pulp and paper generate the largest export values of Portuguese wood product exports and about 99 per cent of the fibre used is sourced domestically. In 2006, sales of eucalypt for pulp and paper production totalled about 4.8 million cubic metres and sales of pine about 1.1 million cubic metres. About 96 per cent of this production was exported to other EU countries and only about 1 per cent consumed domestically (CELPA, 2007).

Portugal leads the world in the production of cork, with an annual production of 120,000 metric tonnes (Fonseca and Parresol, 2001).

Community forestry

Communal forests (known as *baldios*) account for about 7 per cent of Portugal's forestlands (ISA, 2006). The history of these communal properties has been fraught with struggle between different users and between local and central authorities. *Baldios*, which historically covered some four million ha, played an important role in traditional farming systems. Village councils administered the management of these lands for a range of purposes, including the grazing of goats and sheep and the maintenance of woodlands to provide building material and fuel wood (Brouwer, 1995).

The 1933 New State Regime brought major changes to the country's land tenure system, including the privatization of most of the *baldios* lands in all but some remnant areas in the north. Much of the remaining communal land was forested with pine for commercial wood production and placed under the control of a highly centralized Forest Service. Conflicts between the Forest Service and local communities precipitated the abolition of communal property in 1966 (Brouwer, 1995).

The 1974 leftist coup brought further shifts in government priority. In a new alignment between local community groups and the Forest Service, an agreement was reached in 1976 (Law 39167) to reinstate the *baldios* under a new institutional structure. As of 1990, about one-third of the common land area was administered by 132 commoners' assemblies and management councils. These communally managed forests play an important role in the provision of revenue and infrastructure for a number of rural communities (Brouwer, 1995).

Sweden

Forests and forest ownership

60 per cent of Sweden, or about 27.5 million ha, is forested. The majority of this area, 22.1 million ha, is classified as semi-natural forest – the largest such area in Western Europe. Plantations total around 670,000ha (FAO, 2007).

More than 80 per cent of Sweden's forests are privately owned: 57 per cent by small-scale, private owners and 24 per cent by corporations. Nineteen per cent of the forests are state-owned. The share of government owned forestland has fluctuated with the privatization, in 1992, and renationalization, in 2002, of the firm AssiDomän/Sveaskog (Sveaskog, 2009).

Historically, small-scale private forest owners – usually families – in Sweden have drawn their livelihoods from both agriculture and forestry on their properties, but the number of farmers doing so has declined by 30 per cent over the past 50 years. There are now 355,000 family forest owners, 70 per cent of whom live on their properties. The average forest area under family ownership is 46ha (SFA, 2009a).

Small-scale private forestland ownership is predominant in the south of Sweden; industrial ownership is greatest in the north. There are a small number of large non-industrial forest owners – notably the Church of Sweden, which gives each of its eight dioceses authority over forest management on its estates. Industrial ownership is fairly concentrated, with five large Swedish forest companies owning and managing one-third of Sweden's industrial forestlands and controlling approximately 95 per cent of pulp and paper processing capacity (van Kooten et al, 1999, p160). These companies produce about one third of the total harvest volume (van Kooten et al, 1999, p160).

Native forests

Sweden's forest ecosystems are similar to those described above for Finland. Four vegetation zones are recognized: boreal forests, which comprise the largest zone, extend north and northeast into the Arctic Circle. The alpine/sub-alpine zone is mostly located in the Scandinavian mountain range of the northwest coast. In the south are the boreonemoral and nemoral mixed European forests (SFA, 2009a).

The most common tree species are pine (39 per cent of standing volume), in particular Scotch pine (*Pinus silvestris*); spruce (41 per cent of standing volume), in particular Norway spruce (*Picea abies*); birch (*Betula* spp.) (12 per cent); other deciduous trees (3 per cent) and dead trees (3 per cent) (SFIF, 2007). About 4.7 million ha are classified as primary forest and 12.8 million ha as semi-natural forest resulting from assisted regeneration (FAO, 2007).

Many of Sweden's forests were degraded over the course of the 19th century by land clearing, grazing and wood consumption. For several decades in the mid-20th century, industrial wood production emerged as the country's largest industrial sector and a dominant source of forest impacts (SFA, 2009a). After many years of relatively intensive management, the structural diversity of Sweden's forests has been greatly reduced by the widespread removal of standing dead trees, coarse woody debris and large old growth pine and spruce (Boreal Forest, 2008b).

In the late 20th century, Sweden began to place greater emphasis on environmental forest values. At the same time, acid rain originating from other European countries began to pose a serious forest threat. Consequently, Sweden was at the forefront of international efforts to curb acid rain and eutrophication and these latter threats have largely abated (Lorenz et al, 2008; SFA, 2009a).

Planted forests and plantations

'Planted semi-natural forests' comprise about one-third, or 9.3 million ha, of Sweden's total forest area; plantations total 668,000ha (FAO, 2007). Species commonly planted include the non-native North American lodgepole pine (*Pinus contorta*) (FAO, 2006b), as well as native Scots pine (*Pinus sylvestris*) and Norway spruce (*Picea abies*).

Forest governance and policy

The first Forestry Act in Sweden was established in 1903 and has been revised several times since then to reflect changing national priorities. The most recent revision was in 1994, marking a shift towards equal weighting of production and conservation goals. The Forestry Act applies to all forest ownership types. It is complemented by the 1999 Environmental Code, which encompasses biodiversity conservation and environmental protection (SEPA, 2009b).

Sweden's National Forest Policy was most recently revised in 2005, to specify interim quantitative targets for forest conservation and management. These targets complement those specified in the 2001 Environmental Policy (SFA, 2009b): both seek to enhance biodiversity outcomes associated with forest management (Uliczka, 2003).

The Swedish Forest Agency (SFA) (Skogsstyrelsen) within the Ministry of Agriculture Food and Fisheries is responsible for implementation of forest policy and oversight of forest management (SFA, 2009c).

The Swedish Federation of Forest Owners (Skogsägarnas Riksförbund) encompasses seven forest owners' associations, which together account for about 50 per cent of family forestland. Large forest companies have likewise organized themselves under the Swedish Forest Industries Association (Skogsindustrierna) (Boreal Forest, 2008b).

Forest practices system

The forestry Act requires that all forest owners prepare a plan, describing their forest and its natural and cultural values. This plan must be made available to the State Forest Agency on request, but is not otherwise checked. Owners must register any forest management activities affecting an area greater than 0.5ha with the SFA, which then issues permission to conduct those activities for a period of up to three years (Uliczka, 2003). The SFA has developed management guidelines for private owners (KSLA, 2009).

Forest production and trade

Non-industrial forestry is an important contributor to wood production, accounting for over 60 per cent of the fibre harvested annually (van Kooten et al, 1999, p160). Industrial forest owners tend to manage forests more intensively than small-scale operators, through such practices as heavier commercial thinning (Yrjölä, 2002). In a 1997/98 survey, about 60 per cent of the gross felling volume came from final fellings, while some 48 per cent of the thinnings are first thinnings (Yrjölä, 2002). The proportion of deciduous trees was estimated to grow from 10 per cent to 20 per cent of the total fellings by the end of the 20th century (Gustafsson and Thurlesson, 2001). Legislative changes in 1994 reduced the final felling stand age by 25 per cent. The decrease is mostly due to the fellings of younger spruce stands in southern Sweden. Sweden's total growing stock continues to increase, with annual harvests amounting to about 70 per cent of growth (Boreal Forest, 2008b).

The Swedish forest industry is similar in a number of ways to that of Finland, including its major importance to the national economy, its very high technical capacity, as well as its high level of globalization. In Sweden, the industry reportedly accounts for 11–12 per cent of total industrial employment and for 11 per cent of exports. Sweden is Europe's third largest pulp and paper producer, after Germany and Finland, and the world's third largest exporter of paper and sawn timber. The top importers of Swedish forest products are Germany, the United Kingdom, the Netherlands and Denmark (SFIF, 2007).

Indigenous peoples, reindeer herding and forestry

The situation of indigenous forestry in Sweden is similar to that described previously for Finland. Perhaps a quarter of all Saami people live in Sweden; their rights over and access to their traditional forestlands are the subject of a number of governmental and judicial processes (Baer, 1996; Barklund, 2006). Forest owners have adopted a suite of measures to minimize the conflict between forest management for wood production and the reindeer herding practices of Saami people (Barklund, 2006).

A COMPARISON OF BIODIVERSITY CONSERVATION MEASURES

All of the case study countries covered in this chapter are members of the European Union, which has enacted a number of key pieces of legislation that apply within member countries and facilitate cross-boundary collaboration for biodiversity conservation. Among these are the Birds Directive, first passed in 1979, and the Habitats Directive, enacted in 1992. The latter Directive created the Natura 2000 network of protected areas, a system designed to comply with the United Nations Convention on Biological Diversity. An overarching objective of these EU initiatives is to 'meet the European Council's goal of halting biodiversity decline within the EU by 2010' (EC, 2003). In addition to Europe-centred conservation efforts, the EU has also developed legislation to address the impacts of its role in global trade, including the EC Regulation No. 338/97. This latter regulation formalizes agreements under the Convention on International Trade in Endangered Species of Wild Fauna and Flora (CITES).

Protected areas[3]

At the EU level, the Natura 2000 network has established two major categories of protected areas: Special Areas of Conservation (SAC) to address the Habitats Directive, and Special Protection Areas (SPA) to address the Birds Directive. Inclusion in the network does not preclude the commercial extraction of resources, so long as these activities are conducted in a 'sustainable' manner.

As of 2008, 9 per cent of Finland's total land area was under protected area status. This includes 3.2 per cent under Categories I–IV (which generally do not allow extraction of natural resources) and 5.8 per cent under Categories V–VI and unclassified (see Chapter 2, Figure 2.15).

With regard specifically to forestlands, strictly protected forests (including both forest and scrub land) cover 9 per cent of Finland's forest area. Most of these areas are located in the northern parts of the country, with 16 per cent of northern forests under strict protection compared to 2 per cent of southern forests (FFRI, 2008). Protected forests and forests under restricted forestry use cover 13 per cent of the forest area (forest and scrub land), including 22 per cent of northern forests and 4 per cent of southern forests (Finland MAF, 2008b). 'The Forest Biodiversity

Table 4.2 *Percentage of total forest area in Germany in different reserve types and summary of forestry activities allowed (some forest stands fall into more than one reserve category)*

Reserve type	Percentage of forest in reserve	Limitations on forestry
Landschaftsbestandteile	0.05	Unclear
Naturwaldreservate	0.26	No forestry allowed
Nationalparke	1.27	No forestry allowed in 'core zone'
Biosphaerenreservate	2.27	Sustainable forestry practised for educational purposes
Naturschutzgebiete	3.34	No forestry allowed
Wasserschutzgebiete	5.74	Unclear
Bannwald	7.23	Conversion to non-forest use prohibited
Schutzwald	7.71	Conversion to non-forest uses and clearcutting require permits
Landschaftsschutzgebiete	19.83	Unclear
Naturparke	37.51	Very limited forestry allowed
Total	85.21	

Sources: Bayerisches Staatsministerium für Ernährung Landwirtschaft und Forsten, 1999; Pan European Forest Certification Bavarian Working Group, 2000

Programme for Southern Finland 2008–2016 (METSO)' aims to acquire the most ecologically valuable forest sites in southern Finland for temporary conservation or permanent protection. The participation of forest owners in this programme is voluntary (Finland MAF, 2008a).

The Ministry of the Environment oversees the Finnish Environment Institute and 13 regional Environmental Centres, each responsible for promoting conservation at the regional level. The Ministry of Environment also oversees Metsähallitus as the entity responsible for managing Finland's state-owned nature reserves, and other protected areas, wilderness areas and outdoor recreational areas (MOE, 2009a).

As of 2008, Germany reported that 41.2 per cent of its land area was set apart for protection. Only about 3.5 per cent of this area was in IUCN Categories I–IV, with the rest under the less strictly protected IUCN Categories V and VI (see Chapter 2, Figure 2.15).

At the state level, Bavaria has established ten different classifications of forest reserve, now encompassing about 85 per cent of its total forest area. These range from full protection to areas allowing limited commercial forestry activities. Forests adjacent to some reserve types must be managed in a way that avoids influencing or damaging the reserve. Table 4.2 below summarizes the percentage of land area under each of the ten reserve categories.

In Portugal, Law Decree No. 19/93 outlines seven different classes of protected area. These are natural monuments, natural reserves, natural parks, national parks, regional or local protected areas, classified sites and private protected areas (ICN, 2003). In 2008, 6.6 per cent of the country was classified under the IUCN protected areas system, with 1.4 per cent under the stricter IUCN Categories I–IV (see Chapter 2, Figure 2.15). Twenty per cent of mainland Portugal is listed as Natura

2000 sites (ICNB, 2009). However, the proportion of forest protected specifically for biodiversity is very low (MCPFE, 2007: Figure 46).

In Sweden in 2008, a total of 10.2 per cent of the country was in protected areas. This includes 8.8 per cent under the strictest IUCN Categories I–IV (see Chapter 2, Figure 2.15).

According to the Swedish classification system, protected areas include national parks, nature reserves, culture reserves, nature management areas and wildlife sanctuaries (SEPA, 2004). Some 6 million ha, 15 per cent of Sweden's area, are listed as Natura 2000 sites; 60 per cent of these are under strict protection (SEPA, 2009a). These include about 1 million ha, corresponding to 4 per cent, of productive forest-land (SFA, 2009d). Montane areas are the best represented of Sweden's ecosystems and account for 90 per cent of the total reserve system (SEPA, 2004).

Protection of species at risk

The Natura 2000 network, as well as the EC Birds and Habitats Directives (EU, 2009), all state the protection of threatened species as a primary goal. These policies address both the protection of the plants and animals themselves, as well as their key habitat. The Directives include annexes with prioritized listings of species at risk and habitat types of Community interest. For some habitat types it is required to establish special areas of conservation. For animal and plant species 'in need of strict protection', as listed in Annex 4(a) of the Habitats Directive, it is required to restore them to 'favourable conservation status'. In 2005, these requirements were further strengthened by the adoption of an EU framework that defines methods for assessing conservation status. While these EC policies do hold legal weight, the level to which they are implemented depends to a large degree on the individual governments responsible for their implementation.

The 1997 Finnish Nature Conservation Decree (160/1997, as amended by 916/1997, 14/2002 and 913/2005) lists protected species, threatened species, species needing special protection and species which need strict protection according to the EU Habitats Directive (MOE, 2009b). The Finnish classification for endangered species is based on that of IUCN and the type of habitat and threat factors, adjusted to conditions for Finland. The law prohibits the destruction of habitats necessary for survival of endangered species, or any action that might impair their existence (Mikkelä et al, 2001). Metsähallitus is responsible for the monitoring and protection of many threatened species (MOE 2009a).

In Germany, the central piece of legislation addressing endangered species is the Federal Protection of Species Ordinance (BArtSchV). The Federal Agency for Nature Conservation maintains an online database of threatened species and their protection status. The direct taking of endangered species is prohibited (Bundesartenschutzverordnung). Within Bavaria, there are no standardized state-level guidelines for protecting endangered species' habitats, although activities that could lead to the destruction of certain special biotopes such as bogs and riparian forests are prohibited (Bayerisches Naturschutzgesetz). In addition, certain endangered tree and bush species are afforded special protection and the Bavarian state forestry agency has a programme which encourages the expansion of existing stands of these species (Bayerisches Staatsministerium für Ernährung Landwirtschaft and Forsten, 1999).

In Portugal, there is a requirement to conserve habitats classified under the EU Habitat Directive (Decreto-Lei No. 93/90). In addition, 'massive' trees and shrubs and exemplary endemic species are protected from harvest (Art. 10 of Regulatory Decree 55/81, Law Decree 28468). Legislation has also been developed that specifically targets the conservation of oak trees. This includes Decree Law 11/97 that prohibits all land uses that affect the natural regeneration of cork groves. Likewise, EEC Regulation 2078/92 provides a measure to assist in the regeneration of holm oak groves (MOE and NCI, 1998).

The Swedish Forestry Board classifies key habitats or 'biotypes' that must be subject to extra environmental protection measures. As of 1998, about 0.8 per cent of the Swedish forestlands were officially designated as special management zones (Yrjölä, 2002).

FOREST PRACTICE REGULATIONS: NATIVE FORESTS

Riparian zone management (Indicator: Riparian buffer zone rules)

Finland's riparian zone policies differentiate between 'natural' streams and other watercourses. Specifically, the 1996 Forest Act states that the management of 'natural' or 'semi-natural' streams must be carried out in a manner that 'preserves the special features of the habitats' (Finland, 1996). We consider this a 'mixed' policy approach in that the policies specify a substantive conservation outcome but do not include mandatory, standardized thresholds (Parvainen, 2004; Saarenmaa, 2004). In addition to these qualitative requirements, Finland has also developed quantitative voluntary guidelines or 'best management practices' that recommend specific buffer widths for streams and springs based on slope and stage of forest harvest (see Figure 4.1).

German federal law prohibits activities on private land that lead to the 'destruction' of wetlands, riparian forests and 'natural and near-natural' streams, rivers, lakes and ponds (Germany 1998, Article 20c). No forestry activities occur in riparian areas on state-owned lands and state restoration programmes have been instituted to bring back the natural species composition where necessary. On both private and state lands, riparian buffer zone widths and management are handled on a case-by-case basis (Pan European Forest Certification Bavarian Working Group, 2000). As in the case of Finland, we classify this as a 'mixed' policy approach.

In Portugal, best management practice guidelines recommend riparian buffer zones of ten metres. These voluntary substantive guidelines suggest that buffer management 'may require planting low lying species along riparian zones to control for erosion' (Direccao-Geral das Florestas, 2003).

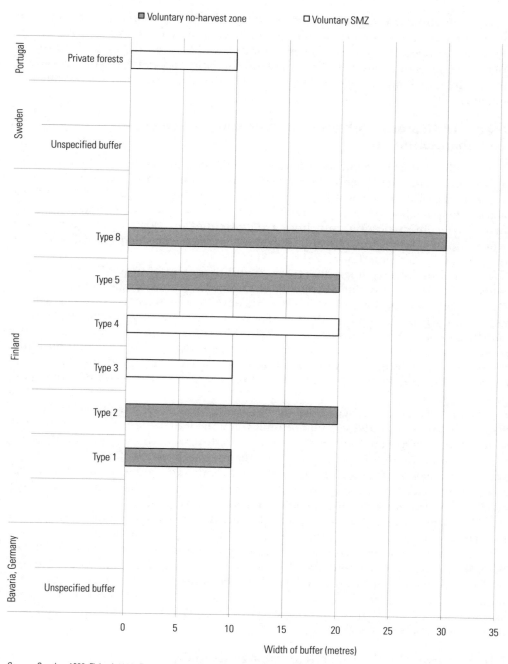

Sources: Sweden, 1993; Finland, 1996; Germany, 1998; Direccao-Geral das Florestas, 2003

Figure 4.1 *Riparian buffer zone policies in Western European case study jurisdictions*

Sweden also has no standardized requirements for riparian buffer zones. The Swedish Forestry Act instead includes generalized and mandatory requirements to 'leave protective buffer zones adjacent to water' and to 'plan felling and transport operations so as to avoid or limit damage to the land and water course' (Sweden, 1993). These qualitative policies, like those of Finland and Bavaria, are classified as 'mixed'.

Roads (Indicators: Culvert size at stream crossings, road abandonment)

In Finland, technical specifications and environmental principles for dimensioning of forest roads are described in forest road construction norms. The size of culverts at stream crossings is derived from the extent of the catchment area and there are tables to help to choose the appropriate culvert size. There are no standardized rules for road decommissioning, but there are guidelines for closing roads for nature protection or bird nesting. In sum, therefore, we have classified Finnish policy as 'mixed' in its treatment of our road policy indicators.

The size of culverts at stream crossings and treatment of unused roads on state lands in Bavaria are governed by mandatory, non-discretionary rules outlined in the 'Richtlinien fuer die Erschliessung des Staatswaldes in Bayern' (Bavaria, 1982). Regarding private forestlands, Die ForstFoeP-RL (Bavaria, 1995) outlines some road building details. For example: 'Nr. 4.2.3. Skid trails should not be turned into permanent roads' and 'Nr. 2.3.6. The landscape ecology effects of road building should be considered.' No prescribed thresholds or voluntary guidelines, however, have been identified for culverts at stream crossings or for road decommissioning on private lands.

Portuguese best practices caution that roads should not alter the direction of water flow, and that bridges should be built when the water is lowest and constructed so that the river/stream margins are not disturbed. It is recommended that there be a programme of regular check-up and maintenance particularly for drainage systems, water crossings and periods of intense rain. It is also recommended that roads near waterways, valleys and summits be priorities for decommissioning. We classify these various guidelines as voluntary substantive policies.

The Swedish Forestry Act stipulates that the government holds authority to regulate the routing of forest roads (Section 30) for the purposes of nature conservation and cultural heritage conservation (1993). Furthermore, under Section 17.2 of the same Act, the government may decline approval of harvest plans 'where the cost of the construction of forest access roads exceeds the benefits for forest management, or where the road cannot be incorporated into a road network plan'. No specific management prescriptions are provided. We classify this approach as a mandatory procedural policy.

Clearcutting (Indicator: Clearcut size limits)

No size limits for clearcuts are specified for private and public forests in Finland or private forests in Bavaria, although the small size of most private forest ownerships precludes large clearcuts. For example, the average final cut size in Finnish private forests is 1.7ha (Saarenmaa, 2003). Bavarian state forests have committed to a 'natural forest' management regime that precludes plantations and clearcutting (BSLF, 2006; Tschacha, 2007).

Portugal represents yet a different context, where most wood production derives from private plantations. There are no legal limits placed on clearcut size in either plantation or natural production forests. Legal restrictions are placed on the harvest of young trees, however. For maritime pines (*Pinus pinaster*) in harvest areas equal to or larger than 2ha, the following cut regulations are in effect: 75 per cent of the trees must be more than or equal to 17cm in DBH (diameter at base height) or a 'perimeter' greater than 53cm. For eucalyptus trees in harvest areas of one hectare or more, 75 per cent of the trees must be more than or equal to 12cm at DBH or a 'perimeter' higher than 37.5cm (Decreto-Lei No. 173/88).

In Sweden, clearcutting is commonly referred to as 'regeneration felling'. State permission is required for regeneration fellings exceeding 0.5ha in size. In addition, clearcuts are limited to 20ha in alpine areas (Sweden, 1993). Assuming permission is obtained, there are no upper limits set for the size of regeneration fellings in lowland areas.

Reforestation (Indicator: Requirements for reforestation, including specified time frames and stocking levels)

Reforestation in Finland is required on all ownership types. The Ministry of Agriculture and Forestry has given an order based on the Forest Act regarding the stocking levels and time limits. The stocking levels prescribed vary from south to north. If the minimum stocking level is not reached, fill-in planting is required. The optimum stocking levels for regeneration vary depending on species, habitat type and region. The difference between the optimum and minimum level varies (e.g. for Norway spruce the optimum is 1600–1800 and the minimum 1200 seedlings/ha, and for broadleaf dominated the optimum is 1600–2000 and the minimum 1000 (Finland MAF, 2009d; Tapio, 2009).

According to the Bavarian Forest Law, BayWaldG Article 15, clearcut or naturally disturbed, unstocked forest areas must be reforested within three years on both public and private lands. In forest areas where regeneration is not successful, it must be supplemented within five years. Natural regeneration is now practised on approximately 40 per cent of forestlands (Bundesministerium für Verbraucherschutz Ernährung und Landwirtschaft, 2001).

In Portuguese natural forests, best management practices encourage silvicultural treatments that promote natural regeneration. However, apart from general requirements to maintain certain forest types (see 'Protection of Species at Risk' above), there are no mandatory policies prescribing reforestation practices.

The Swedish Forestry Act (1993) requires that forestlands be reforested within three years after regeneration felling. Stocking levels are prescribed at the national level (Westland Resources Group, 1995). The Act also sets minimum limits on stand rotation ages between regeneration fellings, ranging from 45 to 100 years for coniferous forests. For forest areas larger than 50ha, no more than half of the forest management unit may be of a stand age younger than 20 years. Forest areas larger than 1000ha are subject to further restrictions.

Annual allowable cut (AAC) (Indicator: Cut limits based on sustained yield)

In Finland, calculations of annual cut volumes are required at the following levels: nationwide, by regional forest centres (there are 13 in the country), by forest management associations' operative areas and, voluntarily, by forest owners. AAC is calculated at the national level every ten years in connection with the national forest inventory conducted by the Finnish Forest Research Institute. These figures are used in implementing forest policy instruments. During the last 40 years, annual increment has constantly exceeded annual removals. Metsähallitus calculates regional-level AACs for state lands in its natural resources plans every ten years; these AACs are followed very strictly. There are no formal requirements regarding AAC for private lands.

In Bavaria, the annual allowable cut (AAC) on both public and private lands is determined at the level of the individual forest management unit. Forest managers must submit harvest plans that detail how the cut area will be reforested. We classify this policy approach as mandatory and procedural.

Best management practices in Portugal suggest the planning of harvest volumes around a number of ecological constraints. These include guidelines on thinning and retention cycles as follows: for long rotation cycles (30–40 years), selective logging should be conducted every four to five years for cleaning and ensuring the maximum production potential of the stand. For even longer cycles, selective cuts should be made every seven to eight years. These selective cuts should allow for 50 to 350 future trees for logging per hectare to remain. As mentioned in the above section on clearcutting, there are legal restrictions on the harvest of immature pine and eucalypt.

The Portuguese General Directorate of Forests requires 30 days' advance notice of harvest and reserves the right to verify that all information disclosed is correct (Art. 7). Records of harvest volumes are kept in order to maintain national statistics of annual volume extracted. However, there are no requirements or guidelines in regard to the calculation of overall cut volumes and hence we classify Portugal as having 'no policy' on AAC.

Sweden does not require calculation of AAC and hence we classify Sweden as having 'no policy' for this indicator. However, it is of tangential relevance that there are requirements to reforest and rules on minimum rotation ages (45 to 100 years). There are also limitations on the percentage of forest properties that can fall below 20 years of age.

FOREST PRACTICE REGULATIONS: PLANTATIONS

No Finnish forests are currently classified as plantations and there are no separate forest practice rules for plantations in Finland. The Finnish Forest Act requires the use of native species for reforestation purposes. It is possible, however, to apply to the Regional Forest Centre for permission to reforest with exotics. In practice, the use of exotics, as well as the practice of intensive plantation management, is rare in Finland (FAO, 2003).

There are no special rules for plantation management in Bavaria but, as noted above, the intent of forest policy is to discourage the conversion of natural forest to plantation.

In Portugal, the rules for plantation and natural forests are identical with regard to our five forest practice criteria. There are, however, additional guidelines that safeguard plantation productivity, which address: soil and climatic conditions; species selection; use of reproductive plant material, and resistance to damage from resident wildlife and from fire.

There are no specific forest practice rules for plantation management in Sweden. However, the Swedish Forestry Act requires 'notification' for the use of exotic (non-native) species (Sweden, 1993).

ENFORCEMENT

In Finland, 13 regional forestry centres, under the authority of the Ministry of Agriculture and Forestry, are responsible for enforcement, auditing and inspections of forestry activities on both public and private lands.

Within the Bavarian State Ministry for Agriculture and Forests (MLF), Department V Forest Administration is responsible for the supervision of forest management on state forests. This Department is divided into four directorates and 127 forest offices (Roering, 2004). MLF's role on private lands is advisory.

In Portugal, the National Forest Authority is responsible for the fulfilment of regulations.

In Sweden, ten regional forestry boards are responsible for enforcing forest management regulations. Forest operations are routinely audited for rule compliance.

FOREST CERTIFICATION

Since the 1990s, Europe has been described as a central driver of forest certification in the international market-place (Hansen, 1999; Hansen et al, 1999; Lindstrom et al, 1999; Rametsteiner, 1999, 2002; Auld, 2001; Cashore et al, 2001, 2004, 2005; Newsom, 2001; Raunetsalo et al, 2002; Sasser, 2002; Rametsteiner and Simula, 2003). While the politics of certification have evolved differently in each country, all European case study countries have become closely engaged in the certification movement in various forms. There is a history of heated debate between supporters of the Forest Stewardship Council (FSC) on the one hand and supporters of the European-initiated Programme for the Endorsement of Forest Certification schemes (PEFC), on the other. Figures 2.18 and 2.19 show the relative growth of each scheme in our case study countries.

In Finland, most woodlot owners have sought and received certification from the 'Finnish Forest Certification Scheme' (FFCS), which is recognized by the PEFC. The total PEFC certified area in January 2008 was 20.7 million ha (PEFC, 2008). Based on FAO data on forest cover (FAO, 2007), this amounts to 92 per cent of the country's forests. The FFCS has proceeded without support from environmental groups such as the World Wide Fund for Nature, whose office in Finland eventually withdrew from Finnish certification negotiations. By January 2008, only three Finnish operations had been FSC certified, consisting of the small, 93ha Family Jalas Forest, Stora Enso Wood Supply Finland at 9397ha and the large Stora Enso Oyj Wood Supply Russia, totalling 424,262ha.

The first FSC German regional standards were created in 1998 and were backed by most environmental groups, but few forest producers. The few exceptions were the Working Group on Natural Silviculture and a handful of municipal and state forestry agencies. The PEFC (which was originally known as the 'Pan-European Forest Certification Programme') emerged as a competitor in 1999 and was promoted by forest owners and most state forestry agencies as a more appropriate programme for Germany. Among the reasons stated for this preference was that the PEFC certified regions instead of individual ownerships and recognized the long-standing German silvicultural tradition (Cashore et al, 2004). The competition between the FSC and PEFC in Germany has been fierce and has been made more vehement by the long history of conflict between the country's forestry and conservation interests. Currently, most state forestry agencies have elected to certify their forests through the PEFC. Private landowner associations also, for the most part, have supported the PEFC. As of January 2008, 7.2 million ha, or 66 per cent, of German forests were certified under the PEFC. The certified area under FSC totalled 418,773ha, or 4 per cent of Germany's forests, and was divided among 57 forest management certificates.

In Portugal there are currently three forest management operations and one non-timber forest product operation certified by the FSC, together totalling about 73,500ha of both plantation and semi-natural forest. This equals about 2 per cent of the total forestlands. As of January 2008, there were no forests certified to a PEFC-endorsed standard.

Table 4.3 *Matrix of policy approaches to the five forest practice criteria in natural forests*

Case Study	1) Riparian	2) Roads	3) Clearcuts	4) Reforestation	5) AAC
Bavaria (Public)	Mandatory mixed	Mandatory mixed	Mandatory substantive	Mandatory substantive	Mandatory procedural
Finland (Public)	Mandatory mixed	Mandatory mixed	Mandatory substantive	Mandatory substantive	Mandatory substantive
Bavaria (Private)	Mandatory mixed	Mandatory mixed	Mandatory substantive	Mandatory substantive	Mandatory procedural
Sweden (Private)	Mandatory mixed	Mandatory procedural	Mandatory mixed	Mandatory substantive	No policy
Finland (Private)	Mandatory mixed	Mandatory mixed	No policy	Voluntary	No policy
Portugal (Private)	Voluntary	Voluntary	No policy	Voluntary	No policy

Note: Mandatory mixed policies require some form of management action but lack standardized, quantified thresholds of performance and/or apply to only a portion of the forest landscape.

Key:

Mandatory substantive
Mandatory procedural
Mandatory mixed
Voluntary
No policy

In Sweden, all major industrial players have sought and become certified under the FSC, while family forest associations have adopted the PEFC (Elliott, 2000; Elliott and Schlaepfer, 2001; Levy, 2003; Cashore et al, 2004). Sweden leads the European case studies in area certified under the FSC, with 20 certificates totalling 11,327,576ha certified, covering 41 per cent of the country's forestlands; PEFC certification accounted for 26 per cent of forestlands.

SUMMARY

Finland, Bavaria and Sweden have all established mandatory rules for many of the policy indicators. However, in a number of cases the approach taken is qualitative and there are no standardized quantitative performance thresholds (e.g. numeric riparian buffer size requirements or clearcut size limits). Portugal, in contrast, takes a more voluntary approach coupled in some cases with quantitative guidelines. In general, for those case studies where public forest policies were assessed due to relatively large areas of public forest, rules governing public forest management were relatively prescriptive.

Plantation production is most significant in Portugal. Forest practice policies governing plantations are the same as those for natural forests in regard to our five standardized forest practice criteria. However, Portugal has also developed best management guidelines that specifically address plantation management.

Forest governance, including compliance and enforcement, is generally considered strong in all of these countries (Esty and Cornelius, 2002). Poor forest governance in other world regions, however, has been a driving force behind

European demand for certified forest products. This demand, in turn, has spurred growth in certified forest area within our European case study countries.

Certification's growth in the region has been fraught with conflict among environmental groups and forest producers regarding what constitutes a legitimate certification scheme. The PEFC system arose out of European producers' objections to the environmentalist-backed FSC. Family forest producers, in particular, have resisted the cost of pursuing FSC certificates in favour of regional-scale group certification under the PEFC.

NOTES

1   With the contributions of Professor Dr Ulrich Schraml, University of Freiburg, Germany and Ms Terhi Koskela, Finnish Forest Research Institute.
2   According to the 2005/2006 national forest inventory of the Portuguese Direccao Geral de Recursos Florestais, forests covered 3.4 million ha or 38.4 per cent of the country, an increase of 93,000ha from the previous 1995/98 inventory (CELPA, 2007).
3   The Ministerial Council for Protection of Forests in Europe (MCPFE, 2007, p68) notes that the particular conditions of protected forests in Europe, in which areas are often small and fragmented, means that minor differences in interpretation of guidelines can generate wide variations in results. It therefore reports protected forest areas according to a harmonized definition. The results of this assessment (MCPFE, 2007, Indicator 4.9) are similar to those presented in our analysis.

REFERENCES

ACPWP (2006) 'Country report: Germany', Rome: Advisory Committee on Paper and Wood Products
AFN (2009a) 'Gestão Florestal', www.dgrf.min-agricultura.pt/portal > Gestão Florestal, accessed May 2009
AFN (2009b) 'Política e planeamento', Autoridade Florestal Nacional, www.dgrf.min-agricultura.pt/portal > Política e Planeamento, accessed May 2009
AFN (2009c) 'Regime florestal', autoridade florestal nacional, www.dgrf.min-agricultura.pt/portal > Gestão Florestal > Regime Florestal, accessed May 2009
Alves, Rosario (no date) 'The role of private forest in Portugal: Forestis, a private owners federation', Porto: Forestis
Auld, Graeme (2001) 'Explaining certification legitimacy: An examination of forest sector support for forest certification programs in the United States Pacific Coast, the United Kingdom, and British Columbia, Canada', MSc Thesis, Auburn, AL: School of Forestry and Wildlife Sciences, Auburn University
Baer, L. A. (1996) 'Boreal forest dweller: The Saami in Sweden', *Unasylva*, 47, 186, www.fao.org/docrep/w1033e/w1033e05.htm#boreal%20forest%20dwellers:%20the%20saami%20in%20sweden, accessed November 2009
Barklund, A. (2006) 'Introductory notes: Policy for balancing the forestry and the reindeer herding interests in Swedish PEFC', www.pefc.se, accessed May 2009

Bavaria (1982) *Richtlinien fuer die Erschliessung des Staatswaldes in Bayern*, Government of Bavaria

Bavaria (1995) *ForstFoeP-RL 1995*, Government of Bavaria

Bayerisches Staatsministerium für Ernährung Landwirtschaft und Forsten (1999) 'Zukunft wald: Nachhaltigkeit in bayerns wäldern', Munich: Bayerische Staatsforstverwaltung

Beck, Roland and Ullrike Krafft (2001) 'Analysis of extension efforts in Bavaria: Extensionists perception', paper read at Forestry Extension: Assisting Forest Owner, Farmers and Stakeholder Decision-Making, 29 October–2 November, Lorne, Australia

Boreal Forest (2008a) 'Finland – forests and forestry', www.borealforest.org/world/world_finland.htm (accessed November 2008)

Boreal Forest (2008b) 'Sweden – forests and forestry', www.borealforest.org/world/world_sweden.htm, accessed November 2008

Brouwer, Roland (1995) '*Baldios* and common property resource management in Portugal', *Unasylva*, 46, 180, www.fao.org/docrep/v3960e/v3960e07.htm, accessed November 2009

BSLF (2006) 'Agriculture and forestry in Bavaria: Facts and figures 2006', Munich: Bayerisches Staatsministerium für Landwirtschaft und Forsten

Bundesministerium für Verbraucherschutz Ernährung und Landwirtschaft (2001) 'Gesamtwaldbericht der Bundesregierung', Dessau-Roßlau: Bundesministerium für Verbraucherschutz Ernährung und Landwirtschaft

Cashore, Benjamin (1997) 'Governing forestry: Environmental group influence in British Columbia and the US Pacific Northwest', PhD, Political Science, Toronto: University of Toronto

Cashore, Benjamin, George Hoberg, Michael Howlett, Jeremy Rayner and Jeremy Wilson (2001) *In Search of Sustainability: British Columbia Forest Policy in the 1990s*, Vancouver: University of British Columbia Press

Cashore, Benjamin, Graeme Auld and Deanna Newsom (2004) *Governing Through Markets: Forest Certification and the Emergence of Non-State Authority*, New Haven, CT: Yale University Press

Cashore, Benjamin, G. Cornelis van Kooten, Ilan Vertinsky, Graeme Auld and Julia Affolderbach (2005) 'Private or self-regulation? A comparative study of forest certification choices in Canada, the United States and Germany', *Forest Policy and Economics*, 7, 1, pp53–69

CELPA (2007) 'Statistics reports 2006', Lisbon: CELPA – Associacao da Industria Papel

Chatham House (2009) 'Finland', www.illegal-logging.info > Countries and regions > Europe > Finland, accessed May 2009

Comissão Técnica de Normalização (2004) 'Norma Portuguesa (NP 4406/ 2003) – Sistemas de Gestão Florestal Sustentável. Aplicação dos critérios pan-europeus para a gestão florestal sustentável', Lisbon: Confederação dos Agricultores de Portugal

CPI (2008) 'Fact sheet: Paper and the forest', Swindon: Confederation of Paper Industries

Direccao-Geral das Florestas (1991) 'Perfil florestal – Portugal' (mimeo), Lisbon

Direccao-Geral das Florestas (2001) 'Inventário florestal nacional. Portugal continental 3a. Revisão, 1995–1998 Relatório Final', Lisbon

Direccao-Geral das Florestas (2003) 'Principios de boas practicas florestais, Ministerio da Agricultura, desenvolvemento rural e pescas', Lisbon

EC (2003) 'MEMO on commission strategy to protect Europe's most important wildlife areas – frequently asked questions about NATURA 2000', European Commission, http://ec.europa.eu/environment/nature/info/pubs/docs/nat2000/2003_memo_natura.pdf, accessed November 2009

EFC (2000) 'Forest policy, law and institutions', Rome: European Forestry Commission, Timber Committee

Elliott, Christopher (2000) 'Forest certification: A policy perspective', Bogor: Center for International Forestry Research

Elliott, C. and R. Schlaepfer (2001) 'The advocacy coalition framework: Application to the policy process for the development of forest certification in Sweden', *Journal of European Public Policy*, 8, 4, pp642–661

Esty, Daniel C. and Peter Cornelius (eds) (2002) *Environmental Performance Measurement: The Global Report 2001–2002*, New York: Oxford University Press

EU (2009) 'Protection of nature and biodiversity', http://europa.eu/legislation_summaries/ environment/nature_and_biodiversity/index_en.htm

European Commission (2009) 'FLEGT/FLEG', last updated 20 November 2008 http://ec.europa.eu/environment/forests/flegt.htm, accessed May 2009

FAO (2003) State of the World's Forests, Rome: Food and Agriculture Organization of the United Nations

FAO (2006a) 'Global forest resources assessment 2005; Progress towards sustainable forest managment', in *FAO Forestry Paper 147*, Rome: Food and Agriculture Organization of the United Nations

FAO (2006b) 'Global planted forests thematic study: Results and analysis', A. Del Lungo, J. Ball and J. Carle (eds) *Responsible Management of Planted Forests: Voluntary Guidelines*, Rome: United Nations Food and Agriculture Organization

FAO (2007) *State of the World's Forests – 2007*, Rome: United Nations Food and Agriculture Organization

FAOSTAT (2008) 'ForesSTAT', Rome: United Nations Food and Agriculture Organization

FDC Tapio (2007) 'Tapion vuositilastot: Tapios årsstatistik', Tapio: Forest Development Centre

Fernandes, Paulo M. and Eric Rigolot (2007) 'The fire ecology and management of maritime pine (*Pinus pinaster* Ait.)', *Forest Ecology and Management*, 241, p13

FFA (2004) 'Forestry in Finland and in Europe', Finnish Forest Association, www.forest.fi/yr2004/eng/2_1.html, accessed November 2009

FFRI (2006) 'Statistical yearbook of forestry', in *Official Statistics of Finland: Agriculture, Forestry and Fisheries*, Helsinki: Finnish Forest Research Institute

FFRI (2007) 'Forest Finland in Brief', Helsinki: Finnish Forest Research Institute (METLA)

FFRI (2008) 'Statistical yearbook of forestry 2008', Helsinki: Finnish Forest Research Institute

Finland (1996) Forest Act, Government of Finland 1093/1996; amendments up to 552/2004 included, unofficial translation, www.finlex.fi/en/laki/kaannokset/1996/en19961093, accessed November 2009

Finland MAF (2001) 'Annual report 2000', Helsinki: Ministry of Agriculture and Forestry, Finland

Finland MAF (2008a) 'Government resolution on the forest biodiversity programme for Southern Finland 2008–2016 (Metso)', Helsinki: Ministry of Agriculture and Forestry, Finland, www.mmm.fi/attachments/metsat/5yckfcmWR/METSOResolution2008-2016_ENGL.pdf, accessed November 2009

Finland MAF (2008b) 'Forests: Nature conservation', www.mmm.fi/en/index/frontpage/ forests/nature_conservation.html, accessed May 2008

Finland MAF (2009a) 'Forest policy', www.mmm.fi/en/index/frontpage/forests/ forest_policy.html, accessed May 2009

Finland MAF (2009b) 'Legislation', www.mmm.fi/en/index/frontpage/forests/forest_policy/legislation.html, accessed May 2009

Finland MAF (2009c) 'Reaching forest use objectives through advice and planning', www.mmm.fi/en/index/frontpage/forests/commercial_forests/forest_planning_advice.html, accessed May 2009

Finland MAF (2009d) 'Maa- ja metsätalousministeriön päätös metsälain soveltamisesta', Helsinki: Ministry of Agriculture and Forestry, Finland, www.mtk.fi/metsa/metsapolitiikka/lakitieto/fi_FI/metsalaki/_files/11623114870095089/default/MMM%20paatos%20metsalain%20soveltamisesta.pdf, accessed November 2009

Finnish Forest Certification Council (1999) 'Caring for our forests', Helsinki: Finnish Forest Certification Council

Finnish Forest Industries Federation (2003a) 'The Finnish forest industries facts and figures 2003; statistics 2002', www.forestindustries.fi, accessed December 2003

Finnish Forest Industries Federation (2003b) 'The Finnish Forest Industries Federation', http://english.forestindustries.fi/, accessed May 2003

Finnish Forest Industries Federation (2009) 'Metsäteollisuuden tuotanto väheni Suomessa 10 prosenttia vuonna 2008: Kustannustason nousu on saatava kuriin', www.metsateollisuus.fi/tiedotteet/arkisto/Metsateollisuudenajankohtaiskatsaus22009.htm, accessed March 2009

Fonseca, Teresa J. F. and Bernard R. Parresol (2001) 'A new model for cork weight estimation in Northern Portugal with methodology for construction of confidence intervals', *Forest Ecology and Management*, 152, pp131–139

Germany (1998) Federal Nature Conservation Act (Bundesnaturschutzgesetz, BNatSchG). Government of Germany. English translation: www.iuscomp.org/gla/statutes/BNatSchG.htm, accessed November 2009

Grey, G. (1988) 'Finland–forestry and technology', *Journal of Forestry*, 86, 7, pp23–26

Gustafsson, Karl, and Tomas Thurlesson (2001) 'Forest impact analyses 1999: Possibilities for forest utilization in the 21st century', Skogsstyrelsen

Hansen, Eric (1999) 'Environmental marketing: Opportunities and strategies for the forest products industry', paper read at Environmental Marketing: Opportunities and Strategies for the Forest Products Industry, 26–28 September, Portland, Oregon

Hansen, Eric, Keith Forsyth and Heikki Juslin (1999) 'A forest certification update for the ECE region', Geneva: United Nations Economic Commission for Europe and Food and Agriculture Organization of the United Nations, Timber Section

Hofmann, Frank, Jutta Kill, Roland Meder, Harald Plachter and Karl-Rheinhard Volz (2000) Waldnutzung in Deutschland: Bestandsaufnahme, Handlungsbedarf und Maßnahmen zur Umsetzung des Leitbildes einer nachhaltigen Entwicklung, D. R. v. S. f. Umweltfragen (ed.), vol 35, Metzler-Poeschel, Stuttgart

Hytönen, Marjatta (2002) 'Public participation in forestry in Finland: An overview', Finnish Forest Research Institute 2002, www.metla.fi/pp/mhyt/ppoverview.htm, accessed May 2003

ICN (2003) 'Thematic report on protected areas or areas where special measures need to be taken to conserve biologica', Lisbon: Institute of Nature Conservation

ICNB (2009) 'Nature 2000', Institute for Nature Conservation and Biodiversity (Portugal), http://portal.icnb.pt > Natura 2000, accessed March 2009

INE (1997) 'A floresta nas explorações agrícolas 1995', Lisbon: Instituto Nacional de Estatística

ISA (2006) 'Report for Cost E42 from Portugal', Lisbon: Instituto Superior de Agronomia

JRC-IES (2007) 'Forest fires in Europe: Report No. 7/ 2006', Luxembourg: European Commission Joint Research Centre & Institute for Environment and Sustainability / Land Management & Hazards Unit

Kooten, G. C. van, Bill Wilson and Ilan Vertinsky (1999) 'Sweden', in B. Wilson, G. C. van Kooten, I. Vertinsky and L. Arthur (eds) *Forest Policy: International Case Studies*, Wallingford, IK: CABI Publishing

KSLA (2009) 'The Swedish forestry model', produced by the Royal Swedish Academy of Agriculture and Forestry (KSLA), Stockholm, www.ipef.br/eventos/2009/graduatecourse/21-KSLA_2009.pdf, accessed November 2009

Levy, Marcelo (2003) 'Comparative analysis of boreal Canadian, Sweden and Finnish standards', Ottawa: The Forest Stewardship Council, Canada

Lindstrom, T., E. Hansen and H. Juslin (1999) 'Forest certification: The view from Europe's NIPFs', *Journal of Forestry*, 97, 3, pp25–30

Lorenz, Martin, Hans-Dieter Nagel, Oliver Granke and Philipp Kraft (2008) 'Critical loads and their exceedances at intensive forest monitoring sites in Europe', *Environmental Pollution*, 155, pp426–435

MCPFE (2003) 'Europe's forests in the spotlight', Vienna, 4th Ministerial Conference on the Protection of Forests in Europe

MCPFE (2007) 'State of Europe's forests 2007', Ministerial Council on Protection of Forests in Europe, www.mcpfe.org/publications/pdf, accessed April 2009

Mendes, Américo M. S. Carvalho and Rafael A. R. da Silva Dias (2002) 'Financial instruments of forest policy in Portugal in the 80s and 90s', paper produced for Working Group 2 of the COST E19 Action under supervision of the Cooperation in the Field of Scientific and Technical Research (COST) Forests and Forestry Products Technical Committee. As of 2009, COST members included 35 European states, www.metla.fi/eu/cost/e19/wg2papers.htm, accessed November 2009

Metsähallitus (2009a) 'Management of commercial forests', 28 November 2008, www.metsa.fi/sivustot/metsa/en/Forestry/Forestmanagement/Sivut/Managementof CommercialForests.aspx

Metsähallitus (2009b) 'Questions and answers', last updated 28 November 2008, www.metsa.fi/sivustot/metsa/en/Forestry/Forestmanagement/Environmentalissues/QuestionsandAnswers/Sivut/QuestionsandAnswers.aspx, accessed November 2009

Mikkelä, Heli, Susanna Sampo and Jaana Kaipainen (2001) 'The state of forestry in Finland 2000: Criteria and indicators for sustainable forest management in Finland', Helsinki: Ministry of Agriculture and Forestry

Ministry of Agriculture and Forestry (2001) 'Ministry of Agriculture and Forestry: Annual report 2000', Helsinki: Ministry of Agriculture and Forestry

MOE (2009a) 'Administration', Ministry of Environment, Finland, www.ymparisto.fi/default.asp?node=6089&lan=en, accessed April 2009

MOE (2009b) 'Protected species under the Nature Conservation Decree', Ministry of the Environment, Finland, www.ymparisto.fi/default.asp?node=12511&lan=en, accessed April 2009

MOE and NCI (1998) 'First Portuguese national report: To be submitted to the Conference of the Parties to the Convention on Biological Diversity', Ministry of the Environment, The Nature Conservation Institute

MTK (2009) 'The Role of Forest Management Associations', Central Union of Agricultural Producers and Forest Owners (MTK), last updated 2 January 2006, www.mtk.fi/MTK_briefly/forestry/en_GB/forestry_associations/, accessed November 2009

Nabuurs, Gert-Jan, Risto Paivinen, Ari Pussinen and Mart-Jan Schelhaas (2003) 'Development of European forests until 2050', *European Forest Institute Research Report 15*, Leiden: Koninklijke

Newsom, Deanna (2001) 'Achieving legitimacy? Exploring competing forest certification programs' actions to gain forest manager support in the U.S. Southeast, Germany, and British Columbia, Canada', Master's of Science, , Auburn, AL: School of Forestry and Wildlife Sciences, Auburn University

Nilsson, Sten (2007) 'Mobilizing wood resources – What's the big deal?', paper read at 'Mobilizing wood resources: Can Europe's forests satisfy the increasing demand for raw material and energy under sustainable forest management?', 11–12 January, Geneva

Ottitsch, A. and M. Palahi (2001) 'European forest policies', in M. Palo, J. Uusivuori, G. Mery (eds) *World Forests, Markets and Policies*, Dordrecht: Kluwer Academic Publishers

Pan European Forest Certification Bavarian Working Group (2000) 'Regionalbericht Bayern', Munich: Paneuropdiacräisches Forestzertifikat (PEFC)

Parvainen, Jari (2004) Personal communication with Jari Parvainen, Director of METLA in Joensuu, Chair of the Finnish Forest Certification Council, 8 January 2004

PEFC (2008) 'PEFC Council information register', http://register.pefc.cz/, accessed February 2008

Portucel, Soporcel (2004) 'The Group Profile', Grupo Portucel Soporcel, www.portucelsoporcel.com/en/group, accessed February 2004

Punttila, P. and A. Ihalainen. (2006) 'Luonnontilaisen kaltaiset metsät suojelu- ja ei-suojelluilla alueilla', in Paula Horne, Terhi Koskela, Mikko Kuusinen, Antti Otsamo, Kimmo Syrjänen (eds) *METSOnjäljillä- Etelä-Suomen metsien monimuotoisuusohjelman tutkimusraportti*, available at wwwb.mmm.fi/metso/asiakirjat/1_Luvut_I-III_s1-52.pdf (last accessed December 2009)

PwC (2006) 'PricewaterhouseCoopers global forest, paper & packaging industry survey – 2006 edition', Vancouver: PricewaterhouseCoopers

Rametsteiner, Ewald (1999) 'The attitude of European consumers toward forests and forestry', *Unasylva*, 50, p196

Rametsteiner, Ewald (2002) 'The role of governments in forest certification – A normative analysis based on new institutional economics theories', *Forest Policy and Economics*, 4, 3, pp163–173

Rametsteiner, Ewald and Markku Simula (2003) 'Forest certification – An instrument to promote sustainable forest management?', *Journal of Environmental Management*, 67, pp87–98

Raunetsalo, Jenni, Eric Hansen, Keith Forsyth and Heikki Juslin (2002) 'A forest certification update for the ECE region', Geneva: United Nations Economic Commission for Europe and Food and Agriculture Organization of the United Nations, Timber Section

Roering, Hans-Walter (2004) 'Study on forestry in Germany', Hamburg: Institute for Economics, Federal Research Centre for Forestry and Wood Products and Department of Wood Science, University of Hamburg

Saami Council (2006) 'Finland ignores human rights issues', press release from Saami Council and Finland's Sámi Central Organization, www.arcticpeoples.org/2006/10/20/finland-ignores-human-rights-issues-press-release/, accessed November 2009

Saarenmaa, Liisa (2003) Personal communication with Liisa Saarenmaa, Counsellor of Forestry, Ministry of Agriculture and Forestry, 26 August 2003

Saarenmaa, Liisa (2004) Personal communication with Liisa Saarenmaa, Counsellor of Forestry, Ministry of Agriculture and Forestry, 8 January 2004

Sasser, Erika N. (2002) 'The certification solution: NGO promotion of private, voluntary self-regulation', paper read at 74th Annual Meeting of the Canadian Political Science Association, 29–31 May, Toronto, Ontario

Schraml, Ulrich (2009) Email communication with Dr Ulrich Schraml, Institute of Forest and Environmental Policy, University of Freiburg, 28 May

Schraml, U. and H. Thode (2000) 'Forstpolitik im nichtbaeuerlichen kleinprivatwald – einstellung und verhalten der waldbesitzer und konsequenzen für die forstpolitik der länder', paper read at Forstwissenschaftliche Tagung 2000, 11–15 October, Freiburg, Germany

Schraml, U. and G. Winkel (1999) 'Germany', in P. Pelkonen, A. Pitkänen, P. Schmidt, G. Oesten, P. Piussi and E. Rojas (eds) *Forestry in Changing Societies in Europe*, Joensuu, Finland: Silva Network

SEPA (2004) 'Sweden's National Parks', Swedish Environmental Protection Agency, www.internat.environ.se/index.php3?main=/documents/nature/engpark/enpstart.htm, accessed March 2004

SEPA (2009a) 'National parks', Swedish Environmental Protection Agency, www.naturvardsverket.se/en/In-English/Menu/Nature-conservation_and_wildlife_management/Nature-conservation-and-species-protection/National-parks-and-other-ways-to-protect-nature/, accessed May 2009

SEPA (2009b) 'The environmental code', Swedish Environmental Protection Agency, available from www.naturvardsverket.se > Legislation and other policy instruments > The Environmental Code, accessed May 2009

SFA (2009a) 'Forests in Sweden', Swedish Forestry Agency (Skogsstyrelsen), available from www.svo.se > International > Forests in Sweden, accessed May 2009

SFA (2009b) 'Policy objectives and quantative targets', Swedish Forestry Agency (Skogsstyrelsen), available from www.svo.se > International > Swedish forest policy, accessed May 2009

SFA (2009c) 'Swedish Forest Agency' Swedish Forestry Agency (Skogsstyrelsen), www.svo.se, accessed May 2009

SFA (2009d) 'Site protection', Swedish Forestry Agency, www.svo.se > International > Site protection, accessed May 2009

SFIF (2007) 'The Swedish forest industries: Facts and figures 2006', Stockholm: Swedish Forest Industries Federation

Stanbury, William T. (2000) 'Environmental groups and the international conflict over the forest of British Columbia 1990 to 2000', Vancouver: Simon Fraser University, University of British Columbia Centre for the Study of Government and Business

Stanbury, W. T., I. Vertinsky and J. L. Howard (1991) 'Europe 1992 – Threat or opportunity for the Canadian Forest Products Industry and implications for public policy', *Forestry Chronicle*, 67, 6, pp674–690

Sveaskog (2009) 'History', Sveaskog 2009, www.sveaskog.se/en/About-Sveaskog/Our-operations/History/, accessed May 2009

Sweden (1993) Forestry Act, Government of Sweden

Tapio (2009) 'Finnish forest management practice recommendations', Forestry Development Centre Tapio, www.tapio.fi/finnish_forest_management_practice_recom, accessed April 2009

Thelander, Goran (2000) 'Sweden's new forest policy', in F. Schmithusen, P. Herbst and D. C. L. Master (eds) *Forging a New Framework for Sustainable Forestry: Recent Developments in European Forest Law*, Sopron, Hungary: EuroLAN(c) Kft

Tschacha, Karl (2007) 'Biomass strategy of the Bavarian State Forest Enterprise based on natural forestry', powerpoint presentation at the 8th Conference of European State Forestry Organizations, Dunkeld, Scotland, 5–8 June 2007

Uliczka, H. (2003) 'Nature conservation efforts by forest owners – Intentions and practice in a Swedish case study', *Silva Fennica*, 47, 4, pp459–475

Uuttera, J., M. Maltamo and J. P. Hotanen (1997) 'The structure of forest stands in virgin and managed peatlands: A comparison between Finnish and Russian Karelia', *Forest Ecology and Management*, 96, pp125–138

Volz, K.-R. and A. Bieling (1998) 'Zur soziologie des kleinprivatwaldes', *Forst und Holz*, 3, pp67–71

Westland Resource Group (1995) 'A review of the Forest Practices Code of British Columbia and fourteen other jurisdictions: Background report – 1995', Victoria, Crown Publications, Inc

Yrjölä, Tiia (2002) 'Forest management guidelines and practices in Finland, Sweden and Norway', Joensuu, Finland: European Forest Institute

Chapter 5

Asia: China, India, Indonesia and Japan[1]

INTRODUCTION

Forests and the forest industry in South, Southeast and East Asia are globally significant in many respects. These regions contain around 55 per cent of the world's population and include the world's two most populous countries. Some Asian economies are the world's richest and most rapidly expanding, while others remain persistently poor. Rapid economic growth in South and East Asia, combined with increasingly strict domestic environmental forest policies, have made these regions leading hubs in the international trade of both Siberian and tropical timber (Zhu et al, 2004; ITTO, 2006b; McDermott et al, 2008a).

Our four regional case studies – China, India, Indonesia and Japan – together account for 10 per cent of the world's forests. These forests include some that are among the world's most biodiverse and others, the most commercially valuable. In Southeast Asia the rates of forest loss and degradation are high, as are corresponding contributions to global greenhouse gas emissions (Brown and Durst, 2003, FAO, 2007). Plantation forests are important in most Asian countries and the Asian region is the locus of their expansion globally (Kanowski and Murray, 2008).

All four Asian case study countries vary greatly in forest endowment, level of economic development, role in global trade and primary objectives for forest management. Indonesia is the only one of the Asian case studies with large areas of primary forest remaining, while India, China and Japan rank first, third and fourth lowest of all case study countries in per capita forest area. China, India and Indonesia each have large rural populations dependent, to varying degrees, on forests for livelihood needs.

Indonesia is a net producer and exporter of wood and wood products. In contrast, China, India and Japan are the world's first, second and third largest importers of tropical timber (ITTO, 2006a), and China and Japan are the second and fourth largest importers of all primary processed forest products by value (FAOSTAT, 2008). The reliance of these countries on wood imports, both to meet their domestic needs and for processing and re-export, has major impacts on forests outside their borders.

Table 5.1 *Distribution of forest ownership*

	Public	Private	Collective
China (forests, not forestland)*	42%	20%	38%
Madhya Pradesh**	90%	10%	
Indonesia***	100%		
Japan	42%	58%	

Notes: * The government owns all forestlands in China, but has granted ownership of the forests growing on portions of those lands to individuals and collectives.
** In Madhya Pradesh all lands officially designated as forestlands are owned by the state, while private lands with forest cover account for approximately 10 per cent of the total forest area.
*** In Indonesia, forests on private lands are not considered part of the Forest Zone and thus not counted as 'forest'.

The priorities for forest management within the four Asian case study countries reflect their highly diverse environmental, social and economic conditions. Much of China's focus since the 1970s has been on plantation establishment, for protection and production; it now has the world's largest area of plantation forest and has largely curtailed production from its native forests. In contrast, since the 1980s, India has focused increasingly on community-based forest management, for both basic livelihood needs and the commercial production of non-timber forest products, and on plantation production for industrial needs. Forest-based development has been central to Indonesian economic policy for the past 40 years, initially focusing on native forests and subsequently on plantations. In Japan, plantation-based reforestation expanded greatly after World War II; while many of these plantations were established with commercial species, they now serve principally protective functions. Japan now has the largest area of protective plantations worldwide.

Our policy comparison focuses at the national level for China, Indonesia and Japan. While Indonesia has recently undergone a process of decentralization, local regulations still lack the clarity necessary for sub-national analysis. In India, considerable authority is vested in the country's 28 states and seven territories. We have therefore chosen a sub-national case study, the state of Madhya Pradesh, as our unit of analysis for Indian policy. Madhya Pradesh (MP) leads the country in forest area (about 12.4 per cent of India's total forest cover (MPFD, 2008)) and volume of wood production.

CHINA

An overview of forests and forestry

Forests cover about 21 per cent (197.3 million ha) of China's land area (FAO, 2007). Included in these figures is the largest area of planted forests and plantations worldwide – a consequence of major reforestation efforts over the past 50 years. China is the world's most populated country, with about 1.33 billion people and – consequently – a very low per capita forest area of 0.15 hectares (FAO, 2007). All forestlands are owned by the state: 42 per cent directly and 58 per cent through collectives. Individual households have use rights of various forms over 80 per cent of collectively owned forests (Miao and West, 2004).

The loss and degradation of China's native forests has been a long-standing challenge, and one that has prompted strong government action. In 1998, in response to severe flooding of the Yangtze and Yellow Rivers, China instituted a logging ban across much of its natural forest area. At the same time, it enacted national strategies for large-scale forestation and reforestation for both protective and productive purposes (Wenming et al, 2002).

China is both a major producer of wood domestically and a major actor in the global forest products trade. Its annual consumption of nearly 300 million m^3 of roundwood or roundwood equivalent is sourced approximately equally from domestic production and imports, but a range of factors mean it is relying increasingly on the latter. China is now the world's second largest importer of wood, the world's second largest pulp and paper producer and exports the equivalent of 70 per cent of its import volume (White et al, 2006).

Native forests

China's native forests are concentrated in the northeast, south and southwest. Forest types are 59 per cent subtropical, 29 per cent temperate and 3 per cent tropical (FAO, 2005). Approximately half are coniferous and half deciduous, with tropical forests occurring on Hainan Island and neighbouring areas of the southern mainland. About 6 per cent of China's forests are classified as primary and 59 per cent as modified natural.

Dominant native species include oak (*Quercus* spp.), Masson pine (*Pinus massoniana*), Chinese fir (*Cunninghamia lanceolata*), birch (*Betula* spp.) and larch (*Larix* spp.), which together cover about 50 per cent of China's forest area (USDA FAS, 2005). China has the largest bamboo resource in the world and these forests are important both symbolically and materially (Ruiz Pérez et al, 2004).

Planted forests and plantations

China's planted forests comprise over 54.1 million hectares designated primarily for production and 17.2 million hectares primarily for protection. Of these, 28.5 million hectares are classified as productive plantation and 2.8 million hectares as protective plantation (FAO, 2006, 2007). Many of these forests are the result of major national forestation programmes with ambitious targets: for example, it is estimated that some 30 million ha were established under various national plans in the 30 years following the establishment of the communist state in 1949 (Smil, 1993). The Three-North Shelterbelt Programme (including the 'Great Green Wall' project) was initiated in 1978 to address desertification in northern China and aims to establish 35 million ha by 2050. The Sloping Land Conversion Programme, initiated soon after the 1998 floods, aims to establish 15 million ha by 2010 (Ma, 2004; Xu et al, 2004). In total, the Six Key Forestry Programmes have targetted 76 million ha for afforestation (Wang et al, 2007).

The leading provinces for productive plantation development are those in southeastern China – Guangxi, Guandong, Hunan, Fujian and Sichuan. Major established plantation species, accounting for 60 per cent of the plantation area, are Chinese fir (*Cunninghamia lanceolata*), Masson pine (*Pinus massoniana*) and poplar (*Populus* spp.) (USDA FAS, 2006). Recent industrial plantation expansion in southern China has been based on *Eucalyptus* species (Bull and Nilsson, 2004).

Notwithstanding the substantial expansion in the extent of China's planted forests, success rates per unit area have not been high, reflecting a mix of political and environmental factors (Smil, 1993). Between 1949 and 2003, only about 37 per cent of the afforested area was successfully established. Over the same period, intensive demands for timber meant that most stands were not allowed to grow to a mature stage, limiting their value for habitat as well as their potential contribution to production (Zhang and Song, 2006).

Forest governance and policy

The framework for forest governance in China is provided by the Forest Law of 1998, amending that of 1985, and the associated Regulations for Implementation (2000). Article 9 of the Forest Law assigns ownership of all forestlands to the state other than those that belong to a collective. However, ambiguities in the definition of the latter term have led to subsequent ownership and policy conflicts. The new provisions were superimposed on a variety of other tenure arrangements that had been implemented since the establishment of the People's Republic of China in 1949. Collective ownership in China is a form of state (typically local government), rather than community, control.

However, since the process of tenure reform began in the early 1980s, individual households have been granted access to collectively held forestlands on the basis of 15-year contracts under a 'Contract Responsibility System' (FAO, 1993). These households were granted ownership of the trees they planted on these lands. In the 1990s, the duration of contracts on collective lands was extended to 30 years or more and in 2000, the Rural Contracting Law enabled the extension of forestland contracts to 70 years or more. Consequently, more than 80 per cent of collective forests are now administered by individual households, which may also choose to enter cooperative arrangements, for example to supply wood to state or private companies.

Fifty-eight per cent of China's forests are under collective ownership and the majority of these are found in southern China (Miao and West, 2004). State forestry farms, similar to state-owned companies, are dominant in the north (Bruce et al, 1995). About 75 per cent of state forestry farms are run by county governments, 15 per cent by prefectures and 10 per cent by provincial governments (SFA, 2005a).

The Forest Law also defines five classes of forest with designated uses – protection, timber (including bamboo), economic (including orchards and other non-timber products), fuel wood and special use forests (including forests to address pollution and those of historical and aesthetic value) (Miao and West, 2004), and precludes the conversion of forestland to non-forest land.

Around the same time as the Forest Law was amended, the Yellow and Yangtze River floods prompted China to place a moratorium on logging in natural forests in 12 provinces and autonomous regions. This was extended in 2000 to 18 provinces, totalling about 92 million hectares, or 56 per cent of the total forest cover in the country. This ban included a complete prohibition on commercial logging in natural forests of the Yangtze and Yellow River watersheds and a reduction of commercial logging in state-owned natural forests in northeast China, Inner Mongolia, Xinjiang and Hainan (Zhu, 2001; Xu and White, 2002). The logging ban was instituted under the 'Natural Forest Protection Programme' which, together with the 'Sloping Land Conversion Programme', was aimed at integrating 'water and soil conservation with agricultural restructuring, poverty reduction, and sustainable development' (Xu and White, 2002). Subsequently, there have been both recommendations (Xu and White, 2002) and policy initiatives to replace the complete ban on harvesting with sustainable forest management in some areas. These experiments have focused particularly on collectively owned forests, reflecting some of the adverse impacts – on employment, fuel wood supply, and grazing access – consequent to the ban (Xu and White, 2002). The third policy initiative dating from the same period was the provision of substantial government subsidies for fast-growing plantation establishment, as part of a strategy to boost resource supply (White et al, 2006).

Forest practices system

Implementation of China's forest governance framework involves a complex hierarchy of national, provincial, prefectural, county and local entities. The State Forest Agency (SFA) is the national authority responsible for all forests. The responsibilities for forest planning, implementation and regulatory enforcement have been in a state of flux since the 1980s, with a general trend towards decentralization (Hyde et al, 2003). The Regulations for the Implementation of the Forest Law of 1998 (Articles 29–34) specify substantial planning and reporting requirements associated with forest harvesting, except for the case of individually owned trees felled for incidental household use and fuel wood (MEP, 2009b).

For much of the period since 1949, timber harvest, pricing, and sale were centrally handled by the state. More recently, local governments and communities have gained increasing control over these matters and pricing is determined by the market. One result of decentralization and a shift to market pricing was a rapid and sometimes crippling increase in taxation at various levels of government, adding up to 70 per cent of sales in some provinces. This reportedly discouraged forestry investment and fuelled illegal logging. In response, many provinces have since enacted legislation to reduce the tax burden (Dai and Jiang, 2006).

Forest production and trade

China's enormous population and its spectacular economic growth have made substantial demands on forest production domestically and internationally. Due to exponential growth in trade and production over the past decade, China has become the world's second largest importer of forest products and second largest pulp and paper producer. China re-exports a volume equivalent to 70 per cent of its

wood imports (White et al, 2006), much of it to Europe and the US. Domestic industrial roundwood harvest and roundwood equivalent imports were of comparable
magnitude in 2005, at 150 million m³ and 134 million m³ respectively, but are trend-
ing in opposite directions. Domestic production declined by nearly one-third in the
decade to 2005, whereas the roundwood equivalent volume of imported wood
tripled over the same period (White et al, 2006).

The decline in domestic production is a consequence of a number of factors: the
overexploitation of China's forests; the 1998 logging ban and other measures to
enhance environmental protection; the effects of regulations requiring forest owners
to obtain permits for forest harvesting; and heavy taxation. The level of illegal
harvesting in Chinese forests is thought to have risen as a result (White et al, 2006).

In 2003, China's top timber producing provinces were Heilongjiang, Inner
Mongolia and Jillin in the northeast, and Fujian and Guangxi in the south. As the
role of plantations continues to increase, the relative contribution of the southern
provinces is expected to increase. For example, under the 2006–2010 logging quota,
plantations account for 63 per cent of production, up from 38 per cent in 2001–2005
(USDA FAS, 2006).

Imports – which include round and sawn wood, plywood, pulp, and paper –
originate principally from Russia, which provides around 50 per cent of the total
import volume, Malaysia (8 per cent) and Indonesia (6 per cent). However, China's
wood imports draw widely from both native and plantation forests globally and
there are many concerns about the impacts of its demand growth in exacerbating
unsustainable harvesting and illegal logging in producer countries (Zhu, 2001;
White et al, 2006; McDermott et al, 2008b).

China now leads the world as an exporter of wooden furniture, with the EU
followed by the US as the largest buyers. Other key products for export include
additional value-added items such as flooring (in particular laminate flooring), as
well as hardwood, plywood and paper (ITTO, 2006a).

The majority of China's pulp and paper industry, like that of India, was
historically based on non-wood fibres produced by small-scale mills. However, the
expansion of modern large-scale pulp and paper production capacity has been such
that China is now the world's second largest producer and is expected to dominate
global growth in production capacity in most major pulp and paper grades (He and
Barr, 2004; Stafford, 2007).

Indigenous and community forestry

There are 55 officially recognized ethnic minority groups in China and the country's
policy on minorities is stated in the Nationality Regional Autonomy Law. This law
includes the provision of regional autonomy in regions where ethnic minorities live
in 'compact communities'. The 'autonomy' granted in this legislation includes rights
to manage and protect local natural resources in accordance with law. About half of
China's natural forest protection programme and land conversion (forestation)
programmes are located primarily in ethnic minority regions (World Bank, 2005).

China's unique system of 'collective forests' is a dominant tenure arrangement,
comprising 58 per cent of China's forest area. Miao and West (2004) describe the

complex history and arrangements of rights and responsibilities associated with these forests, management of 80 per cent of which has now been devolved to individual households under a variety of contractual arrangements. In general, these arrangements give rights to trees that are planted to the individuals or cooperatives who planted them. In a minority of cases, such as those reported by Lui and Edmunds (2003) in southern China, benefit-sharing arrangements approximate forms of community forestry prevalent elsewhere. However, collective forests remain, in the eyes of the state if not the households managing them, a state resource (Miao and West, 2004).

India and Madhya Pradesh

An overview of forests and forest ownership

Forests cover approximately 20 per cent (64.1 million hectares) of India (FAO, 2007), the second largest land use after agriculture. An estimated 49 per cent is considered 'natural' forest and/or agroforestry. The remaining 33 million ha of planted forest ranks second worldwide (ITTO, 2006b).

India's per capita forest area is only 0.06 hectares (FAO, 2007). Despite rapid economic growth and urbanization, three-quarters of India's population are rural villagers, heavily reliant on subsistence forest use. In 2000, the government of India estimated that about 41 per cent of the forests were in a degraded condition and that about 70 per cent of these degraded forests had lost the capacity for natural regeneration (GOI, 2000). Extensive reforestation and forestation projects have led to net growth in forest cover, but mask ongoing natural forest loss (ITTO, 2006b).

Thirty per cent of Madhya Pradesh, some 7.1 million ha, is classified as forest, and 50 million people depend on these forests for their livelihoods. About half of the forest-dependent population consists of tribal or scheduled castes, who historically are among the poorest people (Kumar et al, 2000).

Almost all land officially classified as forest in India is owned and managed by the government, 93 per cent of it by the forest departments of each state and 4 per cent by the revenue department. Corporate entities, communities and small-scale private owners share ownership of the remaining 3 per cent (Saigal et al, 2002).

Two-thirds of Madhya Pradesh's legally designated state forestlands are formally classified as Reserved Forest, and one-third as Protected Forest. Approximately 63 per cent of state forests are under Joint Forest Management and about 11 per cent are formally designated as National Parks or Sanctuaries. Private lands with planted forests, and communally owned lands, comprise 10 per cent of the state's forest area (FSI, 2005).

Native forests

Indian native forest types reflect the diversity of the subcontinent's geography – ranging from tropical moist forests in the south and northeast to Himalayan alpine forests, and from desert thorn forests to extensive coastal mangroves. Tropical evergreen forests and wet and dry deciduous monsoon forests account for the largest areas. Approximately half of India's natural forests are classified as modified natural, and half as semi-natural (FAO, 2007). The proportions classified as dense and open forest are similar, at around 30 per cent (FSI, 2005). India's forests are globally significant for their biological diversity, contributing to India's status as one of the world's 12 mega-diverse countries (ITTO, 2006b).

The major forest types of Madhya Pradesh are dry thorn, dry and moist deciduous, sub-tropical evergreen and tropical moist evergreen. Common tree species include teak (*Tectona grandis*) and sal (*Shorea robusta*) and bamboo is also widespread (FSI, 2005).

Planted forests and plantations

Planted forests and plantations are of central importance to Indian forestry. However, it is difficult to provide an accurate profile of this resource, in part because of evolving classifications that shifted most of what was previously recorded as 'plantation' into the category of planted forest. According to FAO's 2007 State of the Forests report, the area of planted forest is about 33 million hectares. Fifty-seven per cent of this area was planted primarily for productive purposes and 43 per cent primarily for protective purposes (FAO, 2007). Agroforestry is playing an increasing role in planted forest systems and an estimated 25 per cent of planted forests are located on private, communal and non-forest public land. The rate of private planting is believed now to exceed the rate of public planting (ITTO, 2006b).

Intensively managed production plantations comprise about 1 million ha and about 2.2 million hectares are classified as protective plantation (FAO, 2006). Nearly half of India's plantations are eucalypt and acacia species grown on short rotations; 8 per cent are teak and 10 per cent are pines and other conifers. However, many plantation forests have performed poorly (ITTO, 2006b).

The estimates above include planted forests greater than 1ha. The area of smaller-scale tree planting outside designated forestlands has been estimated separately. The total area of such planting in India is estimated at 9.2 million ha and in Madhya Pradesh 0.6 million ha – around 3 per cent and 2 per cent, respectively, of the nation and the case study state (FSI, 2005).

Forest governance and policy

The national and state governments share authority over state forests in India. The agency in charge at the national level is the Ministry of Forests and Environment (FAO, 2007).

The basis of Indian forest law is the 1927 Indian Forest Act, amended in 1951. Other key pieces of national legislation include the Wildlife Protection Act, 1972 (amended 1991); the Forest Conservation Act, 1980 (amended 2003); the Environmental Protection Act, 1986, and the Biological Diversity Act, 2002. Together, these Acts regulate access and use of forests and forest products and support conservation (FAO, 2005; ITTO, 2006b).

India's National Forest Policy was released in 1988. This policy outlines major priorities for forest management, including conservation of forest, soil and water resources; the expansion of forest cover; meeting livelihood needs; improving forest products utilization; reducing pressure on forest resources, and increasing the use of farm forestry to meet industrial wood needs.

State forests in Madhya Pradesh are managed by the Madhya Pradesh Forestry Department. The Department operates on a geographic basis, with the state divided into three regions and 16 Working Plan divisions (MPFD, 2008). Given the extent of forest under joint management in Madhya Pradesh, Joint Forest Management Committees are now of fundamental importance to forest governance (see 'Indigenous and community forestry' below).

Madhya Pradesh has enacted a number of state-level Forest Acts. In addition to Acts focused on state-owned forests, the Lok Vaniki (People's Forestry) Act seeks to facilitate private tree growing by creating supportive legal, institutional and market environments (Saigal et al, 2002). The 2005 State Forest Policy focuses on addressing forest encroachment and forest-related crime, on forestation and on the distribution of revenues (The Hindu Staff Correspondent, 2005).

Forest practices system

The foundation of state forest management throughout India is the Working Plan, first instituted under British Colonial rule (Kant, 2001). Working Plans are prepared every five years at the state and sub-state levels. These plans outline the forestry budget and set prescriptions and targets for management. In Madhya Pradesh, the Lok Vaniki Act provides the basis for oversight of forest practices on private lands (Saigal et al, 2002).

Forest production and trade

Forest production in India is dominated by subsistence use, with over 94 per cent of wood harvest consumed as fuel wood (FAO, 2007). Approximately 50 per cent of India's wood supply comes from 'non-forested' private and communal lands. Agroforestry and farm forestry, including outgrower schemes, are also playing an increasing role in commercial solid wood and pulp production (Saigal et al, 2002; ITTO, 2006b).

All of India's forested states have established forest development corporations as government-owned businesses responsible for forest production on public lands. In Madhya Pradesh, timber harvested from public lands is auctioned at commercial depots or sold at government prices from *'nistar'* (community use) depots (MPFD, 2008).

In addition to timber uses, natural and semi-natural forests are a key source of non-timber forest products (NTFPs). NTFPs are important to the livelihoods of 30 million Indians and contribute over 75 per cent of India's forest-based export revenue (ITTO, 2006b).

India now ranks third worldwide as an importer of tropical logs, most of which originate from Malaysia and Myanmar, with growing volumes from Africa (ITTO, 2006a).

Indigenous and community forestry

The 1988 Indian National Forest Policy places a high priority on rural and indigenous subsistence needs. It provides the legal foundation for the system of Joint Forest Management (JFM), whereby village user groups manage forest resources in conjunction with state, district and local authorities.

Joint Forest Management (JFM) has since taken a variety of forms, ranging from community control to shared management with state forestry agencies. According to Kant (2001), India's JFM system resembles both a return to old traditions and an innovative approach to addressing modern natural resource management challenges. Perhaps due to the combination of historical precedent and popular support, JFM has grown very quickly. Between its establishment in 1988 and January 2000, 22 Indian states initiated programmes of Joint Forest Management, covering about 10.24 million hectares and involving 36,075 village committees (GOI, 2000). Despite the significant land area associated with JFM, though, the majority of forestlands in India still remain under the direct control of the state. Furthermore, state and community objectives for JFM often do not mesh well, with state priorities focusing on environmental rehabilitation as well as commercial timber production, and local villagers concerned with meeting their subsistence needs for fuel wood, fodder, and other non-commercial forest products (Khare et al, 2000; Rao et al, 2005).

There are 645 indigenous groups in India that are recognized in the Indian Constitution and referred to as 'Scheduled Tribes'. These people are also referred to as *Adivasi* or 'First Inhabitants'. While *Adivasi* are commonly considered of low caste, they are culturally, linguistically and economically distinct from the broader Indian caste system. Many are forest-dependent, and – since 1990 – have been allowed first claim on forest produce (Kumar et al, 2000). The controversial Scheduled Tribes and Other Traditional Forest Dwellers (Recognition of Forest Rights) Act 2006 grants rights to traditional forest dwellers to occupy and cultivate forested ancestral lands. The Act has been criticized by indigenous organizations for granting rights to 'other' traditional forest dwellers that are equal to the rights granted to tribal groups (AITPN, 2006). It has been also been strongly criticized by some environmentalist groups as contributing to forest destruction (Basu, 2007). Implementation of the bill is reported to have been impeded by the level of controversy surrounding it (Basu, 2007).

Madhya Pradesh has been at the forefront of the development and implementation of JFM. In 2005, there were more than 14,000 JFM Committees managing some 6 million ha (63 per cent) of state forest. The Madhya Pradesh JFM programme involves some 1.7 million families, of whom 0.8 million are from Scheduled Tribes (FSI, 2005). Management objectives and arrangements vary according to whether the forest is zoned as Protected Area, Dense Forest or Degraded Forest. The state has also adopted a variety of additional mechanisms for granting communal rights, including the system of *nistar* (royalty-free subsistence) rights. Under this system, 'unoccupied' state land, as defined by the Madhya Pradesh Land Revenue Code of 1959, is to be provided for every village for the provision of *nistar* rights, including timber and/or fuel reserves. *Nistar* land is considered communal land, subject to a set of terms and conditions. Communities may also be given access to 'wasteland' as well as to land and resources in neighbouring villages (Ramanathan, 2000).

Nistar may also be provided from state reserve forests via *nistar* depots. However state production is reportedly not adequate for local needs, resulting in considerable forest 'encroachment'. The approach of providing only partial, state-controlled community rights has been criticized for creating numerous contradictions and tensions over the legitimacy and legality of forest production (Prasad, 1999; Ramanathan, 2000).

Indonesia

An overview of forests and forest ownership

Indonesia is the world's largest archipelago: a nation of over 17,000 islands straddling the equator and stretching over 5000km from the northwest tip of Sumatra to the eastern border with Papua New Guinea. Its geographic breadth and diversity are reflected in both its people and its forests; Indonesia's population of 220 million (2005 data) includes over 500 ethnic and 600 language groups and its forests are of global significance for both biodiversity and commercial values. Its tropical forest has been second in size only to the Amazon, but rapid substantial deforestation over the past 40 years has reduced its forest cover to only 60 per cent of its 1950 extent (ITTO, 2006b; Singer, 2009).

Most forest loss has occurred in the 'Outer Islands' – Kalimantan, Papua, Sumatra and Sulawesi – as a consequence of forest-based economic and social development policies of Suharto's New Order regime of 1967–1998. These were founded on the nationalization in 1967 of Indonesia's forests by the creation of the Forest Estate (*Kawasan Hutan*), government sponsored transmigration of populations to the Outer Islands and the revocation of the customary (*adat*) rights of local communities. Indonesia has also pursued aggressive industrial wood plantation development policies. Precise figures on the actual area under plantation are lacking. However, some ITTO and government sources have estimated that about 3 million ha of mostly corporate-owned plantation concessions have been established, principally on land converted directly from native forest. There is also an estimated 1.5 million ha of long-established plantations of high value trees, principally teak in Java, with both the state forestry corporation (Perum Perhutani) and farmers owning substantial resources (ITTO, 2006b; Manurung et al, 2007; Singer, 2009).

The period of *reformasi* which followed Suharto's downfall in 1998 has emphasized the devolution of forest governance and seen a resurgence of *adat* claims. The state has responded to these by recognizing customary forests and initiating a number of community forestry schemes (Singer, 2009). Formally, however, almost all forests in the Outer Islands continue to be classified as public forests. Meanwhile, fluid, ambiguous and inconsistent policies and implementation have facilitated continuing forest exploitation, including illegal logging (Resosudarmo, 2007). Although there have been some recent signs of improvement (Manurung et al,

2007), forest management continues to suffer from cronyism and a lack of adequate enforcement (Dauvergne, 2001; Curran et al, 2004). As a result, and greatly facilitated by the rapid expansion of plantation crops, particularly oil palm (Koh and Wilcove, 2008), forest exploitation and conversion accelerated in the period 2000–2005 (Singer, 2008, 2009).

Native forests

Nearly 90 per cent of Indonesia's native forests are classified as tropical moist forest. Six types are distinguished for management purposes: mixed hill forests (representing 65 per cent of the total); freshwater swamp forests; mangroves; peat swamp forests; savanna, bamboo, deciduous and monsoon forests; and sub-montane, montane and alpine forests (ITTO, 2006b). In 2005, Indonesia was ranked second among our tropical case study countries in the percentage of forests classified as primary (see Figure 2.11). An estimated 10 million of Indonesia's poorest people depend on these forests for their livelihoods (World Bank, 2006).

Indonesia is also ranked first among the case study countries, and amongst the highest globally, in its rate of deforestation, at about 2 per cent (approximating to 2 million ha) annually (ITTO, 2006b; World Bank, 2006). The reasons for Indonesia's high rate of forest loss and degradation have been explored by many authors (Dauvergne, 2001; World Bank, 2006; Tacconi, 2007) and include the country's political economy and governance, its long-standing policy of forest-based and agricultural plantation development, both planned and unsanctioned forest conversion and harvesting, the gap between industrial wood demand and sustainable supply, and the inequitable distribution of benefits from forests. Fire, usually associated with land conversion practices, is also a major threat. As a result of this forest loss, the majority – more than two-thirds – of Indonesia's natural forests are now those remaining in Kalimantan and Papua.

Planted forests and plantations

Indonesia classifies most of its non-primary forest, about 36 million hectares, as semi-natural forests of assisted natural regeneration. The area classified as 'planted' consists entirely of plantations. According to 2005 FAO data, 3.4 million hectares of tree plantations were established, all designated primarily for production purposes (FAO, 2007). The principal species planted are *Tectona grandis* (teak), *Pinus* spp., *Acacia* spp. and *Eucalyptus* spp. Teak plantations are largely confined to Java, while acacia and eucalypt plantations are grown on the Outer Islands. These latter plantations, primarily aimed at pulp production, have shown the most rapid growth and now total around 1.8 million ha (ITTO, 2006b; World Bank, 2006). The area classified as 'tree plantations' is dwarfed in size by plantations of oil palm (*Elaeis guineensis*), coconut (*Cocos nucifera*) and rubber (*Hevea brasiliensis*), totalling some 11 million ha (World Bank, 2006; Manurung et al, 2007).

Government policy has strongly encouraged plantation development, including targets for conversion of natural forests to commercial tree plantations and other crops. The Ministry of Forestry plans to expand Indonesia's forest plantations by a further 9 million ha, with a 40:60 ratio of large- to small-scale ownership (World Bank, 2006; Manurung et al, 2007). The policy of allocating native forest for conversion to forest plantations is known to be problematic for various reasons. First, it has greatly expanded the area of degraded land, since many concession areas allocated for plantation development are never successfully reforested (World Bank, 2006; Manurung et al, 2007). Second, critics argue that the policy has been exploited as a means to log native forests relatively free of environmental constraints (WWF Germany, 2005).

Forest governance and policy

The foundations of contemporary Indonesian forest governance were established by the Basic Forestry Law of 1967, which asserted state ownership and control of all land and forests designated as the Forest Estate; at the time, this corresponded to 74 per cent of Indonesia's area, including lands which were, or subsequently became, not forested. The Basic Forestry Law did not recognize any prior customary (*adat*) rights. Instead, it established four new forest use categories: conservation, conversion, production and protection. Large scale concessions were issued to private entrepreneurs, usually associates of the ruling elite, to exploit production and conversion forests and to establish plantations. The national Ministry of Forests was established to administer the Forest Estate and forestry sector.

The most recent Indonesian Long-Term Forestry Development Plan identifies 127 million ha (67 per cent of Indonesia's land area) as forestland. It allocates 30 per cent of these forests to production, 17 per cent to protection, 12 per cent to conservation, 8 per cent to conversion and 33 per cent to 'other land'; the latter category includes a large proportion of forestland formerly classified as conversion forest (World Bank, 2006).

The Basic Forestry Law was revised in 1999, as part of the process of *reformasi*, and Forestry Law 41/1999 is now the central piece of forestry legislation. It continues the essential principles of the 1967 law, in terms of state ownership and the categorization of forests for assigned uses. However, this law introduced key changes in terms of processes for decentralization and community participation. Law Number 22/1999 on regional autonomy establishes a non-hierarchical distribution of authority among the federal, regional, district and local levels, in some contradiction to the strong central government role asserted in Forestry Law 41/1999. A series of regulations gave effect to both decentralizing and centralizing intentions, leading to considerable conflict among the various levels and branches of government, and resulting in various incongruencies such as the granting of overlapping forest concessions and development of conflicting forest practice laws. For the purposes of consistent analysis, we assess forest practice policies that have been established at the national level, principally those embodied in Regulation 34/2002, Forestry and the Formulation of Plans on the Management, Exploitation and Use of Forest Land, and several other relevant policy documents.

The major provisions of Regulation 34/2002 were to classify the Forest Estate in three categories: conservation, protection and production; to reassert the authority of the national Ministry of Forests, as the sole entity able to issue Commercial Timber

Utilization Permits (IUPHHK) and limit earlier local government authority to do so, and to extend the Ministry's authority over domestic timber transport and marketing and over domestic wood processing industries (Singer, 2009).

Indonesia's forest governance is weak in many respects, reflecting a legacy of policies that concentrated control of the wood processing industries in a few politically powerful hands, centralized forest administration in ways which result in little effective management or enforcement capacity in the field and perpetuated corrupt and collusive practices (World Bank, 2006). These dynamics have shifted, but have not been resolved, through decentralization over the past decade (Resosudarmo, 2007). Meanwhile illegal logging practices continue to cause major problems.

Forest practices system

Forest practices in natural production forests in Indonesia are required to comply with to the Indonesian Criteria and Indicators for Sustainable Forest Management and with the Indonesian Selective Cutting and Planting System (Tebang Pilih Tanam Indonesia, TPTI),[2] established by decrees issued by the Ministry of Forestry. TPTI focuses on silviculture pre- and post-harvest, rather than on harvesting and extraction (Klassen, 2002). Concessionaires are responsible for preparing annual and five-year management plans and an overall management plan for the life of the concession (FAO, 2004). The Ministry of Forestry oversees management plans and assesses concessionaires' compliance with the Criteria and Indicators (C&I) and the TPTI regulations (ITTO, 2006b).

Forest production and trade

Indonesia has been the world's largest producer of tropical industrial roundwood since the 1980s, when wood production from native forests grew exponentially. Since then, solid wood (plywood and sawn wood) production has declined steadily. The number and area of native forest concessions halved in the decade to 2004 and industrial wood intake into processing facilities in 2004 was only about 60 per cent of 1990 levels. In contrast, pulp production drawing from a mix of native forest and plantation wood almost trebled over the decade to 2005 (World Bank, 2006). This is largely as a result of pulp mill development on the island of Sumatra (Manurung et al, 2007). Forest concession and wood and pulp production are highly concentrated in the hands of major conglomerates (ITTO, 2006b; World Bank, 2006). The state forestry corporation, Perum Perhutani, manages a relatively small area, primarily consisting of teak plantations in Java.

Estimates of total wood production are confounded by the extent of illegal harvesting. Consumption of domestically produced wood by processing facilities in 2004 was estimated at 50.5 million m^3, but official log harvest was recorded as only 13.5 million m^3 (World Bank, 2006). ITTO estimated 2003 industrial log production at 25 million m^3, but notes that illegal production 'may exceed the official cut' (ITTO, 2006b). According to Ministry of Forestry task force figures, in 2002 about 82 per cent of wood production was illegal. By 2005 this figure had dropped to about 45 per cent and is believed to have continued its decline thereafter

(Manurung et al, 2007). Eighty-five per cent of 2003 industrial wood production originated from native forests and 15 per cent from plantations (ITTO, 2006b). Fuel wood production is estimated at around 86 million m³ annually (ITTO, 2006b).

The wood product sectors vary considerably in market structure, shaped, in part, by Indonesia's forest trade policies. The export of unprocessed logs was banned in 1985, followed by a ban on rough-sawn timber in 1992. However, despite the ban on raw log exports, substantial volumes of wood are smuggled, unprocessed, through neighbouring countries. Sawn wood, including rough-sawn and other products, is produced largely by medium- and small-scale operators, and is sold in nearly equal volumes for domestic and export consumption. In contrast, the major-ity of plywood exports originate from large companies. Pulp and paper production is the most heavily consolidated with three companies currently dominating produc-tion. In 2005, Indonesia was the world's ninth largest pulp producer and twelfth largest paper producer. Value-added furniture and handicrafts, produced largely by medium- and small-scale companies, are also showing growth in export revenues (Manurung et al, 2007).

Indonesia is a significant net exporter of forest products, which generated around US$5.3 billion in 2002. Principal export destinations are Japan, Taiwan, China and South Korea. A further $3 billion annually may be forgone through illegal harvesting and trade (ITTO, 2006b; World Bank, 2006).

Issues of illegal logging in Indonesia have received considerable international attention. A number of countries, including China, Japan, the UK and the US have developed bilateral Memoranda of Understanding (MOUs) to help address the issue. In addition, Indonesia has been very active in the Asian Forest Law Enforcement and Governance (Asia FLEG) process, and is far along in the process of developing a Voluntary Partnership Agreement (VPA) with the European Union under the European Union Forest Law Enforcement, Governance and Trade (EU FLEGT) initiative (Chatham House and DFID, 2008).

Indigenous and community forestry

There was little formal recognition of traditional (*adat*) rights over forests until *refor-masi* and the 1999 Basic Forestry Law. This 1999 law established a category of community-based customary forest (*hutan adat*) and expressly required that the rights of 'customary law communities' be respected. However, the mechanism for determining legitimate customary rights was not well elaborated (FWI, 2002, p64). There has been considerable debate in Indonesia about how the term 'indigenous' should be defined and its interpretation has varied considerably, ranging from being used to describe all long-term residents of the archipelago, to only 'isolated' settled communities or nomadic tribes (ADB, 2002; Singer, 2009). A national Alliance of Adat Peoples (AMAN) was established in 1999 and has pressured government to grant customary owners greater access to forestland and resources; however, the Indonesian government has been reluctant to do this for both political and economic reasons (Singer, 2009).

Various forms of community forestry have been initiated by the Indonesian government since *reformasi*, following earlier exploration of benefit-sharing arrangements between local communities and the state-owned Perum Perhutani. One suite of initiatives, focused on local communities' rights to access and manage state forests for defined periods, followed decrees and laws between 1998 and 2001. In practice, there were many administrative constraints and loopholes that restricted uptake of these initiatives, or led to perverse outcomes (Singer, 2009). More recently, Government Regulation 6/2007 created three forms of community forest – village forests, people's forests and partnerships – each of which allowed local communities access to and benefits from natural forests (Singer, 2009). A third suite of initiatives (the community-based plantation forest programme, HTR) focuses on plantation development by communities, independently or in partnership with forestry companies. These offer a means of better formalizing and advancing arrangements for benefit-sharing between some forestry companies and local communities (Nawir and ComForLink, 2007). The formation of cooperatives to represent communities' interests is a feature of many of these initiatives.

These various indigenous and community forestry arrangements have been seen by government and business as important for addressing social unrest associated with the exclusion of local communities. In practice, however, their outcomes have been limited (Singer, 2009), in large part due to the reluctance of key Indonesian forest sector actors to cede authority over, and access and use rights to, Indonesia's forests.

Japan

An overview of forests and forestry

Japan is unusual because it is a densely populated country that is also densely forested. Approximately 67 per cent (25 million ha) of Japan's land base is forested; 40 per cent of this is plantation forests. Japan has a long history of managing and establishing forests for primarily protective purposes, reflecting its mountainous topography and exposure to natural hazards such as earthquakes and typhoons. More than one-third of Japan's forests are managed primarily for protection.

Japan's forestlands are 58 per cent privately owned, 31 per cent nationally owned and 11 per cent owned by sub-national governments (prefectures) or communities (cities, towns and villages) (Shinrin-ringyo hakusyo, 2002). The average area of private forest holdings is 2.5ha (Forestry Agency Japan, 2006b).

Wooden buildings and other uses of wood have a strong place in traditional Japanese culture (JOFCA, 2002). However, for a number of economic and social reasons, domestic wood production has been declining steadily since the 1960s. Correspondingly, Japan has assumed a major role globally as an importer of forest products – particularly of tropical hardwoods, but also of wood fibre for its pulp and paper industry (JOFCA, 2002; Forestry Agency Japan, 2006b).

Native forests

Native forest types range north to south from 'sub-polar' coniferous (*Abies* spp. (fir), *Picea* spp. (spruce)), to cool temperate deciduous broadleaf forests (*Fagus* spp.

(beech), *Quercus* spp. (oak) and *Poaceae* spp. (bamboo)) to warm temperate evergreen broadleaf (*Lauraceae* (laurel family)), to semi-tropical and mangrove rainforests (Karan and Gilbreath, 2005).

About 18 per cent of Japan's forest cover is classified as primary forest and 40 per cent as 'modified natural'. As explained in the following section, Japan does not classify any of its forests as 'semi-natural'.

Planted forests and plantation

Japan ranks fifth in the world in terms of the total area of tree plantations, which comprise 40 per cent of its forests. All of these plantations are classified as primarily for protective functions (FAO, 2003). Forests are considered 'man-made' if they are planted or seeded, rather than naturally regenerated (Blandon, 1999). Native coniferous species, such as Japanese cedar (*Cryptomeria japonica*) and Japanese cypress (*Chamaecyparis obtusa*), are the most commonly planted (Forestry Agency Japan, 2003). Some 10 million ha of Japanese plantations have been established since the end of World War II (JOFCA, 2002).

Forest governance and policy

Japan's first national forest law, establishing a national system of protective forests, was promulgated in 1897 (JOFCA, 2002). The key legislation now governing forest management on both public and private lands in Japan is the 2001 Basic Law on Forest and Forestry, which was revised in 2004. The 2001 law establishes new guiding principles for forest management, emphasizing the multiple uses of forests for sustainable resource management, as well as conservation, recreation and other values (Imaizumi, 2001).

An integrated national forest planning system comprises national-level plans for all forests and for natural forests and associated implementation plans; regional-level plans for both national and non-national forests; district plans for sub-national public and community forests, and management plans for private forests (JOFCA, 2002).

National forests, *kokuyuurin*, are generally subject to more stringent regulatory requirements than other public or privately owned forests, *minyuurin* (Blandon, 1999). All forests, whether public or private, are further classified in one of two basic categories: (1) forests with harvesting and/or management restrictions known as 'restricted forests' (including 'protection forests'), or (2) 'ordinary forests' (Imaizumi 2004). By definition, 'restricted' or 'protected' forests, whether public or private, are the most stringently regulated. Some 35 per cent (9.4 million ha) of Japan's forests are in this category; around half these forests are nationally owned and half by other public, community and private owners (JOFCA, 2002). An increasing number (currently about 25 per cent) of private forest owners are non-resident, reflecting broader changes in Japanese society and presenting challenges to established forest management practices (Blandon, 1999; Forestry Agency Japan, 2006b).

A further classification of forestlands is made at the level of local forest plans. At this level, forests are classified according to their primary function, i.e. as:

1   'forests for water and soil conservation';
2   'forests for the people' (i.e. forests for conservation, landscape, and recreation); and
3   'forests for cyclic use' (i.e. forests for efficient and sustainable wood use).

Nationally, the proportions of forest area in each category are, respectively, 54 per cent, 27 per cent and 19 per cent (JOFCA, 2002). Most 'protection forests' fall within one of the first two categories, depending on their specific management objectives (Imaizumi, 2004).

Forest practices system
The range of forest practices permitted in any specific forest depends on its classification, as discussed above. Primary responsibility for those practices and their oversight depends on ownership. Practices for national forests are developed by the national Forestry Agency and specified in five-year Regional Management Plans; those for sub-national (prefecture) forests are developed by prefecture government agencies and specified in ten-year Regional Plans; those for community forests are developed by the local authority, within the context of the relevant ten-year Regional Plan, and those for private forests are voluntarily developed by their owners as five-year Forest Management Plans, also within the context of the relevant Regional Plan. The Forest Agency provides guidelines for the development of private plans and local authorities provide any necessary approval (JOFCA, 2002).

Forest production and trade
Legally, 95 per cent of Japan's forest area is available for timber harvest (IFCO, 2003). Nevertheless, Japan's domestic timber production has been in decline, falling to about one-half of the peak harvest levels reached in the 1960s. The value of forest production has declined by 40 per cent since 1980 (Forestry Agency Japan, 2006b). Reasons for the drop in production include a decline in the price of wood products, relatively high labour costs and changing social values. Private forestlands account for over two-thirds of total production. Around 20 per cent of Japanese wood consumption originates from domestic production (Forestry Agency Japan, 2006b).

It has been estimated that Japan has sufficient growing stock to achieve sustained self-sufficiency within the next 20 years, but there is much debate as to whether Japan will significantly increase its domestic production, given current and likely future socio-economic conditions. For example, the changing demographic of private forest owners is shifting the emphasis in private forest management to values other than production (Forestry Agency Japan, 2006b). The general trend in public forests is also towards less production and more emphasis on protection (Blandon, 1999).

Japan plays a major role in global forestry as a forest products importer, principally to supply domestic demand. This demand, however, has declined by around 20 per cent over the past 20 years (Forestry Agency Japan, 2006b). Until recently, Japan was the second largest importer of industrial roundwood, after the United States, and the largest importer of tropical logs (Dauvergne, 1997, 2001; Blandon, 1999; FAO, 2003). While it remains the largest importer of tropical plywood, it has been overtaken by China and Finland as importers of industrial roundwood, and by China and India as the leading importers of tropical logs. Japan remains a major importer of wood fibre for its pulp and paper industry, and a minor exporter. About half of the timber exports go to China and the United States, with increases in log exports largely going to China (Forestry Agency Japan, 2006a).

Indigenous and community forestry

The Ainu people of Japan's northern and central islands are the only recognized indigenous people in Japan and were the subject of assimilation policies until 2008 (Fogarty, 2008). Consequently, there is little concept of 'indigenous forestry' in Japan.

In the Japanese context, 'community forestry' refers to the management of forests owned by cities, towns and villages. These forests, which are under the management of local authorities, comprise about 6 per cent of Japan's forest area. Their management has to be consistent with the classification of the forest, as discussed above. Historically, these forests have been important in sustaining the livelihoods of those who live in 'mountain villages', providing both wood and non-timber forest products. Demographic and economic changes are challenging the viability of these communities and the revitalization of mountain villages is one of the priorities of Japan's Forestry Agency (JOFCA, 2002; Ishihara and Oka, 2005; Forestry Agency Japan, 2006b).

FOREST USE AND BIODIVERSITY CONSERVATION

Protected areas

At 16.7 per cent of the land area, China's network of protected areas is relatively large. However, the vast majority falls under the less strictly protected IUCN Categories V–VI (Chapter 2, Figure 2.15). The establishment of forest reserves, management and administration of forests and prevention of hunting in forests is covered in Articles 4, 11, 20 and 21 of the Forest Law of 1979. Provisions for nature conservation are also incorporated in Articles 9, 10, 22 and 26 of the new 1982 Constitution. China's Environmental Protection Law (1986) provides a legal framework for the conservation of rare ecosystems, flora and fauna. In addition, reserve management is governed by the Regulation on Natural Reserve Management and Regulation on Wild Plants Protection (Miao, 2004).

India has designated about 5.4 per cent of its land area as protected under the six IUCN categories. About 4.9 per cent of the land area falls under the stricter IUCN Categories I–IV, which generally do not allow extraction of natural resources (Chapter 2, Figure 2.15).

The 1995–2000 World Bank Forestry Project in Madhya Pradesh led to an expanded network of parks and preserves throughout the state. In 2003, the MP Forest Department reported about 10,862 square kilometres of national parks and sanctuaries covering 11.4 per cent of the forest area and 3.52 per cent of the state's total land area. The Forest Department has meanwhile set a goal to increase its protected areas to cover 15 per cent of the forest area and 5 per cent of the total land area, as recommended by the State Wildlife Board (MPFD, 2008).

Reflecting the need for local cooperation in protecting these reserve areas, 502 'Ecodevelopment Committees' have been established among communities lining the reserve areas. In addition, Madhya Pradesh has instituted village relocation projects, relying on the 'voluntary' transfer of rural communities from land areas slated for parks. A compensation programme has also been instituted for losses related to tigers, leopards, wolves, bears, elephants, wild pigs, gaurs and hyenas. A fixed rate of 50,000 rupees (about US $1184.01 as of August 2008, up from US $443.00 in 2004) is paid to the successors of persons killed by these species, treatment valued up to 10,000 rupees is paid for injuries and no more than 5000 rupees is paid in compensation for lost livestock (MPFD, 2008). All of these measures serve to place in sharp relief the conflicts between the needs of local communities and broader conservation goals.

Indonesia has designated 14.6 per cent of its land base as protected areas, with 8.3 per cent under IUCN Categories I–IV (Chapter 2, Figure 2.15). The Indonesia Biodiversity Strategy and Action Plan (BSAP) 2003–2020, developed under the United Nations Convention on Biodiversity (CBD), called for the financing of priority protected areas (PAs) and expansion of the PA system, and specified several new conservation areas (World Bank, 2001). The 2006 Long-Term Forestry Development Plan classes some 22 million ha of forest as conservation forest, namely, forest managed primarily for biodiversity conservation (World Bank, 2006).

However, the process of decentralization has proven particularly problematic with regard to protected areas. The management of protected areas has not been devolved to district governments and therefore still falls under the authority of the Ministry of Forestry. Hence, district governments have little, if any, incentive to conserve them. At the same time, all levels of government lack the financial and human resources to manage them properly. Consequently, many protected areas are considered to be in a degraded condition (ITTO, 2006b; Gaveau et al, 2007).

Japan has nearly 16.5 per cent of its land area under protected status in 2006, with 4.1 per cent under IUCN Categories I–IV (Chapter 2, Figure 2.15). A recent gap analysis indicates that these protected areas are relatively well distributed in terms of forest type, but do not capture the full range of geographic context (Kamei and Nakagoshi, 2006).

Protection of species at risk

The protection of biodiversity in India is particularly challenging given the country's dense rural population and its direct dependence on subsistence uses for forest resources. India has taken a number of different approaches to addressing the challenge. These include the establishment of a Wildlife Act in 1972, which provides wildlife protection measures, including prohibitions on the taking of endangered species. There is no specific piece of Indian legislation, however, dedicated to habitat protection. The new Biodiversity Act (crafted in 2002), if enacted, would serve to fill this gap.

A number of species-specific projects have been implemented in Madhya Pradesh, including several projects aimed at the protection of tigers and their habitats. In addition, the Asiatic lion has been recently reintroduced into the state in one special habitat area. Biodiversity protection outside reserve areas is covered in the state guidelines for working plan preparation (MPFD, 2008).

Indonesia refers to the CITES treaty's red databook for its list of threatened and endangered species. The central government holds authority for species and habitat protection. Law 5/1990, 'Conservation of Biodiversity and Its Ecosystems', provides a regulatory framework for biodiversity protection. Habitat protection is based on habitat scarcity, as well as species' habitat dependence.

The protection of 'rare animals and plants' is enshrined in Article 9 of the Chinese Constitution and China has enacted a number of laws aimed specifically at the protection of species at risk. These include the law on wildlife protection, the regulation on natural reserves and the regulation on protection of wild plants (Miao, 2004; MEP, 2009a).

In Japan, major pieces of legislation covering the protection of species at risk are the Protection of Endangered Species of Wild Fauna and Flora Law, No. 75, 1992 and the Wildlife Hunting and Protection Law, No. 32, 1918. The National Guidelines for the Conservation of Endangered Species outline requirements for the protection of individual organisms, habitats and viable populations. The hunting law lists species that may be lawfully hunted, as well as restrictions on the dates, location and method of hunting (MOE, 2008).

FOREST PRACTICE REGULATIONS

Riparian zone management (Indicator: Riparian buffer zone rules)

In China, quantitative no-harvest zones apply, with size varying according to the width of the stream (MEP, 2009c). As previously discussed, no logging is allowed in a number of major watersheds. In Madhya Pradesh, the protection of streamside riparian zones is spelled out in each Division Working Plan. In Indonesia, no harvest is allowed within 50 metres of tributaries, or 100 metres of major rivers (Law No. 41/1999). In some cases, regional and/or district regulations may also apply. In Japan, clearcutting in protection forest riparian zones on public and private land may be prohibited on sites where forest regeneration may be difficult (Imanaga, 1988; Shinrinhozen kenkyukai, 1997; Norinsuisanroppo henshushitsu, 2003). No standardized buffer zone widths, however, are specified by law.

Roads (Indicators: Culvert size at stream crossings, road abandonment)

In China, no national requirements were identified relating to culvert sizing at stream crossings. Grade Four roads must be decommissioned within ten years (SFA 2005b).

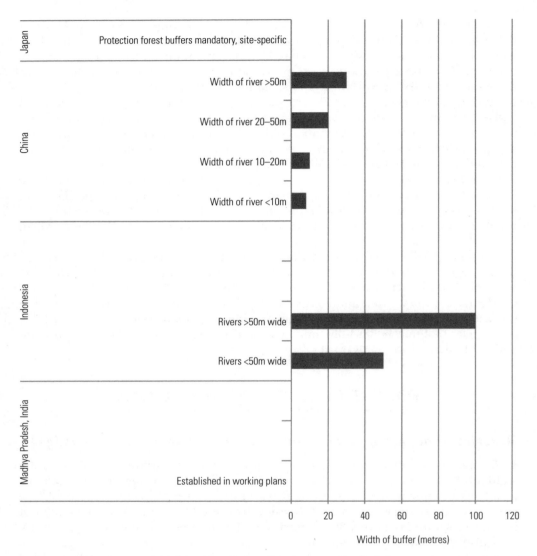

Figure 5.1 *Riparian buffer zone policies**

Note: * The above riparian zones are all 'no-harvest' zones, i.e. zones where all timber harvest is prohibited.
Sources: Imanaga, 1988; Indonesia, 1999; MEP, 2009c; Norinsuisanroppo henshushitsu, 2003; Shinrinhozen kenkyukai, 1997

Forest road building in Madhya Pradesh is covered in the Division Working Plans. Road maintenance is covered by separate budget allocations (MPFD, 2008).

In Indonesia, culvert size at stream crossings is based on peak flow. Forest concessionaires are responsible for road maintenance for the life of the road and are required to reforest roads no longer in use (Indonesia, 1989, 1999).

In Japan, the Forest Road Regulation governs stream crossings and decommissioning on national lands and roads on private lands that have received government subsidies (Nihon Rindo Kyokai, 2002). Since almost all road building on private forestlands in Japan receives state funding, government road building standards apply *de facto* to almost all forest roads (Imaizumi, 2004). In general, Japan's road network is well developed and maintained in a fairly stable state. Due to a lack of substantive rules specifically aimed at road building on private lands, however, we classify Japan's private forest road building policies as effectively a 'mixed' approach.

Cutting rules (Indicator: Clearcutting or cutting rules relevant to tropical forestry)

China has instituted a maximum clearcut size limit of 20 hectares (China, 1987).

In Madhya Pradesh, minimum forest management requirements are listed in Working Plan prescriptions and then specified in each of the Working Plans for the 16 state forestry divisions (MPFD, 2008).

In Indonesia, no clearcuts are allowed except in conversion forests. Selectively harvested areas must be limited to 100 hectares per 'compartment' (i.e. harvesting block) (Forestry Minister's decree No. 485/Kpts-II/1989 on Indonesian Selective Cutting and Planting System (Tebang Pilih Tanam Indonesia)). Indonesian cutting rules are based on minimum diameter limits, a technique commonly used in Southeast Asia. Specifically, Indonesian regulation dictates that 'all commercial trees above 60 cm. dbh (diameter at breast height) can be felled with[in] a 35-year cutting cycle (Forestry Minister's decree No. 485/Kpts-II/1989 on Indonesian Selective Cutting (Tebang Pilih Tanam Indonesia)) (Sist et al, 2003).

There is considerable debate over the appropriateness of minimum diameter cutting limits as a means of managing Southeast Asian mixed dipterocarp forests. A major basis for criticism is that cutting rules do not take adequate account of the environmental conditions necessary for successful forest regeneration and the maintenance of biodiversity (Curran et al, 1999; Sist et al, 2003).

In Japan, the maximum clearcut size for all national forests and for protection forests on other landownership types is 20 hectares (Japan, 1999). In addition, there are ten hectare limits for 'erosion, sand and drought' forests and five hectare limits for 'water conservation forests'. There is no prescribed maximum for privately owned, non-protection forests (although the average private property size is only 2.5ha) (Forestry Agency Japan, 2006b). We classify Japan's policies for private forests as 'mixed' because they do not apply in all cases.

Reforestation (Indicators: Requirements for reforestation, including specified time frames and stocking levels)

In China, reforestation is required within one year. Stocking levels through planting and/or natural regeneration are specified over three years (China, 1987; MEP, 2009c).

In Madhya Pradesh, Working Plan prescriptions require the setting of targets for the defined forest area that will be subject to reforestation 'treatments'. In addition,

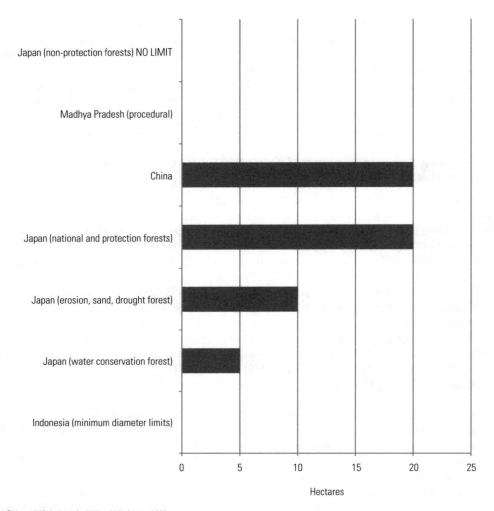

Sources: China, 1987; Indonesia 1989, 1999; Japan, 1999

Figure 5.2 *Clearcut size limits*

all planned harvests must also be regenerated. Working plans set targets for land area to be 'rehabilitated' through reforestation (MPFD, 2008). Livestock grazing and fires are generally the biggest barriers to successful regeneration. Since livestock are generally not fenced, protection from livestock grazing requires that the forests be actively guarded. There are no defined stocking levels or time frames for achieving reforestation.

There are no prescribed stocking levels or time frames for reforestation in Indonesia. Instead, there is a requirement to reforest at least 500 hectares per year (Forestry Minister's decree No. 6652/Kpts-II/2002).

In Japan, reforestation is required on national lands and all protection forests. Where planting is the prescribed method of reforestation in these forests, required time frames and stocking levels have been established (Shinrinhozen kenkyukai, 1997; Norinsuisanroppo henshushitsu, 2003).

Annual allowable cut (AAC) (Indicator: Cut limits based on sustained yield)

In China, the central government determines and approves the national-scale annual allowable cut every five years. The limit is determined according to a variety of factors, but the Forest Law 1998 (Article 29) states it must not exceed annual growth. The process for calculating AAC involves a complex procedure both up and down the chain of command from the national to the local level. State-owned forestry companies, collective forestry farms and individual forest owners must prepare harvest plans and submit them to the SFA. The State Council determines and approves the national-scale AAC and allocates cut levels to the provinces which, in turn, distribute this AAC among individual forest units (China 1987; MEP 2009a).

Annual allowable cuts are calculated in Madhya Pradesh as part of Division Working Plans. The amount that Indian states may harvest each year is subject to federal approval and is based on a budgetary calculation of revenues from the last year's volume of timber harvested, as well as the amount spent on reforestation, forest rehabilitation and other forest protection activities. The National Forest Policy of India (1988) states that timber needs should not eclipse the needs of local subsistence users and that forest processing industries must provide local employment and involve local people 'fully in raising trees'.

Indonesia's AAC is determined at the national level by the Ministry of Forestry and then distributed among the 300 or so forest concessions currently in operation (Forestry Minister's decree No. 156/Kpts-II/2003). The AAC has fluctuated greatly over time: from a high of 22 million m^3 in the 1990s, it was reduced to 5.6 million m^3 in 2004 (ITTO, 2006b), and subsequently raised to 8.15 million m^3 in 2006 and to 9.1 million m^3 in 2007 (USDA FAS, 2007).

In Japan, the volume to be harvested is calculated in regional forest plans (in 158 forest plan areas), both for national and private forests. The calculation is based on multiple-use goals (Norinhoki Kenkyuiinkai, 1999). Rules governing national forests and private protection forests prohibit cutting over and above planned harvest volumes (Imaizumi, 2004). Private forestland owners of non-protection forests are encouraged to develop individual forest management plans that include specified harvest levels.

FOREST PRACTICE REGULATIONS: PLANTATIONS

Plantation forest management in China is subject to the national forest law. There are no regulatory differences between native and plantation forests with regard to our five forest practice criteria.

In India, plantations are managed using the same general policy approach as for natural forests, i.e. through prescriptions specified in sub-state level Working Plans.

Indonesian forest policy and land use classification allows natural forests to be converted to forest plantations; the most recent Indonesian Long-Term Forestry Development Plan assigns 8 per cent of Indonesia's forest to conversion (World Bank, 2006). Although grasslands and unproductive forest areas are identified as priority areas for plantation establishment, in practice, the conversion of natural forest has been preferred because of the financial returns derived from harvesting the standing forest (Dauvergne, 2001).

Forest practices in industrial plantation forests are governed by numerous decrees which specify the proportion of concession area to be used for particular purposes, conserve protected animals and plants and protect cultural and other special feature sites and riparian zones (Nasi et al, 2008). They are also informed by C&I for plantation forests developed by CIFOR (Muhtaman et al, 2000). Establishment of the principal plantation species is limited to 70 per cent of the concession area and specified proportions of the area that should be retained for conservation (10 per cent), enrichment with high value timber species (10 per cent), livelihood crops (5 per cent) and infrastructure (5 per cent). However, lack of clarity about the status and interpretation of relevant requirements, and about the respective responsibilities of the Ministry of Forestry and provincial and district governments, means that implementation of forest practice requirements is inconsistent (Nasi et al, 2008).

In Japan, reforestation requirements (on national and protection forests) and/or voluntary guidelines (on non-national, non-protection forests) are different for planted versus natural forests. Furthermore, planted protection forests have different thinning requirements to natural protection forests. In general, however, there is no difference in the prescriptiveness of policies applying to planted forests and natural forests.

ENFORCEMENT

In China, Forestry Planning and Designing Institutes at national, regional, provincial and prefectural levels are in charge of inspecting forestation, reforestation and harvest activities. The Institutes submit their inspection reports to their respective forestry management authorities. Penalties for illegal harvest are well defined and can be severe. The many legal provisions provide powerful tools for the forestry departments in forest protection. When the logging ban was introduced in 1998, forestry authorities in 30 provinces (autonomous regions and municipal units) established forest security agencies with about 60,000 forest police personnel (Wang et al, 2007).

In Madhya Pradesh, the Forest Department is responsible for auditing compliance with forestry regulations. The Department has a pyramidal structure, delegating responsibilities down the line of command from the state to divisions, districts and local level 'beats'. Villagers are also responsible for policing Joint Forest Management Areas.

Limited state capacity for forest protection has been a serious problem. The following quote from the Madhya Pradesh Forest Department outlines some of the challenges faced (emphases added):

> *Forest protection is beset with myriad problems. For the forestry personnel, it is becoming increasingly difficult to combine the role of a policeman with that of a development agency. As a result of the* phenomenal increase in welfare, development, production, commercial, administrative and co-ordination responsibilities, *the time available for patrolling or field supervision has seriously declined over the years ... Therefore, it is difficult to hold the staff responsible for illicit felling even in cases of dereliction of duty.* (MPFD, 2008)

The above quote suggests there is a conflict inherent in the Joint Forest Management Model regarding the appropriate role of the forest service, i.e. as 'policeman' or 'development agency'. It is unclear how much of the solution, in this case, lies in strengthening state authority or in further sharing authority with local communities.

For Indonesia, ITTO (2006b) concludes that 'prescriptions for the management of production forests are conceptually sound but implementation has been weak'. This long-standing situation has been exacerbated by the decentralization of governance under *reformasi* and by overlapping responsibilities between both national and sub-national government agencies (Rhee, 2003; Rhee et al, 2006; Resosudarmo, 2007). Similarly, in the case of protected areas, national parks are under national authority and thus district authorities have little incentive to cooperate in their protection (Contreras-Hermosilla and Fay, 2005; Fox et al, 2006). Both the Indonesian government and donor agencies have identified strategies to strengthen enforcement, and – while some progress has been made – there remains 'a gap between the written law and the rule of law' (World Bank, 2006).

Japan's national forest planning system assigns primary responsibility for overseeing compliance with forest planning and practices requirements with the national Forest Agency, for national forests, with prefectures for their forests and with local authorities for city, town, village and private forests.

FOREST CERTIFICATION

In China, FSC forest management certificates as of January 2008 totalled a little over 576,000 hectares. Much more significant, thus far, has been the exponential growth in FSC Chain of Custody (CoC) certificates, awarded to over 400 certified Chinese wood product producers by January 2008 and over 500 by August 2008. These CoC certificates indicate the growth in global demand for certified products from major buyers of Chinese wood products. Parallel to the growth in FSC certificates, the Chinese government has passed legislation establishing national standards. These standards are similar in format to the FSC standards but it is as yet unclear whether they will be recognized by the FSC and/or endorsed under the PEFC system.

Certification has been slow to develop in India. This is to be expected given the relatively low level of forest product exports to countries with developed markets for certified products. As of January 2008, a total of 644 hectares of plantation area had received an FSC forest management certificate. India's exports of non-timber forest products exceed those of timber products. Unlike neighbouring Nepal, however, as of January 2008 no Indian company had yet received an FSC certificate for NTFP production.

Indonesia's state-owned Perum Perhatani was the first company worldwide to receive FSC-accredited certification. Its first certification, completed by a certifying body now accredited to the FSC, occurred in 1990 prior to the FSC's establishment. However, governance challenges continue to plague certified operators in the region, sometimes leading to loss of certificates. As mentioned in the introduction to this book, Indonesia has also established a national certification initiative, the Lembaga Ekolabel Indonesia (LEI). LEI has developed an MOU for joint certification with the FSC. Accounting for overlapping certificates under the FSC and LEI schemes, there are now a little over 1 million hectares certified within Indonesia.

Japan currently has two forest certification schemes in operation, the FSC and the Japanese national Sustainable Green Ecosystem Council (SGEC). About 127,000 hectares of Japan's forests were FSC certified as of January 2008. As with China, Japan's more significant role in global trade is as an importer and processor of wood products from elsewhere in Asia and worldwide. This is dramatically reflected in the number of FSC CoC certificates. As of January 2008 over 600 CoC certificates had been issued to Japanese businesses, with that number exceeding 750 by August 2008. Many of these certificates are for paper, printing and publishing companies (Owari and Sawanobori, 2007). Given the importance of Southeast Asian pulp production for the Japanese market, this suggests corresponding growth in demand for certified pulp from producing countries throughout the region.

Japan's Sustainable Green Ecosystem Council (SGEC) was formed in 2003 to promote the use of Japanese forest products and in response to a concern that 'Japanese consumers tend to believe that imported products with certification labels are more environment-friendly than domestic products' (SGEC, no date). As of January 2008, there were 707,836ha of forests certified under SGEC, across 55 forest management units (SGEC, 2009).

Table 5.2 *Matrix of policy approaches in natural forests to the five forest practice criteria*

Case Study	1) Riparian	2) Roads	3) Clearcuts	4) Reforestation	5) AAC
China (Public)	Mandatory substantive	Mandatory mixed	Mandatory substantive	Mandatory substantive	Mandatory procedural
Indonesia (Public)	Mandatory substantive	Mandatory mixed	Mandatory substantive	Mandatory procedural	Mandatory mixed
Japan (Private)	Mandatory mixed	Mandatory mixed	Mandatory mixed	Mandatory mixed	No policy
Japan (Public)	Mandatory mixed	Mandatory mixed	Mandatory mixed	Mandatory mixed	Mandatory substantive
Madhya Pradesh (Public)	Mandatory procedural	Mandatory procedural	Mandatory procedural	Mandatory procedural	Mandatory procedural

Legend:
- Mandatory substantive
- Mandatory procedural
- Mandatory mixed
- No policy

SUMMARY

The forestry context of these Asian case studies is highly diverse, as is their approach to forest policy. Madhya Pradesh takes the most procedural approach of all of the case study countries worldwide. Furthermore, these procedures are uniquely organized in a strict hierarchy, proceeding from the federal to state to progressively more local levels of governance. Indonesia and China's national-level policies are relatively prescriptive, although both countries have been undergoing varying degrees of decentralization. Indonesia's ongoing conflict among federal, regional and local authorities renders the precise nature of legal responsibilities unclear. Japan's policy framework perhaps most closely mirrors a Western European mixed approach, involving mandatory substantive requirements without a comprehensive set of standardized thresholds.

NOTES

1   With the contributions of Dr Pingyang Liu, Center for Urban Environmental Management Studies, Fudan University, for China and Dr Takuya Takahashi, School of Environmental Science, University of Shiga Prefecture, for Japan.
2   This decree has since been replaced by Ministry of Forestry Regulation No. 11 2009 on Silviculture Systems in Areas with Forest Product Utilization Permits in Production Forests. The new requirements include modified harvest rotations and cutting diameters.

REFERENCES

ADB (2002) 'Indigenous peoples / ethnic minorities and poverty reduction: Indonesia', Manila: Asian Development Bank (ADB), Regional and Sustainable Development Department, Environment and Social Safeguard Division

AITPN (2006) 'Indigenous issues: India's Forest Rights Act of 2006: Illusion or solution?', New Delhi, Asian Indigenous Tribal Peoples Network (AITPN)

Basu, Paroma (2007) 'Tigers pitted against tribes by Indian forest law', *National Geographic News*, 5 December

Blandon, P. R. (1999) *Japan and World Timber Markets*, New York: CABI Publishing

Brown, C. L., and P. B. Durst (2003) 'State of forestry in Asia and the Pacific', Regional Office for Asia and the Pacific (RAP) publication, Bangkok: Food and Agriculture Organization of the United Nations

Bruce, J. W., S. Rudrappa and Zongmin Liu (1995) 'Experimenting with approaches to community forestry in China', *Unasylva*, 46, 1

Bull, G. Q. and S. Nilsson (2004) 'An assessment of China's forest resources', *International Forestry Review*, 6, pp210–220

Chatham House and DFID (2008) 'Voluntary partnership agreements', Chatham House, Department for International Development, www.illegal-logging.info, accessed August 2008

China (1987) 'Administration measures on forest harvesting and regeneration', Beijing: Government of China

Contreras-Hermosilla, Arnoldo and Chip Fay (2005) 'Strengthening forest management in Indonesia through land tenure reform: Issues and framework for action', Washington, DC: Forest Trends

Curran, L. M., I. Caniago, G. D. Paoli, D. Astianti, M. Kusneti, M. Leighton, C. E. Nirarita,and H. Haeruman (1999) 'Impact of El Niño and logging on canopy tree recruitment in Borneo', *Science*, 286, pp2184–2188

Curran, L. M., S. N. Trigg, A. K. McDonald, D. Astiani, Y. M. Hardiono, P. Siregar, I. Caniago and E. Kasischke (2004) 'Lowland forest loss in protected areas of Indonesian Borneo', *Science*, 303, 5660, pp909–1088

Dai, Xingyi and Xinlu Jiang (2006) 'One result of decentralization', in D. Xingyi and J. Xinlu (eds) *Footprints of the Explorer – Research on the Collective Forestry Property Rights in Yong'an*, Beijing: China Forestry Press

Dauvergne, Peter (1997) *Shadows in the Forest: Japan and the Politics of Timber in Southeast Asia*, Cambridge, MA: MIT Press

Dauvergne, Peter (2001) *Loggers and Degradation in the Asia Pacific: Corporations and Environmental Management*, Cambridge and New York: Cambridge University Press

FAO (1993) 'Forestry policies of selected countries in Asia and the Pacific', Rome: Food and Agricultural Organization of the United Nations

FAO (2003) *State of the World's Forests*, Rome: Food and Agriculture Organization of the United Nations

FAO (2004) 'Indonesia: Harvesting practices', Food and Agriculture Organization of the United Nations, www.fao.org/forestry/25535/en/idn/, accessed April 2009

FAO (2005) *State of the World's Forests*, Rome: Food and Agricultural Organization of the United Nations

FAO (2006) 'Global planted forests thematic study: Results and analysis', A. Del Lungo, J. Ball and J. Carle (eds), Rome: United Nations Food and Agriculture Organization

FAO (2007) *State of the World's Forests – 2007*, Rome: United Nations, Food and Agriculture Organization

FAOSTAT (2008) 'ForesSTAT', Rome: United Nations Food and Agriculture Organization

Fogarty, Philippa (2008) 'Recognition at last for Japan's Ainu', BBC News, 6 June 2008 http://news.bbc.co.uk/1/hi/world/asia-pacific/7437244.stm, accessed November 2009

Forestry Agency Japan (2003) 'Montreal process: First country forest report (2003 report)', Tokyo: International Forestry Cooperation Office, Planning Division, Forestry Agency

Forestry Agency Japan (2006a) '4. Recent trends in forestry and the forest industry', in *Annual Report on Trends in Forests and Forestry: Fiscal Year 2006 (Summary)*, Forestry Agency, Ministry of Agriculture, Forestry and Fisheries of Japan, www.rinya.maff.go.jp/new/hakusyoeigo/english18/textp3.htm, accessed November 2009

Forestry Agency Japan (2006b) *Annual Report on Trends in Forests and Forestry: Fiscal Year 2006 (Summary)*, Forestry Agency, Ministry of Agriculture, Forestry and Fisheries of Japan, www.rinya.maff.go.jp/new/hakusyoeigo/english18/index.html, accessed November 2009

Fox, James, Dedi Supriadi Adhuri and Ida Aju Pradnja Resosudarmo (2006) 'Unfinished edifice or Pandora's box? Decentralization and resource management in Indonesia', in B. P. Resosudarmo (ed) *The Politics and Economics of Indonesia's Natural Resources*, Singapore: Institute of Southeast Asia Studies

FSI (2005) 'State of forest report 2005', Dehradun: Forest Survey of India, Ministry of Environment and Forests, www.fsi.nic.in/sfr_2005.htm, accessed November 2009

FWI (2002) 'The state of the forest: Indonesia', Bogor: Forest Watch Indonesia (FWI) and Global Forest Watch

Gaveau, David L. A., Hagnyo Wandono and Firman Seiabudi (2007) 'Three decades of deforestation in southwest Sumatra: Have protected areas halted forest loss and logging, and promoted re-growth?', *Biological Conservation*, 134, pp495–504

GOI (2000) 'Guidelines for strengthening of Joint Forest Management Programme', Delhi: Government of India, Department of Environment and Forests, Forest Protection Division

He, D. and C. Barr (2004) 'China's pulp and paper sector: An analysis of supply–demand trends and medium projections', *International Forestry Review*, 6, pp254–265

Hyde, William F., Brian Belcher and Jintao Xu (eds) (2003) *China's Forests: Global Lessons from Market Reforms*, Washington, DC and Bogor, Indonesia: Resources for the Future and Center for International Forestry Research

IFCO (2003) 'Montreal process first country forest report [excerpt]: Japan', International Forestry Cooperation Office (IFCO), Planning Division, Forestry Agency, www.rinya.maff.go.jp/mpci/rep-pub/2003/2003japan_e.pdf, accessed November 2009

Imaizumi, Yuji (2001) 'Data and information collection for sustainable forest management in Japan: Japan's country report for the international expert meeting on MAR-SFM (monitoring, assessment and reporting on sustainable forest management), Yokohama: Forestry Agency, Japan

Imaizumi, Yuji (2004) Personal communication with Yuji Imaizumi, Forestry Agency, Japan, 5 May 2004

Imanaga, Masaaki (1988) 'Development and implementation of the protection-forest-system', in R. Handa (ed) *Forest Policy in Japan*, Tokyo: Nippon Ringyo Chosakai

Indonesia (1989) Forestry Minister's decree No. 485/Kpts-II/1989 on Indonesian selective cutting and planting system (Tebang Pilih Tanam Indonesia), Jakarta: Government of Indonesia

Indonesia (1999) Basic Forestry Law UU 41/1999, Jakarta: Government of Indonesia

Ishihara, M. and H. Oka (2005) 'Mushroom production in Japan and the Republic of Korea. Box 13.2', in G. Mery, R. Alfaro, M. Kanninen and M. Lobovikov (eds) *Forests in the Global Balance – Changing Paradigms*, Vienna: IUFRO

ITTO (2006a) 'Annual review and assessment of the world timber situation', Yokohama: International Tropical Timber Organization

ITTO (2006b) 'Status of tropical forest management 2005', Yokohama: International Tropical Timber Organization

Japan (1999) 'Order from the Director of the Forestry Agency, 1999, Kokuyurin'ya no kakukinoruikei niojita kanrikeiei no shishin nitsuite (Principles of Management according to each functional classification)', Tokyo: Government of Japan, Forestry Agency

JOFCA (2002) 'Forestry in Japan, seven items introducing Japanese forests and forestry', Tokyo: Japan Overseas Forestry Consultants Association

Kamei, Mikoi and Nobukazu Nakagoshi (2006) 'Geographic assessment of present protected areas in Japan for representativeness of forest communities', *Biodiversity and Conservation*, 15, pp4583–4600

Kanowski, P. J., and H. Murray (2008) 'TFD review: Intensively-managed planted forests: towards best practice', New Haven, CT: The Forest Dialogue (TFD) http://research.yale.edu/gisf/tfd/pdf/impf/TFD%20IMPF%20Review.pdf, accessed November 2009

Kant, Shashi (2001) 'The evolution of forest regimes in India and China', in M. Palo, J. Uusivuori and G. Mery (eds) *World Forests, Markets and Policies*, Dordrecht: Kluwer Academic Publishers

Karan, Pradyumna Prasad and Dick Gilbreath (2005) *Japan in the 21st Century: Environment, Economy, and Society*, Lexington: University Press of Kentucky

Khare, Avrind, Madhu Sarin, N. C. Saxena, Subhabrata Parit, Seema Bathla, Farhad Vanya and M. Satyanarayana (2000) 'Joint forest management: Policy, practice and prospects: India country study', London: International Institute for Environment and Development (IIED)

Klassen, A. W. (2002) 'Impediments to the adoption of reduced impact logging in the Indonesian corporate sector', Rome: Food and Agriculture Organization of the United Nations

Koh, L. P. and D. S. Wilcove (2008) 'Is oil palm agriculture really destroying tropical biodiversity?', *Conservation Letters*, 1, pp60–64

Kumar, Nalini, Naresh Saxena, Yoginder Alagh and Kinsuk Mitra (2000) 'India: Alleviating poverty through forest development', in *Evaluation Country Case Study Series*, Washington, DC: The World Bank

Lui, D. and D. Edmunds (2003) 'Devolution as a means of expanding local forest management in South China: Lessons from the past 20 years', in W. F. Hyde, B. Belcher and J. Xu (eds) *China's Forests: Global Lessons from the Market Reforms*, Washington, DC and Bogor: Resources for the Future, Center for International Forestry Research.

Ma, Q. (2004) 'Appraisal of tree planting options to control desertification: Experiences from the Three-North Shelterbelt Programme', *International Forestry Review*, 6, pp327–334

Manurung, E. G. Togu, Ch. Bintang Simangunsong, Doddy S. Sukadri, Bambang Widyantoro, Agus Justianto, Syaiful Ramadhan, Lisman Sumardjani, Dede Rochadi, Pipin Permadi, Bambang Mardi Priyono and Bambang Supriyanto (2007) 'A road map for the revitalization of Indonesia's forest industry', Jakarta: The Forest Industry Revitalisation In-House Experts Working Group, Ministry of Forestry of Indonesia

McDermott, Constance, Benjamin Cashore, Lloyd Irland, Elizabeth Gordon, Brian Milakovsky and Camille Rebelo (2008a) 'Forestry driver mapping project: Eight reports on forest trade from six priority ecoregions to China and the US', New Haven, CT: Yale Program on Forest Policy and Governance

McDermott, Constance, Lloyd Irland and Benjamin Cashore (2008b) 'China–USA supply chain report', New Haven, CT: Yale Program on Forest Policy and Governance

MEP (2009a) 'Policies and Regulations', China, Ministry of Environmental Protection, http://english.mep.gov.cn/Policies_Regulations/laws/environmental_laws/200710/t20071009_109914.htm, accessed November 2009

MEP (2009b) 'Regulations for the implementation of forestry law of the People's Republic of China', China, Ministry of Environmental Protection, http://english.mep.gov.cn/Policies_Regulations/laws/envir_elatedlaws/200710/t20071009_109960.htm, accessed November 2009

MEP (2009c) Forestry Law of the People's Republic of China, Amended 1998, available from http://english.mep.gov.cn/Policies_Regulations/laws/envir_elatedlaws/200710/t20071009_109917.htm

Miao, Guangping (2004) Personal communication with Guangping Miao, China National Forestry Economics and Development Research Center (FEDRC) Beijing, 28 April

Miao, G. and R. A. West (2004) 'Chinese collective forestlands: Contributions and constraints', *International Forestry Review*, 6, pp282–298

MOE (2008) 'Wildlife protection system and the hunting law', Government of Japan, Ministry of the Environment, www.env.go.jp/en/nature/biodiv/law.html, accessed August 2008

MPFD (2008) 'Forest Department, Madhya Pradesh', Madhya Pradesh Forest Department, www.forest.mp.gov.in, accessed August 2008

Muhtaman, Dwi R., Chairil Anwar Siregar, and Peter Hopmans (2000) 'Criteria and indicators for sustainable plantation forestry in indonesia', Bogor: Center for International Forest Research (CIFOR)

Nasi, R., P. Koponen, J. G. Poulsen, M. Buitenzorgy and W. Rusmantoro (2008) 'Impact of landscape and corridor design on primates in a large-scale industrial tropical plantation landscape', *Biodiversity Conservation*, 17, pp1105–1126

Nawir, A. A., and ComForLink (2007) 'Forestry companies' perspectives: Improving community roles in plantation forest development through partnerships', Bogor: CIFOR

Nihon Rindo Kyokai (ed) (2002) 'Rindo Kitei: Un'yo to kaisetsu (Forest road regulations: Implementation and guidance). Nihon Rindo Kyokai (Japan Forest Road Association)', Tokyo: Nihon Rindo Kyokai

Norinhoki Kenkyuiinkai (1999) 'Norinhoki Kaisetsu Zenshu: Kokuyurin 'ya hen (Guidebook of laws on agriculture and forestry: National Forests)', Tokyo: Taiseishup panssha, Norinhoki Kenkyuiinkai, Committee for studying laws on agriculture and forestry

Norinsuisanroppo henshushitsu (ed) (2003) 'Norinsuisanroppo (Laws on agriculture, forestry and fisheries)', Norinsuisanroppo henshushitsu (Editorial Office for the Statute Book Regarding Agriculture, Forestry and Fisheries), Tokyo: Gakuyoshobo

Owari, Toshiaki and Yoshihide Sawanobori (2007) 'Analysis of the certified forest products market in Japan', *Holz als Roh- und Werkstoff*, 65, 2, pp113–120

Prasad, Ram (1999) 'Joint forest management in India and the impact of state control over non-wood forest products', *Unasylva*, 50, 198, www.fao.org/docrep/X2450E/x2450e0c.htm#joint%20forest%20management%20in%20india%20and%20the%20impact%20of%20state%20control%20over%20non%20wood%20f, accessed November 2009

Ramanathan, Usha (2000) 'Common land and common property resources in Madhya Pradesh', Geneva: International Environmental Law Research Centre

Rao, K. Raja Mohan, M. Sabesh Manikandan and Walter Leal Filho (2005) 'An overview of the impacts of changes in common property resources management in the context of globalisation: A case study of India', *International Journal of Sustainable Development and World Ecology*, vol 12, pp471–477

Resosudarmo, I. A. P. (2007) 'Has Indonesia's decentralization led to improved forestry governance?', Canberra: The Australian National University

Rhee, Steve (2003) 'De facto decentralization and community conflicts in East Kalimantan, Indonesia: Explanations from local history and implications for community forestry', in K. Abe, W. de Jong and L. Tuck-Po (eds) *The Political Ecology of Tropical Forests in Southeast Asia: Historical Perspectives*, Melbourne and Kyoto: Trans Pacific Press and Kyoto University Press

Rhee, Steve, M. Moelono and E. Wollenberg (2006) 'Impacts on land tenure and livelihoods of forest-dependent communities', in C. Barr, I. A. P. Resosudarmo, J. McCarthy and A. Dermawan (eds) *Decentralization of Forest Administration in Indonesia: Implications for Forest Sustainability, Community Livelihoods, and Economic Development*, Bogor: CIFOR

Ruiz Pérez, M., B. Belcher, M. Fu and X. Yang (2004) 'Looking through the bamboo curtain: An analysis of the changing role of forest and farm income in rural livelihoods in China', *International Forestry Review*, 6, pp306–316

Saigal, S., H. Arora and S. Rizvi (2002) *The New Foresters: The Role of Private Enterprise in the Indian Forestry Sector*, London: International Institute of Environment and Development (IIED)

SFA (2005a) 'Development report of China's forestry (Abstract)', Beijing: State Forestry Administration, Chinese Forestry Press

SFA (2005b) 'Code of forest harvesting', Reference Number: LY/T 1646–2005, Beijing: State Forestry Administration, published 16 August 2005 and enacted on 1 December 2005

SGEC (no date) 'Let's foster Japanese forests! Let's use Japanese wood! SGEC forest certification system is suitable for Japan to nurture beautiful nature and expert foresters', Tokyo: Sustainable Green Ecosystem Council

SGEC (2009) 'Certified forest list', Sustainable Green Ecosystem Council, www.sgec-eco.org/certforest/itiranhyou-synrin.pdf, accessed August 2009

Shinrinhozen kenkyukai (ed) (1997) 'Hoanrin no Jitsumu (Handbook of protection of forests)', Shinrinhozen kenkyukai (Study Group of Forest Conservation), Tokyo: Chikyuusya

Shinrin-ringyo hakusyo (2002) 'Annual report on trends of forest and forestry fiscal year 2002', www.rinya.maff.go.jp/new/hakusyoeigo/english14/english14top.htm, accessed November 2002

Singer, B. (2008) 'Putting the national back into forest-related policies: The international forests regime and national policies in Brazil and Indonesia', *International Forestry Review*, 10, 3, pp523–537

Singer, Benjamin (2009) 'Indonesian forest-related policies', Paris: Institut d'Etudes Politiques

Sist, Plinio, Robert Fimbel, Douglas Sheil, Robert Nasi and Marie-Hélène Chevallier (2003) 'Towards sustainable management of mixed dipterocarp forests of South-east Asia: Moving beyond minimum diameter cutting limits', *Environmental Conservation*, 30, 4, pp364–374

Smil, V. (1993) 'Afforestation in China', in A. Mather (ed) *Afforestation*, London: Belhaven Press

Stafford, Brian (2007) 'Environmental aspects of China's papermaking supply', Washington, DC: Forest Trends

Tacconi, Luca (2007) *Illegal Logging: Law Enforcement, Livelihoods and the Timber Trade*, London: Earthscan

The Hindu Staff Correspondent (2005) 'Crack down on forest encroachments', 4 April 2005, Bhopal: The Hindu (Newspaper), www.hindu.com/2005/04/05/stories/2005040508810500.htm, accessed November 2009

USDA FAS (2005) 'People's Republic of China. Solid wood products. China's sixth forest resource inventory report 2005', Beijing: United States Department of Agriculture, USDA Foreign Agricultural Service

USDA FAS (2006) 'People's Republic of China. Solid wood products. Annual report 2006', Beijing: United States Department of Agriculture, Foreign Agricultural Service

USDA FAS (2007) 'Indonesia: Solid wood products annual report 2007', Beijing: United States Department of Agriculture, Foreign Agricultural Service

Wang, G., J. L. Innes, J. Lei, S. Dai and S. W. Wu (2007) 'China's forestry reforms', *Science*, 318, pp1556–1557

Wenming, Lu, Natasha Landell-Mills, Liu Jinlong, Xu Jintao, and Liu Can (2002) 'Getting the private sector to work for the public good: Instruments for sustainable private sector forestry in China', in Instruments for Sustainable Private Sector Forestry Series, London: IIED

White, Andy, Xiufang Sun, Kerstin Canby, Jintao Xu, Christopher Barr, Eugenia Katsigris, Gary Bull, Christian Cossalter and Sten Nilsson (2006) 'China and the global market for forest products: Transforming trade to benefit forests and livelihoods', Washington DC: Forest Trends, Centre for International Forest Research (CIFOR), Rights and Resources Initiative (RRI)

World Bank (2001) 'Indonesia: Environment and natural resource management in a time of transition', Jakarta: World Bank

World Bank (2005) 'Guangxi integrated forestry development and conservation project: Social assessment report', Jakarta: World Bank

World Bank (2006) 'Strategic options for forest assistance in Indonesia', Jakarta: World Bank

WWF Germany (2005) 'Borneo: Treasure island at risk', Frankfurt: World Wide Fund for Nature, Germany

Xu, J., Katsigris E. White T. A. (eds) (2002) 'Implementing the natural forest protection program and the sloping land conversion program: Lessons and policy recommendations', Council for International Cooperation on Environment and Development (CCICED) Secretariat Canadian Office, www.harbour.sfu.ca/dlam/Taskforce/grassPreface.html., accessed November 2009

Xu, Z., M. T. Bennett, R. Tao and J. Xu (2004) 'China's sloping land conversion programme four years on: Current situation and pending issues', *International Forestry Review*, 6, pp317–326

Zhang, Yuxing and Conghe Song (2006) 'Impacts of afforestation, deforestation, and reforestation on forest cover in China from 1949–2003', *Journal of Forestry*, October/November, pp383–387

Zhu, Chunquan (2001) 'International impacts of Chinese forest policy', paper read at International Conference on Sustainable Forestry Development in China, Huangshan, China

Zhu, Chunquan, Rodney Taylor and Feng Guoqiang (2004) 'China's wood market, trade and the environment', Beijing: WWF International

Chapter 6

Central and Eastern Europe: Latvia, Poland and the Russian Federation[1]

INTRODUCTION

Latvia, Poland and Russia vary greatly in the size of their forest resource as well as their role in global forest trade. Russia contains the largest forest area of any country worldwide, amounting to 20.5 per cent of global forest cover. Latvia and Poland, in contrast, together have only about 0.3 per cent of the world's forests. Forest product exports are important to the forest economy of all three countries, although the relative volumes of these exports vary by several orders of magnitude. Russia is a leading exporter, by volume, of primary wood products to other European countries, and is the largest supplier to China (Zhu et al, 2004).

All three case study countries have undergone dramatic changes across all sectors since *perestroika* and the fall of communist regimes in 1989–1991. Common foci of government reforms have been privatization, decentralization and trade liberalization. The opening of national borders has also resulted in increased regional participation in international environmental agreements, such as the Montreal Process and the Kyoto Protocol, as well as various environmental Directives of the European Union.

This rapid pace of change exacerbates the challenge of balancing economic growth with other social and environmental values. All three Central and Eastern European case study countries have enacted new forest laws within the last 20 years, designed to reflect changing national priorities and changing economic contexts, while at the same time adapting to local economic and political situations.

The different pathways towards privatization these countries have pursued serve as interesting examples of the complexities of transforming policy regimes. As pointed out by numerous theorists (Ostrom, 1990; Kissling-Naf and Bisang, 2001), there are many different degrees of privatization and various forms by which it can be accomplished. These include privatization of land ownership (with greater or lesser associated private property rights), privatization of production (in other words, silviculture, harvesting, processing, etc.) and the liberalization of markets.

All case study countries have liberalized their forest product markets by opening their borders to international competition. Each of these countries, however, has pursued different strategies regarding land ownership and forest product production. In Poland and Russia, traditional socialist-era patterns of public forestland ownership have been maintained, with the majority of forestlands administered by the federal government. Only 17 per cent of Polish forests are held under private forest ownership. In Russia, all lands classified as forest are in the publicly controlled Forest Fund. No major land privatization efforts have been instituted in Poland or Russia following the end of Soviet-era rule. While it appeared that the new Russian Forest Code of 2006 might include enabling language for future privatization (Lobovikov, 2000; Kissling-Naf and Bisang, 2001; Polyakov and Teeter, 2003; Meidinger et al, 2006; USDA FAS, 2006), this reform did not materialize in the final draft of the law.

In Latvia, the privatization of forestlands has proceeded at a rapid rate since democratization. Under Soviet rule, all Latvian private forests were transformed into 'cooperative farms'. After independence, Latvia announced the goal of having roughly 50 per cent of forestland under private ownership (Latvia, 2000a) and had almost achieved this goal by 2005 (our cut-off date for case study selection), with 45.1 per cent of forestlands under private ownership (FAO, 2006).

According to our selection criteria, which consider relative forest area as well as volume of production, our detailed analysis of forest practice policies will therefore address public lands in all three countries, as well as private lands in Latvia.

Private forestland owners in Eastern Europe face a number of management obstacles. While national forest codes are supposed to apply to *all* forestlands (Polyakov and Teeter, 2003), they rarely provide specific mention of private lands, creating ambiguity as to the rights and responsibilities of private forest owners (Cirelli, 1999). In Russia, it remains uncertain whether forests established on abandoned farmland are part of the government-controlled 'Forest Fund,' or are private property like other agricultural lands. Furthermore, average woodlot sizes in Poland and Latvia are very small, at approximately one and eight hectare(s) respectively (Jodlowski, 1998; Latvia State Forest Service, 2004; Actins and Kore, 2006), with potentially adverse consequences for the economic feasibility of forest management. However, the fact that Latvia's private production matches that of its state forests suggests that small property size has not presented an insurmountable obstacle to economic utilization.

With regard to public forests, Russia, Poland and Latvia have all taken steps to further privatize timber production processes. Russia's 1993 Basic Forestry Legislation instituted a system of short- and long-term forest leasing, which most closely resembles tenure systems in Canada (Lobovikov, 2000). Through this mechanism, forest management, silviculture and harvesting rights and responsibilities have been sold to provincial governments and private companies (Mabel, 2000; Sheingauz et al, 2001). The new forest code, enacted in 2006, has moved to an open auction system for all commercial timber licences. In Poland and Latvia, the state-owned forests have been corporatized, with a mandate to achieve economic self-sufficiency.

A process of 'devolution', distinct from 'privatization' and 'democratization', has also been occurring in all three countries, most notably in Russia. In the forestry sector, devolution describes a complex array of policy choices involving trade-offs and negotiations between forestry interests at all levels. The complexity of the process is most clearly illustrated in Russia, where transfer of more authority to provincial levels has effectively decreased authority at local levels (Mabel, 2000; Moiseev, 2000). Theorists have also contended that federal level authority may often more effectively protect the needs and values of citizens than regional and local authorities faced with limited resources and conflicting economic and social pressures (Hoberg, 2000; Moiseev, 2000). In the Russian case, some observers feel that the lack of a clear division between federal and provincial responsibilities in the new Forest Code has created a policy vacuum, which can lead to incomplete funding, implementation and monitoring of key forest management tasks, or the *de facto* delegation of these responsibilities to forest leaseholders (Lesniewska et al, 2007; Pstygha, 2007).

Problems of illegal forest activities have presented a major challenge to many post-Soviet and formerly socialist countries (Cirelli, 1999; Moiseev, 2000). However, the nature and extent of these problems vary considerably. Tax evasion and other illegal financial dealings, rather than violation of forest practice laws, present perhaps the greatest challenge in the Eastern European countries and Russian west (Actins and Kore, 2006; Jakubowicz, 2006; Tysiachniouk, 2006). In Siberia and the Russian Far East, in contrast, enforcement of environmental forestry regulations remains highly inadequate (Lankin, 2005; Tysiachniouk, 2006).

Poland

An overview of forests and forest ownership

Historically, forests covered virtually all of what is now Poland. In 2005, about 29 per cent (9.2 million ha) of the country was classified as forest (FAO, 2006). As is true for much of Europe, after centuries of human-induced deforestation, Poland's forest area has been expanding over the past few decades; in Poland's case this growth has approximated 0.2 per cent per annum over the period 1990–2005.

Poland maintained about 17 per cent of its forests under private ownership both during and after the Soviet era. Forestlands on both public and private property are highly fragmented. Less than one-quarter of State Forest Directorate holdings are of more than five hectares; the average area of private forest holdings is about one hectare.

Native forests

Only 53,000ha of Poland's forests are classified as 'primary'; these include the World Heritage-listed Bailowieza Forest, on Poland's border with Belarus. All Poland's other forests are classified as 'semi-natural' and planted, or as 'plantations'.

About 78 per cent of Poland's forests are coniferous, with Scots pine (*Pinus sylvestris*) accounting for 69 per cent of the growing stock, followed by oak (*Quercus* spp.) as the next most common species (7.3 per cent of growing stock). Norway spruce (*Picea abies*) (5.5 per cent of growing stock) predominates in mountainous areas (USDA FAS, 2007b).

Planted forests and plantations

Semi-natural planted forests of native species account for almost all Poland's forest-lands, with about 5.6 million hectares planted for production purposes and about 3.1 million ha planted primarily for protection (FAO, 2006).

Poland has only 32,000 hectares of forest plantation, corresponding to 0.3 per cent of its total forest area (FAO, 2006).

Forest governance

The Ministry of Environment is responsible for developing Poland's forest policy. Poland's public lands are managed by the National Forest Holding State Forests, a state-owned company. The 1991 Forest Act specifies the general rules of forest management, which apply to both state and private forests.

Forest practices system

Forest management plans are required to conduct forestry activities on both public and private lands. State forest managers must also meet the additional detailed prescriptions of the *Technical Handbook of Forest Management*. The handbook's guide-lines are considered voluntary for private forest owners.

Forest production and trade

Poland harvests about 56 per cent of its annual timber increment (USDA FAS, 2007b), about 94 per cent of which is sourced from state lands (Jakubowicz, 2006). Wood production has been increasing since 2005 and growth is expected to continue at least through 2010. Poland's timber industry is highly export-oriented, with Poland leading Europe as the largest supplier of fibreboard (USDA FAS, 2007b). Almost all wood processed in Poland is sourced domestically (MOE, 2007).

State Forests, the state-owned company responsible for production from state lands, has recently moved to an online wood purchasing system. This new system, which allows equal competition among local and international companies, has reportedly favoured large multinational firms (USDA FAS, 2007b).

Poland is the world's fourth largest exporter of wooden furniture (ITTO, 2006). Within Poland, the furniture industry is the largest consumer of solid wood products and roughly 90 per cent of the finished product is sold for export, primarily to EU countries (USDA FAS, 2007b). Poland's furniture industry consists predominantly of small-scale producers (Jakubowicz, 2006).

Latvia

An overview of forests and forest ownership

Latvia classifies most of its 2.3 million ha of forests as 'modified natural'. Forest extent, at 47.4 per cent (FAO, 2006), has increased by over 50 per cent since 1935 (LEDC, 2000).

Latvia has undergone a major restructuring of landownership since 1991, involving the return of lands collectivized under Soviet rule to their original family owners. At the time of the Soviet Union's dissolution, there were no lands held under private ownership. By the turn of the 21st century, 50 per cent of forestlands had been re-privatized. These private ownerships are small in size, averaging about eight hectares.

Native forests

Of Latvia's 2.3 million ha of modified natural forest, only 14,000ha are classified as 'primary' and some 235,000ha as mature. Dominant species are Scots pine (*Pinus sylvestris*) 41 per cent, birches (*Betula* spp.) 28 per cent and Norway spruce (*Picea abies*), 21 per cent (LEDC, 2000).

Planted forests and plantations

Most of Latvia's 'semi-natural' forest is planted, with 567,000ha established primarily for production and 77,000 planted for protection. As of 2005, only 1000ha of 'plantation' were reported (FAO, 2006).

Forest governance

Latvia has divided forest rule-making, oversight of forest practices and public forest management among three different institutional entities. The Forest Sector of the Agriculture Ministry sets the laws, the State Forest Service (SFS) is responsible for enforcing laws on both public and private lands and the state joint-stock company 'Latvia's State Forests' (LVM Ltd) manages state-owned forests. According to forest policy, the state as forest owner requires that the value of the forest capital not be reduced and 'should be increased' and 'desires to profit from its capital [forest]'. The joint-stock company pays an annual dividend to the state in alliance with these objectives (Silmakelen, 2009).

LVM lands are divided into eight 'profit centres' or regions, which are run as independent economic units (Stasa and Sarmulis, 1998). All forest work is hired out to private contractors.

The 2000 law on forests provides the foundations for forest practice regulation in the country. The law generally applies to both private and public lands and the State Forest Service is responsible for its implementation (Latvia, 2000b).

Forest practices system

Management plans are a prerequisite for harvest on all forest ownerships and a forest inventory must be completed at least once every ten years (Silmakelen, 2009).

Forest production and trade

Forest production in the late 1990s was roughly evenly divided between public (55 per cent of production) and private (45 per cent of production) (Stasa and Sarmulis, 1998). The distribution of forest harvest across private owners appears to be highly variable. According to a 2001 survey by the State Forest Service, only 3 per cent of private owners, all with properties of at least 30ha, conducted 'regular harvests' while 60 per cent reported receiving no income at all from forest harvesting (Latvia State Forest Service, 2004; Actins and Kore, 2006).

Forest processing in Latvia is focused on primary and secondary solid wood and engineered products. Paper production is relatively less developed. In 2005, wood processing was the second largest industrial sector and exports accounted for about 70 per cent of production. The primary buyers are EU countries, but there has been rapid growth in exports elsewhere as well, including neighbouring countries Estonia, Lithuania, and Russia (Latvia, 2005a). Important export products include firewood, hardwood roundwood, softwood lumber (sawn wood) and furniture (Actins and Kore, 2006). Latvia also imports significant volumes of wood and forest products (in 2006, around 40 per cent of the volume of forest products exported), mostly from neighbouring countries and mostly for processing and re-export (Latvia, 2007).

The Russian Federation

An overview of forests and forest ownership

As observed in the introduction to this chapter, Russia has the largest forest area of any country, including the world's greatest expanse of intact boreal forest (Aksenov et al, 2002). Approximately 70 per cent of Russia's forest area is found in the central and eastern regions: 43 per cent in Siberia and 27 per cent in the far east. Because the majority of these forests are remote and inaccessible, high proportions are largely undisturbed and have high biodiversity and wilderness values. In contrast, many of the forests in the west of Russia, where 70 per cent of Russia's population live, are much more fragmented and exploited. This skewed distribution of resource, population and infrastructure means that much of Russia's potential for both large-scale conservation and increased production lies in very remote regions of the country. About 40 per cent of Russia's boreal forests are considered commercially inaccessible (USDA FAS, 2006). Arguably, among the greatest threat to Russian forests is the building of transport infrastructure for wood production, rather than the removal of timber. In Siberia and the Russian Far East, human-generated fires, closely associated with the presence of roads, have been increasing in rate and severity and are commonly cited as the single greatest challenge to forest conservation (Cushman and Wallin, 2002; Groisman et al, 2003; Kondrashov, 2004; USDA FAS, 2007c).

The federal government owns all lands classified as forestland in Russia. These forests are referred to collectively as the 'Forest Fund'.

Illegal logging remains a significant and persistent problem in Russia, although estimates of its extent vary. Official reports suggest rates as low as 0.5 per cent of official production, while many environmental organizations estimate rates of 20–50 per cent. The extent of illegal logging varies substantially by region. Chita Oblast (southeastern Siberia, on the border with China) appears to be one of the most problematic areas, followed by Primorsky Krai and Khabarovsk Krai (Russian Far East, on and near the border with China) (USDA FAS, 2006). Depending on the definition used, between 15 and 70 per cent of all logging operations in the far east are illegal (Lesniewska et al, 2007).

Native forests

Conifers (pine, spruce, fir and larch) account for about 70 per cent of Russia's forest area and 80 per cent of the growing stock (USDA FAS, 2006). Hardwoods are of particular commercial importance in the Russian Far East, particularly in the biodiverse Sikhote-Alin mountain range, and are quickly being depleted due to the explosion of demand from China's export-oriented furniture industry (Lankin, 2005; Milakovsky and McDermott, 2008).

Planted forests and plantations

Russia ranks third worldwide in its area of plantation forests, with roughly 17 million hectares (2.1 per cent of its total forest area) classified as such (FAO, 2006). Conversely, no semi-natural planted forests have been reported (FAO, 2006), suggesting that at least some of the area Russia has reported as plantation might elsewhere be classified as planted forest.

Forest governance

Russia's republics and administrative units cover quite a diverse range of environmental, social and economic conditions, ranging from the densely populated temperate regions of western Russia, to the heavily forested, remote taiga regions of Siberia and the far east. As is apparent from the following analyses, the Russian Federation has developed a highly complex and detailed set of forestry regulations addressing appropriate forest management practices within these different forest categories.

In 2006, Russia enacted a new forest code that continues its path towards decentralization and free market trade. The 2006 code outlines a new system for leasing forests to private companies (*lespromkhozi*) for 10 to 49 year periods. Government oversight is carried out by the Russian Federal Forest Agency (Rosleskhoz) and 81 provincial forest services of the federated 'subjects' (republics, *oblasts*, *krais*, autonomous *oblasts* and *okrugs*) of the Russian Federation.[2]

A central feature of the 2006 code is the further transfer of authority from the Federal Forestry Agency to the regions and to private industry. Two of the code's goals have been identified as: 'To establish a balance of power, that will be market-oriented, between the state powers of the Federation and the powers of its regions', and 'To separate the work of the state and economic management within the forest sector' (Petrov, 2007). Reflecting and reinforcing a strong market emphasis, the code was developed within the Ministry of Economic Development, rather than the Ministry of Natural Resources (Pstygha, 2007). During the same reform period, the Rosleskhoz was transferred from the Ministry of Natural Resources to the Ministry of Agriculture (Lesniewska et al, 2007).

The transition from the top-down forest management model of past forest codes to this more decentralized, privatized model is causing considerable turmoil within Russia's forest industry and civil society. Many observers fear that the newly empowered players are not prepared for their new responsibilities and that the effort to encourage a more market-driven system has weakened environmental protections established in past codes.

The sharing of responsibilities between Rosleskhoz and the regional agencies has been in place since Soviet times, when it was known as 'dual subordination' (Carlsson et al, 1999). However, Rosleskhoz was always the dominant player, controlling forest management activities through thousands of local branches (*leskhozi*), which oversaw the *lespromkhozi* with some assistance from the regional authorities. The fall of the Soviet regime and the introduction of free-market reforms in the 1990s saw the privatization of the *lespromkhozi* and some initial decentralization ('joint jurisdiction'). The authority of the Rosleskhoz was further diluted when President Putin authorized its demotion from an independent federal agency to a branch of the Ministry of Natural Resources, and later, of the Ministry of Agriculture.

The division of power between Roskleskhoz and the provincial forest services remains somewhat ambiguous, but certain key features are apparent. The federal level remains responsible for forest monitoring and inventory, the development of forestry regulations and minimum lease rates, research and education and the distribution of federal funds (Petrov, 2007). Oversight for leasing, forest law enforcement, reforestation and management plan development has passed to the 81 provinces. The federal district *leskhozi* have been denied the formerly lucrative right to carry out 'sanitation harvests' in the regions they oversee (Pavlov, 2007). The *leskhozi* have been replaced with *lesnichestvo*, forest districts restricted to administrative and forest management functions.

Each province is responsible for developing its own administrative structure, usually divided in function among the three categories of management, inspection and commercial exploitation (Lesniewska et al, 2007). As part of this structure, many provinces have established state-owned commercial timber operations, often employing former *leskhozi* staff. These state-owned companies are, at least officially, expected to compete for timber on equal footing with private companies (Lesniewska et al, 2007).

The 2006 code has also introduced significant changes to forest classification and silviculture. The fairly complex classification system in place under the 1997 code has been simplified to three categories: protection, production and reserve forests. The first category includes riparian, montane, urban or other sensitive forests. Clearcutting is generally prohibited in this zone. Production forests are intended for 'sustainable, maximum-efficient production of high-quality wood' and allow for both clearcutting and selective cutting. Reserve forests are those in which timber harvesting is not intended for the next 20 years. These are available for future classification as either protection or production forests, depending on the need. The code also eliminates the distinction between industrial (commercial) logging and intermediate logging, the latter covering operations that are marginally profitable but silviculturally important (Sheingauz, 2006). All volume harvested through intermediate cutting must now be counted as part of the leaseholder's annual allowable cut.

The provincial forestry agencies now have responsibility for the development of management plans and forest law enforcement. However, new forest practice prescriptions, as envisioned under the 2006 forest code, have been slow to develop. In part this is due to uncertainty surrounding the devolution of rule-making authority. As of the census date for this book's analysis, January 2007, the previous requirements still apply and hence are the focus of our analysis of forest practice policies.

In general, a continued lack of clarity in Russia's forest regulatory framework, a long reliance on an 'informal' forest economy and lack of adequate funds for regulatory implementation, make it difficult for even the most legally conscientious forest operator to conduct operations without any violations of the law (Lankin, 2005; Lebedev, 2005).

Forest practices system

The 2006 code launches a new State Forest Inventory (SFI) system. The SFI is designed to improve accuracy and be more compatible with inventory systems in other countries. However, the system is not due to be fully introduced until 2020, creating interim uncertainties regarding the discrepancy between old and new estimates of standing volume and other forest features (Lesniewska et al, 2007).

Drawing ultimately on the SFI, each province is expected to develop a forest plan. The provincial plan is to cover all forest districts (*lesnichestvo*) and municipal forest parks (*lesopark*) and specify zones for use and conservation (Lesniewska et al, 2007).

The next planning level down is the forest district (*lesnichestvo*). A ten-year forestry *Reglament* (management regulation) must be established for each district outlining appropriate forest use and annual allowable cut (Lesniewska et al, 2007).

At the level of the forest management unit, individual area-based licences are sold via open auction. Each successful leaseholder is then responsible for preparing a ten-year forest development plan at their individual or company expense (Lesniewska et al, 2007).

Forest production and trade

Official harvest estimates indicate that about 33 per cent of Russia's growing stock, around 183 million m^3, is harvested annually. Softwood logs and lumber are the largest category of solid wood production and further growth is expected due to rapidly expanding domestic demand, as well as demand from China, Finland and elsewhere (USDA FAS, 2006). As mentioned above, estimates of the extent of illegal logging vary, but many fall within the range of 20–50 per cent of legal harvesting (USDA FAS, 2006).

As of 2005, nearly three-quarters of Russia's solid wood exports consisted of unprocessed roundwood (FAOSTAT, 2008). This lack of processing is most pronounced in the Russian Far East, where until recently about 90 per cent of wood product exports were in the form of roundwood (Sheingauz et al, 2001), most of which is destined for China (Lankin, 2005).

Russia is now taking significant steps to force the development of wood processing. In February 2007, the federal government announced a progressive series of export taxes on unprocessed logs, beginning at 20 per cent in July 2007 and increasing to 80 per cent for softwoods and 50 per cent for hardwoods by January 2009. If fully implemented, this decision would produce major reverberations on wood supply flows to China, Japan, Finland and worldwide. Meanwhile, there is some evidence of an increase in illegal log exports in an effort to evade export taxes (USDA FAS, 2007a). The federal government has acknowledged that the increased taxes and accompanying subsidies for firms willing to develop forest industry have yet to successfully stem the flood of roundwood exports (Sharipova, 2007).

Pulp and paper accounted for only about 16.2 per cent of Russia's wood product exports in 2006, mostly in the form of raw pulp. About 80 per cent of Russia's total pulp production is exported while only 40 per cent of paper production is exported (Akim, 2008).

BIODIVERSITY CONSERVATION

Protected areas

Latvia and Poland joined the European Union in 2004, along with eight other Eastern European countries. Both countries have increased their protected area network through active engagement in EU initiatives, including Natura 2000 and the EC Birds and Habitat Directives (see Chapter 5 for an introduction to these EU policies). Prior to and after EU succession, these countries have also participated in the Emerald Network (created in 1998), a network of 44 European countries that serves to link Natura 2000 with non-European Community countries. The Emerald Network was developed on the basis of the Convention on the Conservation of European Wildlife and Natural Habitats (EECONET, 2008).

In addition to such Europe-wide initiatives, Poland has also pursued the creation of transboundary protected areas with neighbouring countries, including Russia, Lithuania, Belarus, Ukraine, Slovakia and the Carpathian mountains region (Conference of the Parties Bratislava, 1996). In Latvia, 8–10 per cent of the state forest estate is managed primarily for conserving and enhancing biodiversity (Latvia State Forest Service, 2009a).

According to 2008 UNEP-WCMC data, protected areas cover 15.4 per cent of Latvia, 22.1 per cent of Poland and 9 per cent of Russia (Chapter 2, Figure 2.15). In Latvia and Russia the majority of protected areas are classified under IUCN Categories I–IV, which generally do not allow extraction of natural resources; while in Poland the majority are still unclassified (Chapter 2, Figure 2.15).

The enforcement and financing of protected areas is a major challenge in parts of Siberia and the Russian Far East (World Bank, 1996; Morozov, 2000; Nefedyev, 2003).

With regard to more general forest zoning, the 2006 Russian forest code has made several additions to the list of economic activities that are allowed within the Forest Fund. Some observers believe that these changes will make it easier to develop forestland for mineral extraction, energy and perhaps even construction purposes. It appears that such activities can even take place in protection forests (Lesniewska et al, 2007).

Protection of species at risk

Latvian legislation concerning endangered species and their habitats includes: On Protection of Species and Habitats (16.03.2000), On the List of Specially Protected Species and Species with Exploitation Limits (No. 396/14.11.2000) and On the List of Specially Protected Habitats (No. 421/12.05.2000). In addition, Latvia has developed a National Biodiversity Strategy and Action Plan and a National Biodiversity Protection Plan for implementation of Natura 2000 requirements in Latvia (2003-2013) (Latvia, 2005b).

In Poland, the 2004 Act on Nature Conservation and associated legal Acts guide the protection of species at risk. The Decree on Conservation of Plant Species (2001) and the Decree on Conservation of Animal Species (2004) list species warranting special protection. Legal requirements cover protection of both these species and their habitats, including the establishment of protection areas and connecting corridors.

In Russia, biodiversity preservation is addressed in the 'On Environmental Protection' Act (Russia, 2002) and in the 'On Wildlife' Act (Russia, 1995a). Biodiversity is defined as diversity of wildlife within a species, between species, and in ecosystems. The 'On Wildlife' Act (Article 22) addresses habitat protection and requires that any economic activity that impacts wildlife habitat must include mitigating measures. Special natural protected territories have been established to protect wildlife or floral species and their habitats. The 2006 forest code also contains provisions for the protection of rare ecological communities. The 'Red Book' of the Russian Federation lists threatened and endangered species and many Russian regions have also prepared regional 'Red Books'.

Russia's Rules of Standing Timber Release, Article 15 (Russia, 1998) specify 33 tree species (e.g. *Acer pseudoplatanus*, *Quercus dentata*), as well as other listed trees, shrubs and vines that are protected from cutting.

FOREST PRACTICE REGULATIONS: NATURAL FORESTS

Riparian zone management (Indicator: Riparian buffer zone rules for rivers and streams)

In the state forests of Poland, clearcutting is not allowed in public forests within 40 metres of lakes and watercourses, regardless of the size of the water body (Poland, 1999). However, this rule is not always observed along very small streams (Czech, 2007).

In Latvia, riparian protection is outlined in the 'Law on Protected Belts' (5 February 1997 with amendments of law in 2001, 2003) and Regulations of the Cabinet of Ministers No. 284 'Methodology for the Designation of the Protected Belts for Water Bodies and Water Courses' (4 August 1998.). The 'Law on Protected Belts' (Article 7, 35 and 37) defines riparian ('surface water objects') buffer ('protected belt') widths and management restrictions. Protected belt minimum widths are defined in quantitative detail. Harvesting restrictions within the protected belts include a ten-metre 'no-harvest' zone and a 50-metre no clearcutting zone. Beyond those areas, restrictions apply to construction, chemical use and other activities apart from the removal of riparian flora.

In Russia, requirements for riparian buffer strips are established in the Fisheries Management Code and the Forest Code. In addition, Russian regional governments may also establish 'forbidden strips' where site conditions are determined to require special protection. No harvesting is allowed in forbidden strips, except as necessary for improving the condition of vegetation along riparian zones (in other words, only 'stand-tending cuts' and 'sanitation cuts' are allowed and only for protection purposes as opposed to timber harvest). The widths for forbidden strips vary on the basis of bank slope; they are specified as 35m for banks with zero or negative slope, 35–50m for bank slopes up to three degrees and 55–100m for bank slopes greater than three degrees.

In 2008 (after the January 2007 cut-off date for our standardized policy comparison), a new Fisheries Management Code specified riparian buffer zone width based on the length and type of the water body. The standards are as follows:

- for rivers up to 10km: 50 metres along each side;
- for rivers from 10 to 50km: 100 metres along each side; and
- for rivers more than 50km: 200 metres along each side.

The forest code forbids clearcutting and pesticide application within all water-protection zones.

Figure 6.1 below presents the riparian zone requirements for each of the case study countries.

Roads (Indicators: Culvert size at stream crossings, road abandonment)

In Latvia, some forestry experts believe that the road network is not sufficiently dense to facilitate efficient forest product harvesting and transportation. Design specifications for paved roads are listed in Latvia's 'Law of Construction'. The development of unpaved roads is not regulated by minimum culvert size requirements or requirements for decommissioning. This approach is classified as 'mixed' since there are regulations that apply only to a subset of forest roads.

Poland has established general rules for culvert sizing on all roads, but lacks specifications for road decommissioning.

In Russia, a country with vast stretches of unroaded forestlands, road building policies are relatively well developed. The Russian federal government categorizes forest roads based on their cover materials and projected length of use. Forest roads normally fall into categories III–V. Forest roads are also divided into all-year roads, winter roads, branch-line roads (*vetka*: three- to six-year use) and temporary roads (one- to two-year use) (Russia, 1995b, 1996b).

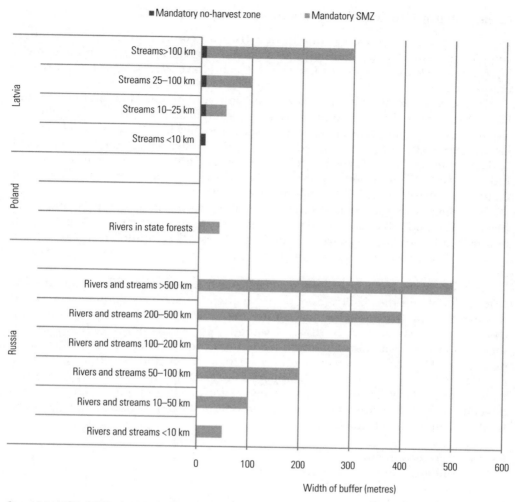

Figure 6.1 *Riparian buffer zone policies in Latvia, Poland and Russia*

Road building is prohibited in the 'forbidden strips' (in other words, no timber harvest zones) of rivers, lakes and other water bodies. In mountainous areas, culverts must be inserted every 100 to 200 metres. Culvert sizes (rather than peak flow capacities) are prescribed and culverts must measure from 0.4 to 0.5m in diameter depending on site conditions (Matveenko, 1981).

The Russian forest code of 2006 as yet lacks specific requirements addressing forest roads culvert design or decommissioning.

Clearcutting (Indicator: Clearcut size limits)

In Latvia and Poland, as is common in European countries, harvesting is often completed in two stages: 'thinning' and 'regeneration felling'. Regeneration felling is equivalent to clearcutting. Both Latvia and Poland have strict restrictions on regeneration felling.

In Latvia, clearcut limits are set on both public and private lands and vary on the basis of soil type. The maximum clearcut size is ten hectares on dry sandy soils and five hectares on other dry soils. At least 40 seed trees per hectare must be left on site. On wet peat soils, if clearcuts exceed two hectares, the strip cut method must be used and cuts must be limited to 50 metres wide on wet peat soils and 100 metres wide on mineral and drained soils (Cabinet of Ministers Regulations No. 152 from 09.04.2002: Tree cutting regulations on forestland).

The maximum size of a regeneration felling on public lands in Poland is six hectares (Poland 1998, 2003). To place this in context, however, it is important to realize that less than one-quarter of the State Forest Directorate forests exceed five hectares in size; this fragmentation clearly precludes extensive use of larger-scale clearcuts.

Forestry in the Russian Federation, particularly in the south, Siberia and the Far east, does not face the challenges of property fragmentation common in central and western Europe. Nevertheless, maximum patch sizes are relatively small in most cases, particularly relative to those of Canada, another country with vast boreal forest resources. Clearcutting is forbidden in protection forests, unless the forests are deemed to have lost their ecological/protective values and selection harvesting cannot restore those values. On 1 January 2007, regulations under the new forest code regarding maximum clearcut patch sizes for production forests had not been developed[3] and the older standards (based on the former Group 1, 2, 3 classification) were in use (see below):

Group 1 (Protection forests)
- coniferous and broad-leaved: 5–10ha
- pioneer hardwoods: 15ha

Group 2 (Populated areas)
- coniferous and broad-leaved: 10–20ha
- pioneer hardwoods: 25ha

Group 3 (Remote, production forest)
- Pine: 25ha, maximum width 250 metres
- Other: 25–50ha

In mountain forests, the clearcut size must be 1.5 times smaller on 11–20° slopes and two times smaller on 21–30° slopes than in corresponding plain forests.
- Hardwood pioneer species in the Far East: 250ha

Figure 6.2 summarizes the above clearcut requirements for Latvia, Poland and Russia.

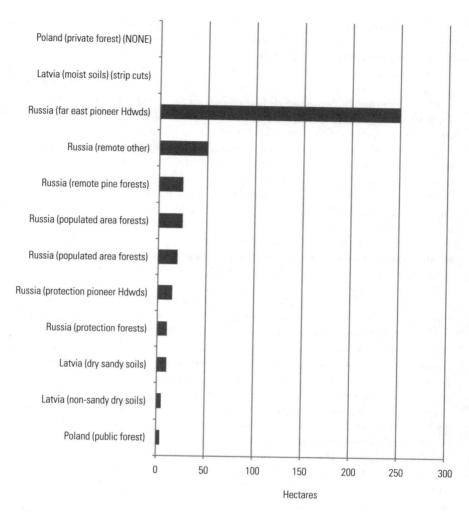

Sources: Poland, 1998, 2003; Russia 1993b, 1993c; Latvia, 2002

Figure 6.2 *Clearcut size limits in Poland, Latvia and Russia*

Reforestation (Indicator: Requirements for reforestation, including specified time frames and stocking levels)

In Latvia, reforestation is required by the Forest Law within three years on both public and private lands. Required stocking levels are determined for different forest regions by the Cabinet of Ministers as mandated in the Nature Protection Regulations in Forest Management (No. 189/05.05.2001).

Under the Polish Forestry Act, reforestation is required within five years and within two years after forest fires and other catastrophic events. For state forests, stocking levels are determined by the *Technical Handbook of Forest Management (Zasady*

Hodowli Lasu) and vary from 1500 seedlings/hectare for larch to 10,000 seedlings/hectare for pine or oak. Stocking levels must be 'adequate' on private lands, but no standardized requirements have been enacted.

The Russian Forest Service has developed numerous mandatory regulations meant to ensure adequate reforestation, among which are time frames and stocking level requirements. Stocking densities of at least 4000 seedlings are required on dry soils and densities of at least 6000 seedlings per hectare are required in the forest-steppe zone (alternating forest and dry, grassy plains).

In Russia, if regeneration is carried out by planting seeds, the planting density must be increased by 20 per cent. In the sub-zone of northern taiga, the planting density can be lowered by 10–15 per cent if advanced regeneration and natural seeding are present. If regeneration is accomplished using transplants,[4] the required planting density is 2500 per hectare, provided that final prescribed stocking levels are reached.

The time to reach reforestation and the required stocking levels vary in Russia according to forest 'group type' and harvest method. In Group 1 forests with clearcutting, adequate stocking should be reached within one to two years. In adjacent cuts, the time frame is two to five years. Group 2 forests must be reforested in one to two years after clearcutting and one to four years in adjacent cuts. In Group 3 forests, conifers must be regenerated within one to three years in adjacent cuts and hardwoods within one year (Russia, 1993d).

Russian federal regulation also dictates the protection of advance regeneration of commercial species during harvest operations. According to Russian law, 75 per cent of advanced regeneration has to be preserved in harvesting operations involving chainsaw logging and tractor skidding. Sixty per cent must be preserved with mechanical equipment logging. Seventy per cent of advanced regeneration must be preserved between skid trails in winter conditions (snow) and 60 per cent must be preserved between skid trails in summer conditions (Russia, 1993e).

In July 2007, the new Declaration of the Regeneration Requirements was released, which contains detailed standards for required seedling stocking for different species and forest types across Russia. Many of the provisions listed above remain in place, with some variation. No clear standard appears to exist regarding the time at which harvested areas must be fully regenerated. The new law still requires the protection of advance regeneration in harvest areas, but lacks specific numerical standards for how much must be protected.

Annual allowable cut (AAC) (Indicator: Cut limits based on sustained yield)

Latvian State Forests calculates annual allowable cut by species for its 1.4 million hectares of public lands. The annual allowable cut on state lands has been set at 8.3 million cubic metres (Stasa and Sarmulis, 1998). AAC calculations are not required on private lands.

According to Poland's Forest Act, each forest district must prepare a forest management plan based on ten-year cycles of production (Poland The State Forests,

2002). The Forest Act states that the cut cannot exceed production capabilities, suggesting that limits to AAC guarantee a non-declining timber supply over both the short and longer term.

According to the Russian Forest Code of 2006 (Article 87), the AAC must be calculated for each *lesnichestvo* (the most important unit of forest organization after the downgrading of the *leskhoz*) or *lesopark* (municipal forest). Responsibility for AAC calculation falls to different agencies and levels of government, depending on the type of forest unit. For the vast majority of *lesnichestva*, it is developed by the provincial forest services. In Moscow Oblast and federally protected forests, Rosleskhoz is responsible. The Ministry of Public Safety is responsible for certain forests under its control. For *lesoparki*, AAC is calculated by municipal authorities. All the agencies are held to certain standards and protocols as defined in the Ministry of Natural Resources Order No. 106. They are required to consult with 'interested parties' in forming the AAC and post it on the internet not less than 30 days from its release (Gagarin, 2007). It should be noted that forest inventory is still under federal control and that a comprehensive inventory will not be completed until 2020. This complicates AAC calculation for all the agencies responsible (Lesniewska et al, 2007).

In all of the case study countries, the actual annual harvest has been well below the annual allowable cut based on annual increment. However, the lack of comprehensive Russian forest inventories could lead to overharvesting in some forest management units and across certain forest types and regions. It is also the case in the Russian Far East that extraction has focused on the most valuable species and the most valuable sections of each tree (in other words, the butt log). This 'high-grading' practice means that, while leaseholders do not harvest more than the volume allotted them, they do not attain that volume from the range of species and timber grades assumed in the AAC, nor do they necessarily fully utilize the trees cut. Large volumes of wood remain on the ground in harvest areas and are often not considered when calculating the volume cut. This discrepancy between 'utilized' AAC and actual volume harvested obviously underestimates the impact of harvesting on the forest and on future standing volumes (Kulikov, 2007).

FOREST PRACTICE REGULATIONS: PLANTATIONS MANAGEMENT

In Latvia, there are a number of distinct differences between the regulations governing plantations and those for natural forests. Article 24 of Latvia's Forest Law, Section 24 states that:

1   a forest stand is considered a plantation if is registered with the state forest as a plantation;
2   provisions for tree felling and forest regeneration shall not apply to plantations;
3   after felling, repeat establishment of a plantation is allowed.

Section 25 then provides requirements for registration as a plantation, including:

1   the use of species consistent with the purpose of the plantation;
2   numerically prescribed minimum tree heights;
3   a prescribed number of trees per hectare based on predominant species;
4   the absence of brush exceeding tree height within a radius of 30 centimetres.

These requirements are supplemented by the Cabinet of Ministers Regulation Number 108, 'Regulations for Forest Establishment and Plantation Forests' (2001). Also relevant is Number 648 'Regulations on Forest Reproductive Material' (2003) (regarding Forest Law Article No. 20). The latter policy includes in-country rules in line with the EU Directive 1999/105/EC on the trade of forest reproductive material within the European Community. The corresponding Latvian regulation bans importing material that has potentially adverse impacts on forests. Finally, the use of 'new' species in Latvia is governed by the 'Law of Protection of Species and Biotopes' (2000), which requires an environmental impact assessment. There is no restriction on using exotic species which have already been introduced.

The forest practice requirements in Poland and Russia, in contrast, apply equally to natural forests and plantations. However, these countries have also developed some regulations specific to plantations. Poland has officially banned the use of exotic (non-native) tree species. In Russia, non-local species may be planted only when proven successful under the given site conditions. In practice, exotic species (for example, Interior Douglas fir) have been used in trials, but have not gained wide industrial use. The Russian Federation has, however, supported plantation projects in the interests of enhancing timber production.

The 2006 Russian forest code includes few legal requirements about plantation establishment or maintenance, but specifically states that plantations can be harvested or tapped for resin without restriction (Article 42). It also authorizes the establishment of plantations both within the Forest Fund and on other lands, such as farmlands. However, the ownership of plantations established on private farmland is not clear. Probably because of this issue, it is difficult to obtain authorization to establish plantations in such areas (Yaroshenko, 2009).

ENFORCEMENT

In Latvia, the State Forest Service (SFS) is responsible for enforcing laws on both public and private lands; it monitors forest practices principally through ranger staff in its district offices (Latvia State Forest Service, 2009b). The SFS reports annually on infringements of forest laws and regulations: around 1300 were reported in 2006, 80 per cent of them on private land. Around a quarter of offences were related to breaches of forest fire safety regulations (Latvia State Forest Service, 2007).

In Poland, the Forest Service (Służba Leśna) is responsible, amongst other things, for combating crimes and any form of harmful activities in state forests. Fifty-six per cent of reported losses in 2005 were associated with timber theft.

The Polish Department of Forestry, Nature Conservation and Landscape Protection of the Ministry of the Environment supervises all forests, regardless of ownership. Audits in state forests are carried out by two units: the first is Inspection of State Forests (subordinated to the General Director) and the second is the division operating under every Regional Directorate: the Department of Control and Property Protection.

There are four types of audits in Poland's State Forests: comprehensive (held every five years), problematic (scrutiny of chosen issue in one or many State Forest Units), verifying (for controlling execution of post-audit decisions) and temporary (short term, connected to current activities or inspecting submitted complaints). Additionally, random audits are performed by various state institutions, including the Inland Revenue, Supreme Chamber of Control or the Ministry of the Environment.

In Russia, responsibility for enforcement of forest regulations has passed from the federal to the provincial forest agencies as per the forest code of 2006. While the work is often done by the same *lesniki* (forest rangers) as before, they now represent the province. Rosleskhoz still collaborates with the provinces on enforcement through its *lesnichestvo* offices, but the extent varies by province.

The transfer of authority and overhaul of earlier forest law challenges Russian forest law enforcement. Some observers cite a rise in violations and illegal logging during this transition period. Blame is sometimes directed at underperforming provincial forest services and sometimes at the federal level for not sufficiently clarifying the law (Lesniewska et al, 2007). The co-existence of new ecological requirements and traditional 'full utilization' standards can create apparent contradictions. *Lesniki* in the province of Karelia use personal judgement to distinguish protected biotopes (legally mandated uncut areas surrounding important ecological features) from illegal *nedarub* ('undercuttings', or patches of leased timber that have not been utilized). Clear standards for distinguishing a biotope from *nedarub* appear to be lacking.

FOREST CERTIFICATION

Forest certification has been of major interest to export-oriented central and eastern European countries, due to its promise of increased market access. Poland was the first European country to achieve FSC certification of its state forests and Russia now ranks second worldwide (next to Canada) in total forest area FSC certified. In Latvia, where almost half of the forests are privately owned, state forests have obtained FSC certification, while certification has been slower to develop on private lands.

All three case study countries have also developed national standards under the PEFC system. Latvia received PEFC endorsement in July 2001, but that endorsement was revoked in February 2008 due to failure to fully harmonize to PEFC's procedural requirements. As of May 2009, Latvia has an application for re-endorsement still pending. The Polish national scheme was endorsed by PEFC in March 2008 and the Russian scheme in March 2009. As of January 2008, our cut-off date for comparison of certified areas, only Latvia had PEFC certified forests and those forests totalled 80,761 hectares (2.7 per cent of total forest area) (PEFC, 2008).

SUMMARY

Russia, Poland and Latvia all favour mandatory, prescriptive forest practice policies. While Russian riparian zones are the largest in size, management restrictions within these zones are limited and subject to loose interpretation. Only Latvia includes no-harvest riparian zones on some watercourses. The maximum clearcut sizes allowed in national forests in Russia, Poland and Latvia are relatively small compared to the North American case studies, with the exception of the Russian Far East pioneer hardwood forest type.

AAC calculations are legislated in all three case studies, with the exception of Latvian private lands. Russia has the most restrictive policies limiting harvest to no more than annual increment. Due to economic, political and geographic factors, the actual rate of timber harvest in Russia is only about one-third of the annual allowable cut. However, this harvest rate is unevenly distributed across the country. It therefore serves to mask overharvest in some key biotopes, such as the hardwood forests of the Russian Far East.

Russia is the only case study country in eastern Europe that has reported large areas of forest plantation, although these forests may more closely resemble semi-natural planted forests in the other case study countries. 'Semi-natural' planted forests are important in both Poland and Latvia. Forest practice laws for plantations and natural forests are similar in Poland and Russia, whereas in Latvia different silvicultural rules apply to plantations.

Table 6.1 *Matrix of policy approaches in natural forests to the five forest practice criteria*

Case Study	1) Riparian	2) Roads	3) Clearcuts	4) Reforestation	5) AAC
Latvia (Private)					
Latvia (Public)					
Poland (Public)					
Russia (Public)					

Mandatory substantive
Mandatory procedural
Mandatory mixed
No policy

NOTES

1   With the contributions of Brian Milakovsky, MSc, Yale School of Forestry and Environmental Studies, for Russia.
2   There are a total of 89 subjects of the Federation, but certain subjects share forest agencies and the federal cities of Moscow and St Petersburg and the subject Moscow Oblast remain under federal control.
3   The new regulations were released in July 2007; they specify patch size for classes based on forest type and geographic zone, for a total of about 25 classes.
4   Transplant (Terminology of Forest Science, Technology Practice and Products Vocabulary, 1971): 'A seedling after it has been lifted and replanted, i.e. "moved", one (and occasionally more) times in the nursery, in contrast to a seedling planted out directly from the seed bed.' Transplants are considered to have better survival rates and this is why the number of transplants per hectare can be smaller.

REFERENCES

Actins, Ansis and Mara Kore (2006) 'Forest certification in Latvia', in B. Cashore, F. Gale, E. Meidinger and D. Newsom (eds) *Confronting Sustainability: Forest Certification in Developing and Transitioning Societies*, New Haven, CT: Yale School of Forestry and Environmental Studies Publication Series

Akim, Eduard L. (2008) 'Russian pulp and paper industry in 2007–2008', Advisory Committee on Paper and Wood Products – 49th session FAO Forest Products and Industries Division, 10 June 2008. Bakubung, South Africa, www.fao.org/forestry/industries/9808/en/, accessed November 2009

Aksenov, Dmitry, Dmitry Dobrynin, Maxim Dubinin, Alexey Egorov, Alexander Isaev, Mikhail Kaprachevskiy, Lars Laestadius, Petr Potapov, Andrey Purekhovskiy, Svetlana Turubanova and Alexey Yaroshenko (2002) 'Atlas of Russia's intact forest landscapes', Moscow: Global Forest Watch

Carlsson, Lars, Nils-Gustav Lundgren and Mats-Olov Olsson (1999) 'Forest enterprises in transition – business behavior in the Tomsk Forest Sector', in *IIASA Interim Report (IR-99-010)*, Laxenburg, Austria: International Institute for Applied Systems Analysis

Cirelli, Maria Teresa (1999) 'Trends in forestry legislation: Central and Eastern Europe', FAO Legal Papers Online # 2, February 1999, www.fao.org/legal/prs-ol/lpo2.pdf, accessed November 2009

Conference of the Parties Bratislava (1996) 'Regional meeting on the convention on biological diversity in Central and Eastern European countries: Implementation of the convention and preparation for the Third Meeting of the Conference of the Parties of Bratislava', Slovakia

Cushman, Samuel A., and David O. Wallin (2002) 'Separating the effects of environmental, spatial and disturbance factors on forest community structure in the Russian far east', *Forest Ecology and Management*, 168, pp201–215

Czech, Andrzej (2007) Personal communication with Andrzej Czech, Director of Nature, Ecology and People Consult (NEPCon), Poland, 1 January 2007

EECONET (2008) 'The European Ecological Network', EECONET: Eeconet Action Fund, European Ecological Network, Nature Net Europe, www.eeconet.org/eeconet/index.html, accessed September 2008

FAO (2006) 'Global forest resources assessment 2005: Progress towards sustainable forest management', in *FAO Forestry Paper 147*, Rome: Food and Agricultural Organization of the United Nations

FAOSTAT (2008) 'ForesSTAT', Rome: United Nations Food and Agricultural Organization of the United Nations

Gagarin, Y. (2007) 'Commentary on the Russian Forest Code', www.rosleshoz.ru/docs/codex, accessed February 2009

Groisman, Pavel Ya., Richard W. Knight, Richard R. Heim Jr, Vyacheslav N. Razuvaev, Boris G. Sherstyukov and Nina A. Aperanskaya (2003) 'Contemporary climate changes in high latitudes of the Northern Hemisphere cause an increasing potential forest fire danger', in *5th Symposium on Fire and Forest Meteorology and 2nd International Wildland Fire Ecology and Fire Management Congress*, Orlando, FL: American Meteorological Society

Hoberg, George (2000) 'How the way we make policy governs the policy we make', in D. Alper and D. Salazar (eds) *Sustaining the Pacific Coast Forests: Forging Truces in the War in the Woods*, Vancouver: UBC Press

ITTO (2006) 'Annual Review and assessment of the world timber situation', Yokohama: International Tropical Timber Organization

Jakubowicz, Piotr Paschalis (2006) 'Forest certification in Poland', in B. Cashore, F. Gale, E. Meidinger and D. Newsom (eds) *Confronting Sustainability: Forest Certification in Developing and Transitioning Societies*, New Haven, CT: Yale School of Forestry and Environmental Studies Publication Series

Jodlowski, Krzysztof (1998) 'Forests and forest contractors in Poland', paper read at the FAO/Austria expert meeting on environmentally sound forest operations for countries in transition to market economies, Ort/Gmunden

Kissling-Naf, I. and K. Bisang (2001) 'Rethinking recent changes of forest regimes in Europe through property-rights theory and policy analysis', *Forest Policy and Economics*, 3, 3–4, pp99–111

Kondrashov, Leonid G. (2004) 'Russian far east forest disturbances and socioeconomic problems of restoration', *Forest Ecology and Management*, 201, pp65–74

Kulikov, A. (2007) Personal communication, with Alexander Nikolaevich Kulikov, Chairman of the Executive Board of The Wildlife Foundation, Khabarovsk, Russia

Lankin, Alexey (2005) 'Forest product exports from the Russian far east and eastern Siberia to China: Status and Trends', Washington, DC: Forest Trends

Latvia (2000a) 'National Report', Rome: European Forestry Commission / Timber Committee Session

Latvia (2000b) 'Law on forests', translated 2007 by the Translation and Terminology Centre, Riga: Republic of Latvia

Latvia (2002) Cabinet of Ministers Regulations No. 152 from 09.04.2002: Tree cutting regulations on forestland, Riga: Republic of Latvia

Latvia (2005a) 'The national economy of Latvia: A macroeconomic review', Embassy of Latvia in Stockholm, Economic Affairs, www.mfa.gov.lv/en/stockholm/bilateral-relations/economic-affairs/, accessed 2009

Latvia (2005b) 'Convention on Biological Diversity: Third national report, Latvia', Riga: Ministry of the Environment, Republic of Latvia

Latvia (2007) 'Market statement', paper read at UNECE Timber Committee 65th Session, 8–12 October 2007, Riga: Republic of Latvia

Latvia (1997) Law on Protected Belts (05.02.1997), with amendments of law in 2001, 2003, Republic of Latvia

Latvia (1998) Regulations of the Cabinet Ministers No. 284, Methodology for the Designation of the Protected Belts for Water bodies and Water courses (04.08.1998), Republic of Latvia

Latvia State Forest Service (2004) 'What does the Latvian forest owner look like?', www.vmd.gov.lv/eng/2/24/246/faq_01.htm, accessed February 2004

Latvia State Forest Service (2007) 'Forest offences going down', www.vmd.gov.lv > News, accessed March 2009

Latvia State Forest Service (2009a) 'The environmental policies of LVM', www.lvm.lv/eng/our_forests/lvm_and_environment/, accessed March 2009

Latvia State Forest Service (2009b) 'Organizational set-up', www.vmd.gov.lv. > About us > Organizational set-up, accessed March 2009

Lebedev, Anatoly (2005) 'Siberian and Russian far east timber for China: Legal and illegal pathways, players and trends', Washington, DC: Forest Trends

LEDC (2000) 'Biodiversity in Latvia: Forests', Latvian Environment Data Centre, http://enrin.grida.no/biodiv/biodiv/national/latvia/ecosys/forests/forests.htm, accessed April 2000

Lesniewska, F., A. Laletic, A. Lebedev and K. Harris (2007) 'Transition in the Taiga: The Russian Forest Code 2006 and its implementation process', Taiga Rescue Network

Lobovikov, Maxim (2000) 'Economic aspects of the Russian Forest Code 1997', paper read at Forging a New Framework for Sustainable Forestry: Recent Developments in European Forest Law, Vienna

Mabel, Marian (2000) 'The flexible domestic state: Institutional transformation and political economic control in the Khabarovsk Krai forest sector', Laxenburg, Austria: International Institute for Applied Systems Analysis

Matveenko, L. S. (1981) *Automobile Roads for Timber-Hauling (Avtomobilnye lesovoznye dorogi)*, Moscow: Lesnaya Promyshlennost

Meidinger, Errol, Ansis Actins, Vilis Brukas, Hando Hain, Gerard Bouttoud, Piotr Paschalis-Jakubowicz, Maria Tysiachniouk, Mara Kore, Rein Ahas and Peep Mardiste (2006) 'Regional overview: Forest certification in Eastern Europe and Russia', in B. Cashore, F. Gale, E. Meidinger and D. Newsom (eds) *Confronting Sustainability: Forest Certification in Developing and Transitioning Societies*, New Haven, CT: Yale School of Forestry and Environmental Studies Publication Series

Milakovsky, Brian and Constance McDermott (2008) 'Amur ecoregional report', New Haven, CT: Yale Program on Forest Policy and Governance

MOE (2007) 'Poland – Statement on the condition of economy and wood market', paper read at UNECE Timber Committee 65th Session, 8–12 October 2007, Geneva, by the Ministry of the Environment, Poland

Moiseev, Nikolai A. (2000) 'Economic and legal aspects of forest management in Russia: Problems and the ways to solve them', paper read at Forging a New Framework for Sustainable Forestry, Recent Developments in European Forest Law, Vienna

Morozov, A. (2000) 'Illegal logging in Russia (Kratkiy obzor nezakonnyh rybok lesa v Rossii (Formy i metody nezakonnyh rubok))', www.forest.ru/rus/publications/illegal/, accessed January 2004

Nefedyev, V. V. (2003) 'Tasks and problems of the Russian forest management (Zadachi i problemy lesopolzovaniya v lesah Rossii) (Report made during the session on Russian Forest Management)', paper read at XII World Forest Congress, Quebec

Ostrom, E. (1990) *Governing the Commons: The Evolution of Institutions for Collective Action*, Cambridge: Cambridge University Press

Pavlov, P. (2007) 'Commentary on the Russian Forest Code', http://rosleshoz.gov.ru/docs/codex, accessed November 2009

PEFC (2008) 'PEFC council information register', http://register.pefc.cz/, accessed February 2008

Petrov, A. P. (2007) 'Forest code of the Russian Federation: State and economic forest management', paper read at Exchange of Experiences on Forest Policies and Institutions in Eastern European Countries, Workshop, 14–18 May, Zamardi, Hungary

Poland (1991) The Forest Act, Warsaw: Government of Poland.

Poland (1998) *Regulation by the Ministry of the Environment in the Matter of Management Plan Compilation, Simplified Management Plan and Forest Inventory*, Warsaw: Government of Poland

Poland (1999) Regulation 11A of the General Director of State Forests, Warsaw: Government of Poland

Poland (2003) *Technical Handbook of Forest Management*, Warsaw: Government of Poland

Poland The State Forests (2002) 'National forest holding annual report', Warsaw: Government of Poland

Polyakov, Maksym and Lawrence Teeter (2003) 'Redesigning forest policy tools under a transitional economy setting', in L. Teeter, B. Cashore and D. Zhang (eds) *Forest Policy for Private Forestry: Global and Regional Challenges*, New York: CABI Publishing

Pstygha, S. (2007) Personal communication with Sergei Pstygha, Deputy Director, Primorski Krai Forest Service, Vladivostok, Russia

Russia (1993a) Manual for stand-tending cuts in the plain forests of the European part of Russia (Nastavlenie po rubkam uhoda v ravninnyh lesah evropeyskoy chasti Rossii), enacted 29 December 1993 # 347, Moscow: Federal Forest Service

Russia (1993b) Main provisions for final harvesting in the forests of the Russian Federation (Osnovnye polozheniya po rubkam glavnogo polzovaniya v lesah Rossiyskoy Federatsii), enacted 30 September 1993 # 260, Moscow: Federal Forest Service

Russia (1993c) Rules of final harvesting in forests of the far east (Pravila rubok glavnogo polzovaniya v lesah Dalnego Vostoka), enacted 30 July 1993 # 201, Moscow: Federal Forest Service

Russia (1993d) Main provisions for reforestation and afforestation in the forest fund of the Russian Federation (Osnovnye polozheniya po lesovosstanovleniyu b lesorazvedeniyu v lesnom fonde Rossiyskoy Federatsii), enacted 27 December 1993 Moscow: Federal Forest Service

Russia (1993e) Main provisions for stand-tending in the forests of the Russian Federation (Osnovnye polozheniya po rubkam uhoda v lesah Rossiyskoy Federatsii), enacted 28 September 1993 # 253, Moscow: Federal Forest Service

Russia (1995a) *On Wildlife (O zhivotnom mire)*, 52-ÔÇ, 24 April 1995, Moscow: State Duma of the Russian Federation

Russia (1995b) *SNiP 2.05.02-85 Automobile Roads (Avtomobilnye dorogi)* Moscow: Ministry of Construction Research Institute (Mistroy Soyuzdornii)

Russia (1996a) Provisions for water protection zones and strips (Polozhenie o bodoohrannyh zonah vodnyh obyektov b bh pribrezhnyh zaschitnyh polosah), enacted 11 November 1996 # 1404, Moscow: Government of the Russian Federation

Russia (1996b) *SNiP 2.05.07-91 Industrial Transport (Promyshlenniy transport)*, Moscow: Ministry of Construction Research Institute (Mistroy Soyuzdornii)

Russia (1998) Rules of standing timber release in the forests of the Russian Federation (Pravila otpuska lesa na kornyu v lesah Rossiyskoy Federatsii), 551, 1 June 1998, Moscow: Federal Forest Service

Russia (2002) *On Environmental Protection (Ob ohrane okruzhayuschey sredy)*, 7-ÔÇ. 10 January 2002, Moscow: Government of the Russian Federation

Sharipova, A. (2007) 'Forest omission (Lesonedorabotka)', *Kommersant*, 234, p3810

Sheingauz, A. (2006) 'Comments on the forest code of the Russian Federation of 2006 and on the Federal Law "On enactment of the forest code of the Russian Federation"', submitted to the World Bank, Washington, DC

Sheingauz, Alexander, Marian Mabel and Natalia Antonova (2001) 'Globalization and the forest sector in the Russian far east', in M. Palo, J. Uusivuori and G. Mery (eds) *World Forests, Markets and Policies*, Boston, Kluwer Academic Publishers

Silamikele, Ilze (2009) Email communication with Ilze Silamikele, Deputy Director, Forest Policy Department, Ministry of Agriculture of the Republic of Latvia, 8 June

Stasa, Janis and Ziedonis Sarmulis (1998) 'Forest operations in Latvia during transition to a market economy', paper read at Proceedings of the FAO / Austria Expert Meeting on Environmentally Sound Forest Operations for Countries in Transition to Market Economies, Ort/Gmunden, Austria

Tysiachniouk, Maria (2006) 'Forest certification in Russia', in B. Cashore, F. Gale, E. Meidinger and D. Newsom (eds) *Confronting Sustainability: Forest Certification in Developing and Transitioning Societies*, New Haven, CT: Yale School of Forestry and Environmental Studies Publication Series

USDA FAS (2006) 'Russian federation solid wood products annual 2006', Washington, DC: United States Department of Agriculture, Foreign Agricultural Service

USDA FAS (2007a) 'Japan solid wood products: Russia announces plans to increase the export tax on logs to 80% in 2007', Washington, DC: United States Department of Agriculture Foreign Agricultural Service

USDA FAS (2007b) 'Poland solid wood products annual 2007', Washington, DC: United States Department of Agriculture, Foreign Agricultural Service

USDA FAS (2007c) 'Russian federation solid wood products annual 2007', Washington, DC: United States Department of Agriculture, Foreign Agricultural Service

World Bank (1996) 'Russian federation forest policy review: Promoting sustainable sector development during transition', Washington, DC: World Bank, Agriculture, Industry and Finance Division

Zhu, Chunquan, Rodney Taylor and Feng Guoqiang (2004) 'China's wood market, trade and the environment', Beijing: WWF International

Chapter 7

Latin America: Brazil, Chile and Mexico

INTRODUCTION

Brazil, Chile and Mexico together account for 14 per cent of total global forest extent. Forests and forest practices within and across these three countries are very diverse, reflecting some of the diversity of Latin American environments and societies. They range from the temperate pine oak forests of Mexico's Sierra Madre Occidental mountains, through the world's largest expanse of lowland, tropical rainforest in the Amazon Basin (the majority of which is in Brazil), to the succession of 12 major forest types located across Chile's 39° of latitude; they also include extensive eucalypt and pine plantations in southern Brazil and central Chile. For much of the 20th century, government policies focused on economic development facilitated by rapid depletion of native forests in these countries. More recently, with the advent of a 'sustainable forest management' paradigm, each country has pursued different trajectories for forest conservation and economic development, adding to their environmental and socio-political diversity.

Forest practices in Brazil vary from indigenous hunter-gatherer systems, to frontier-style exploitation of the Amazon, to some of the world's most technologically advanced plantation forestry. While Brazil is a major forest products exporter, its growing domestic economy consumes most of the wood produced from its native forests. Contested land tenure remains a major issue for stakeholders in both native and plantation forests, although considerable progress has been made in recent years.

In Chile, the rapid depletion of accessible high value native forest timbers, combined with heavy government subsidies for plantation development, have contributed to the emergence of a plantation-based forest economy. Tree plantations now produce 98 per cent of Chile's industrial wood; most are managed by large-scale, high capacity firms oriented towards export markets.

In Mexico, community forest management is the most widespread of all this book's case study countries, accounting for over 50 per cent – and perhaps as much as 80 per cent – of the country's forests and the majority of its wood production

(Antinori and Bray, 2005). This has led to a relatively heterogeneous approach to forest industry development and forest conservation, notwithstanding national constitutional amendments in 1987 and 1999, and ensuing environmental legislation, to enhance environmental protection and biodiversity conservation (Székely et al, 2005).

Brazil and Mexico are amongst the world's most biologically diverse countries and have been ranked first and fourth respectively in these terms (World Bank, 1995; Grupo ARD Inc and Darum, 2003; Huppe, 2007). They also rank very highly in rates of deforestation and forest degradation (see Chapter 2) and illegal logging (Seneca Creek Associates and Wood Resources International, 2004; ITTO, 2008 cited in Chatham House, 2009a). Forest cover in Chile, in contrast, is relatively stable, with a net gain of around 0.05 per cent over the five years to 2005 (see Chapter 2), largely due to the expansion of plantation forests.

Brazil

An overview of forests and forest ownership

Brazil has the largest area of tropical forest, second largest area of forest cover and largest area of primary forest of any country worldwide. Brazil's total forest area is estimated at 477 million ha, including some 5.3 million ha of plantation forests (FAO, 2006). About 95 per cent of Brazil's remaining natural forests are located in the Amazon Basin (ITTO, 2006).[1]

Due to both its ecological and economic importance, the Amazon Basin – defined here as the 'Legal Amazon'[2] – was selected as our unit of analysis for native forest management policies. The Legal Amazon comprises the states of Acre, Amapá, Amazonas, Pará, Rondônia, Roraima, Tocantins; Mato Grosso (centre-west); part of Maranhão and five municipalities of Goiás (IBGE, 2008).

As of 2006, 21 per cent of the Amazon had been designated as indigenous lands, 21 per cent as publicly owned conservation units, 24 per cent as private lands and 9 per cent as 'special areas' including land reform settlements, Afro-Brazilian lands and military land. Twenty-five per cent remains as untitled public land (*terra devoluta*) (Verissimo and Lentini, 2007). The vast majority of forests managed for production are privately owned (ITTO, 2006). There been legal provisions for timber harvesting in National Forests only since 2006 (Verissimo and Lentini, 2007).

There has been little plantation development in the Amazon – partly because it is richly forested, partly as a consequence of its location and limited transport and processing infrastructure, partly because of the requirements to retain native vegetation and partly because of its relative environmental unsuitability for proven plantation species. Consequently, only about 1 per cent (56,000ha) of Brazil's plantations are located in the Amazon (Verissimo and Lentini, 2007). This book's analysis of plantation policies in Brazil, therefore, extends beyond the Amazon to review policies that apply in other states. Ninety per cent of Brazil's plantations are in corporate ownership.

Native forests

Tropical lowland rainforests are the dominant forest type in Brazil, comprising about 50 per cent of its forests; semi-evergreen rainforests comprise a further 10 per cent. Other major forest types are the *cerrado* (savannah) woodlands of the Brazilian highlands, to the south and southeast of the Amazon, and the humid evergreen forests of the Atlantic coast. These biomes originally covered 24 per cent and 13 per cent of Brazil, respectively, but both have been extensively cleared (SBS, 2008). Coniferous forests of *Auracaria* and *Podocarpus* spp. form a transition zone between the Atlantic forests and the *cerrado*. The proportion remaining of these forests from their pre-European extent varies: from only 7 per cent for the Atlantic forest, to 34 per cent for the *cerrado* and 85 per cent for the Amazonian forest (ITTO, 2006; Durigan et al, 2007). Although much of the global focus is on the Amazon forests, the greatest immediate threats to biodiversity due to forest loss are in the *cerrado* (Verissimo and Lentini, 2007).

Deforestation, forest degradation and illegal logging are all serious problems in the Amazon forests, although there is considerable debate as to their precise extent. Between 1990 and 2005 Brazil lost the largest total area of forest cover of any country worldwide. According to the 2005 FAO Forest Resources Assessment, losses across the country as a whole averaged 3.1 million hectares per year (FAO, 2007). Within the Amazon, according to the Brazilian National Institute for Space Research (Instituto Nacional de Pesquisas Espaciais), the annual deforestation rate was 1.8 million ha between 1990 and 2005, with the largest losses occurring along the edges of the basin in Mato Grosso, Pará and Rondônia (INPE, 2008).

The general succession of events that leads to deforestation and forest degradation in the Amazon is well documented (see for example Siqueira et al, 2009). Both state and federal governments promote resource-based development, including the expansion of transportation corridors. Extractive activities such as mining and timber harvest help finance road building. The opening of roads, in turn, encourages migration from more populated parts of Brazil. The subsequent expansion of both commercial farming – principally cattle grazing, sugar cane and soy beans (Geist and Lambin, 2002) – and smallholder agriculture are the principal causes of deforestation. Studies have revealed direct correlations between real prices for beef or mean annual soybean prices and rates of Amazonian deforestation (Morton et al, 2006).

A lack of government capacity contributes to high rates of illegal forest-related activities throughout this process of frontier colonization. The challenges to forest governance are numerous, including unclear tenure rights, conflicts between pro-development, pro-poor, and pro-environmental legislation, lack of enforcement capacity, and the slow pace of enacting legislation that defines and enables legal timber harvest. As a result, while there is a high level of consensus that forest governance is inadequate, there is considerable disagreement as to the relative role of 'illegal logging' in forest loss and degradation. One recent estimate of illegal wood harvest in Brazil places it at 50 per cent of natural forest production (Seneca Creek Associates and Wood Resources International, 2004; Verissimo and Lentini, 2007;

Chatham House, 2008a). A contrasting study suggests 'legal' harvest is cause for at least equal concern: research findings in this case suggest that an estimated 75 per cent of Amazonian timber production comes from legal land clearance by smallholders in new agricultural land settlement schemes (Smeraldi, 2003).

Plantations

The extent of forest plantations in Brazil in 2005, 5.38 million ha, was the greatest of all Latin American countries and ranked seventh worldwide (FAO, 2006). All of these are classified for productive purposes and more than 90 per cent are corporate-owned. The majority are located in the southern and northeastern regions of the country, particularly in São Paulo, Minas Gerais and Bahia states. Only about 56,000ha of plantation have been established in the Amazon region, 60 per cent of which are for pulpwood (Verissimo and Lentini, 2007). As noted above, our analysis of plantation policies therefore extends beyond the Amazon to review federal policies that apply nationally.

Most of Brazil's plantations are of eucalypts (56 per cent) and pines (35 per cent). Brazil has been one of world leaders in the development of highly productive eucalypt plantations and export-oriented industries based on them. Whilst eucalypt-growing has a history of more than 100 years in Brazil, it expanded greatly in response to federal financial incentives during the period 1966–1986. Average productivity for both eucalypt and pine plantations is high by global standards, at round 35 and 28m^3 per hectare per year, respectively, and some eucalypts achieve growth rates of up to 70m^3 per hectare per year (Del Lungo and Ball, 2006; SBS, 2008). Coupled with continuing strong demand for plantation products nationally and internationally, this has led to a continuing high level of investment in both processing facilities and plantations, including through outgrower schemes. As a result, returns from plantation investment in Brazil are very attractive, by both international standards and in comparison to many alternative land uses (Siqueira et al, 2009). Around 60 per cent of plantations are grown primarily for pulp production, 23 per cent for sawnwood and 12 per cent for biomass, principally charcoal for smelting (Del Lungo and Ball, 2006).

Forest governance

Brazil's federal structure comprises 26 states, a federal district and more than 5500 local governments. A new constitution in 1988 devolved much greater authority and resources to the lower levels of government than hitherto, including the rights of states to legislate concurrently with the national government on forestry issues (Gregersen et al, 2004). However, the primary legislation governing forestry activities in Brazil remains the national Forestry Code, Law No. 4.771, enacted in 1965 (Magalhães Lopes, 2000), which applies to all of Brazil's forests.

Two key components of the 1965 Forest Law are the requirement that a proportion of all private forestland must be set aside as a permanent reserve and that management plans be prepared and approved prior to timber harvest. The legal mechanisms to operationalize these requirements, and the capacity to enforce them, have, however, been slow to develop; for example, management plans were not actually required until 1995 (Gregersen et al, 2004). Consequently, regulations governing the forest sector in the Amazon are seen to have been largely ineffective, at least until recently (Seneca Creek Associates and Wood Resources International, 2004).

The level of forest reservation required on private lands is now 80 per cent in the Amazon (increased from 50 per cent in 1996), 35 per cent in the *cerrado* and 20 per cent in other regions (Seneca Creek Associates and Wood Resources International, 2004; ITTO, 2006).

Sustainable forest management was first formally required in 1986, but not defined until Decree No. 1, 282/94 in 1994. This definition was further developed under the 1995 Tarapoto Process for Sustainability of the Amazon Forest and the ITTO Criteria and Indicators for Sustainable Forest Management. In March 2006, the Management of Public Forest Law (11.284/2006) was enacted, recognizing the rights of traditional forest people and establishing a mechanism for granting concessions on public lands (Verissimo and Lentini, 2007). In December 2006, management plan requirements were revised in the 'Decreto Redesenha Normas Para Elaboração De Planos De Manejo Sustentável' (Ambiente Brasil, 2006). The National Forest Programme (PNF) 2010 aims, amongst other objectives, to increase the area of private forest under sustainable management to 20 million ha and to establish 50 million ha of sustainably managed national forests (FLONAs) by 2010 (Macqueen et al, 2003).

The Ministry of Environment (MMA), established in 1992, is the federal agency responsible for environmental policies and their implementation, including protected areas and biodiversity conservation. The Brazilian Institute of Environment and Renewable Resources (IBAMA), established in 1985, is the branch within the MMA responsible for carrying out federal duties to implement national forest policy on non-indigenous lands. The Indian National Foundation (FUNAI) oversees indigenous land management.

Despite the provisions of the 1988 Constitution, the devolution of forest governance has been slow; the reasons for this include the strong national and international interest in the Amazon and the potential local political disadvantages to state and local governments (Macqueen et al, 2003). An important step in this direction was taken in 2004, with the establishment of the Coordinating Commission for the National Forestry Programme (CONAFLOR), comprising federal, state and local government representatives. The Commission has been tasked with multi-scale collaboration in policy-making (ITTO, 2006). In 2006, two years after the establishment of this Commission, the new Public Forest Law (11.284/2006) mandated a major transfer of responsibility and authority from federal- to state-level administrations (see following section on forest practice systems).

Forest practice systems: Amazonian native forests
Because, as noted above, mechanisms for allowing timber concessions in forests on public lands are only very recent, mechanisms for regulating forest practices have been focused on private lands. An approved sustainable forest management plan is

required to harvest in private forests. Permits may also be obtained to convert up to 20 per cent of the total private forest area to non-forest use.

Prior to changes to the Public Forest Law (Lei 11.284 March 2006) and implementing regulations (CONAMA Resoluçao 378, 19 October 2006), IBAMA was responsible for approving both forest management plans and legal deforestation plans. The 2006 changes established a more decentralized regime and transferred that responsibility to state-based environmental agencies (Órgão estadual do meio ambiente, OEMA) under bilateral agreements with IBAMA (Singer, 2006). Exceptions where IBAMA still holds primary responsibility were clarified in a subsequent resolution, as follows:

1   regulated forest activities where there are more than two states involved;
2   when some species of trees (CITES) are to be harvested;
3   in the case of deforestation of native forest of 2000 hectares in the Amazon and 1000 hectares in other regions; and
4   for forest management activities that exceed 50,000 hectares (Resolution 378/2006, Article 1).

The elements required in a management plan were first specified in Normative Instruction 80, 1991; requirements include respect of minimum harvest cycles and a 100 per cent inventory (May 2006). The first step in the approval process is verification of land ownership, which can be problematic because of deliberate or inadvertent multiple titling; the second is the examination of remotely sensed imagery to assess whether any illegal logging has occurred previously; the third is to review the 100 per cent forest inventory and associated proposal for specific trees to either be harvested, identified and retained as seed trees, or retained for other reasons. In addition to specified diameter limits for each species, all trees of 35–45cm dbh must be retained, a maximum of 35 m^3 per hectare can be harvested and a minimum period of 25–30 years is required before re-entry (Seneca Creek Associates and Wood Resources International, 2004).

The requirements of preparing such a detailed and sophisticated management plan are very demanding of forest owners and the relatively onerous requirements for approving it in comparison to the states' and/or IBAMA's staff resources mean that the time required for approval is often very extended (Seneca Creek Associates and Wood Resources International, 2004). As ITTO notes, 'the requirements for the authorization of deforestation are much easier to fulfil than the highly bureaucratic requirements for the approval of forest management plans and annual operation plans' (2006, p211).

Forest practice systems: Plantation forests
Plantation proposals of more than 1000ha are subject to an environmental impact assessment and an operating licence issued by the state environmental protection agency (May 2006). Licences specify operating conditions (which may be restrictive in relation to soil conservation), monitoring of biodiversity and water impacts and riparian zone protection and reclamation; particular requirements vary between

states and for each case. All plantation forestation must respect the minimum requirements for reservation of native forest, noted above, and riparian zone buffers.

Forest production

The Brazilian Amazon is the second largest producer of tropical wood globally, after Indonesia, and wood production is the third-ranked economic activity – after mining and cattle ranching – in the region. In terms of official classification, about 20 per cent of the region is considered production forest and about 70 per cent of this is privately owned production forest. Ninety per cent of timber harvesting in natural forests occurs in the three states of Pará, Mato Grosso and Rondônia. Until 2006, with the enactment of the Management of Public Forest Law, there were no legislative mechanisms for timber concessions on public lands (ITTO, 2006; Verissimo and Lentini, 2007).

Most – nearly two-thirds – of Amazonian production of 32 million m³ (2005 data) is consumed domestically in Brazil, much of it for low value construction materials in the southern Brazilian states. However, both the volume and the added value of wood product exports from the Amazon are growing, the former accounting for 36 per cent of production volume in 2004 (Verissimo and Lentini, 2007). *Swietenia macrophylla* (*mogno*/mahogany) and *Virola surinamensis* (virola) were initially the most prized species, but now an estimated 150 different species are commercially logged (ITTO, 2006). According to official estimates, wood harvested from legally converted forests comprises 25 per cent of Amazonian production; that from forests managed under SFM plans comprises 7 per cent (ITTO, 2006). However, as discussed above, these figures have been the subject of substantial debate.

The 2006 harvest from Brazil's plantations was 184 million m³, 73 per cent of which was eucalypts and 26 per cent pine. While Brazil's plantations also supply a large domestic market, the eucalypt-based pulp and paper industry has had an export orientation from the outset and Brazil is now one of the world's largest producers of pulp and paper. Exports of plantation products, principally pulp and paper, comprise nearly 75 per cent of Brazil's US$10 billion forest products exports and the planted forest sector contributes 4.5 per cent of Brazil's GDP (SBS, 2008).

Indigenous and community forestry

The Amazon has a large and diverse indigenous population, comprising roughly 150 distinct tribes. Some 50 of these groups have no regular contact with the outside world (Hall, 2004; Survival International, 2009). Indigenous forms of engagement with forestry activities are correspondingly diverse. The Brazilian constitution considers all forests to be national assets. In conjunction with the history of colonization of Brazilian Indian lands, this has led to significant and widespread tenure disputes between indigenous and immigrant peoples, and with the state (ITTO, 2006). The principal strategy to address this issue in the Amazon has been to designate indigenous lands, which now comprise 21 per cent of the Brazilian Amazon, and lands for other recognized groups (e.g. the Afro-Brazilian *quilomobola*); a further

25 per cent is currently designated as unclaimed public land, some of which may be transferred to indigenous and community ownership.

A further 21 per cent of the Brazilian Amazon is protected in various forms of conservation tenure. Around 5 per cent of this area (about 5 million ha) has been designated specifically as 'extractive reserves' for the harvest of mostly non-timber forest products, such as latex, nuts, fruits and oils, following the model originally promoted by the rubber-tappers movement (Brown and Rosendo, 2000; Hall, 2004; Ruiz-Pérez et al, 2005). The National Council of Rubber Tappers of Brazil has a goal of securing 10 per cent of the Amazon as extractive reserve (ITTO, 2006).

Thus, the majority of indigenous and community forestry activities conducted in the Amazon focus on non-wood forest products. Community forestry involving timber production is also practised, but has not been widely adopted for a number of economic and institutional reasons (Medina et al, 2008; Pokorny and Johnson, 2008). Likewise, its contribution to wood production is correspondingly minor. Medina et al (2008) reported that less than 2 per cent of communities in four Amazon Basin case study locations participated in community-based forest harvest, compared to the 95 per cent who informally negotiated the sale of timber to loggers.

Chile

Forests and forest ownership

Chile has a total forest area of some 16 million ha, including 2.7 million ha of plantation forests. About 29 per cent of Chile's native forests are in the national protected area system (SNASPE); a further 2 million ha of private native forest is protected. Small forest owners claim ownership of approximately 67 per cent of native forest and larger forest owners 4 per cent. In total, some 6 million ha of private native forest is available for production (Morales, 2006; Espinosa and Acuña, 2007). Chile's indigenous peoples hold recognized legal rights to traditional territories; however, in many cases, indigenous tenure claims remain ill-defined and under dispute (Neira et al, 2002).

Plantation ownership is dominated by large, internationally oriented firms, which own more than 70 per cent of each of the radiata pine and eucalypt estates; two companies alone hold 50 per cent of the plantation estate (Mery et al, 2001). All of Chile's plantations are managed primarily for production – nearly 50 per cent primarily for solid wood production and 35 per cent primarily for fibre (Del Lungo and Ball, 2006).

Land ownership disputes involving indigenous people also extend to plantation forests and there are long-running conflicts between some indigenous groups. In particular, there are long-standing conflicts between the Mapuche people and plantation owners in the plantation-intensive central-southern region (Espinosa and Acuña, 2007).

Native forests

Chile's latitudinal and topographic range contribute to a wide diversity of forest types, dispersed among eight distinct bioregions. Chile extends 4300km from north to south and its average east-west width is only 170km. Its topography is dominated on the eastern border by the Andes Mountains, paralleled for much of their length by the coastal mountains, and separated by the central valley. Climatic zones range from the sub-Antarctic in the far south, to humid temperate zones in the southern and central regions, to desert in the north.

Eighty-two per cent of the 13.4 million ha of Chile's native forest grows south of 39° latitude, about midway in Chile's longitudinal range. Twelve principal forest types are recognized in legislation. Among these, key types include Siempreverde, Lenga, Coihue de Magallanes, Ciprés de las Guatecas and Alerce (Espinosa and Acuña, 2007).

Chile's temperate forests represent one-third of the world's relatively intact temperate forests (Bryant et al, 1997) and its temperate coastal rainforests are second in extent only to those of the Pacific Northwest of North America (Wilcox, 1996).

Plantations

Chile has a long history of tree plantation management using exotic species that dates back to the early decades of the 19th century. Incentives for plantation development were first introduced in 1931, and by 1974 some 300,000ha had been established across both private and public sectors. Under the Pinochet government, a new forest policy placed major emphasis on the development of a globally competitive plantation-based private forestry sector. This led to the privatization of large areas of plantations and the development of subsidized processing facilities. The 1974 Forest Development Law (Decree Law 701) provided an enabling legal framework for plantation expansion and included substantial incentives for the further establishment of intensively managed plantations (Mery et al, 2001). Plantations expanded rapidly as a result: by 2005, plantations totalled about 2.7 million hectares (FAO, 2006), around 13 per cent of the country's total forest area.

Most plantations are located between latitudes 36° and 39° south; about two-thirds are radiata pine (*Pinus radiata*) and most of the remainder are eucalypt (*Eucalyptus* spp.) (Espinosa and Acuña, 2007). Although the conversion of native forests was not intended by the 1974 law promoting plantations, a significant proportion – perhaps half – of the plantations established since 1974 have been on sites converted from native forest (Espinosa and Acuña, 2007; WWF, 2009).

Forest governance

Authority over environment and natural resource issues in Chile is vested in the national government. Chile's first significant forest legislation, the Forest Law (Decree Law 265), was implemented in 1931. It included provisions to promote plantations and protect riparian vegetation and remains a central piece of forestry legislation. It is complemented by the 1974 Forest Development Law (DL 701) which, as noted above, focused on plantation expansion and industry development. The 1974 law also required the establishment of management norms for specific forest types and the preparation and submission of management plans for forest operations.

Law 701 was amended in 1998, by Decree Law 1956, which focused forestation incentives on degraded sites and on small- rather than large-scale owners (Silva,

2004). A native forest law, first introduced to parliament in 1990 as one of the first environmental initiatives after Chile's redemocratization, initially met strong opposition and subsequently languished for more than a decade. It was revived by the government in 2006 and, after further protracted negotiation, became law in December 2008 (Silva, 2004; CONAF, 2009), subsequent to the census date for this book.

The Chilean National Forestry Corporation, CONAF, was established in 1973 under the Ministry of Agriculture, to contribute 'to the protection, conservation, increase, management and improvement of renewable forest resources ... with the end of achieving rapid growth and development of productive forest activity in concert with the protection and conservation of forests and the environment' (RENACE, 2002). CONAF has responsibility for the management of Chile's national system of protected areas (SNASPE) as well as for oversight of forests managed for production. The National Environment Commission (CONAMA) is responsible for evaluating environmental impact assessments, including those that apply to proposed forest operations.

Forest practice systems

Since the enactment of DL 701 in 1974, operations in Chilean native forests and in plantations which receive a government subsidy require a forest management plan prepared by a forester or agronomist and approved by CONAF. This law also prohibited harvesting of forests on slopes greater than 45 per cent (although in fact harvesting continues on slopes of up to 60 per cent), and banned logging of particular forest types (*Araucaria, alerce*) and threatened species. It also specified silvicultural systems that vary by forest type and slope. The 1994 Environmental Law (DL 19300) requires an environmental impact assessment for proposed forest operations of greater than 20ha in northern Chile, greater than 500ha in central-southern Chile, and greater than 1000ha in far southern Chile (Morales, 2006; Espinosa and Acuña, 2007).

There is no requirement for a management plan, or any other type of permit, for plantations that are not established under the subsidy provisions of DL 701 and which are less than the area thresholds for environmental impact assessments. Voluntary best management practices (BMPs) have been developed to guide plantation management.

Forest production

All production originates from private forests. The forestry sector is the second-ranked in the Chilean economy (Espinosa and Acuña, 2007), with plantations responsible for 98 per cent of the country's commercial wood production of about 33 million m^3. Much of this production is exported. In 2005, total wood product exports were 512,000 metric tons, with a value of roughly US$26 million (FAOSTAT, 2008), representing some 11 per cent of total national export income. Eighty-five per cent, some 7.5 million m^3, of wood harvested from native forests is used for firewood (Espinosa and Acuña, 2007).

Indigenous and community forestry

Chile is the territory of 13 regional groups of indigenous peoples (Being Indigenous, 2009b); the Mapuche, who now number about 500,000, comprise the majority, both historically and currently (Being Indigenous, 2009a). As noted above, the traditional lands of the Mapuche coincide with the region of greatest plantation development and this has led to continuing conflict (Espinosa and Acuña, 2007).

Chile has sought to address some of the conflicts over land ownership through land acquisitions overseen by the National Corporation of Indigenous Development (CONADI). More than 200,000ha have been acquired and transferred to local indigenous communities (Espinosa and Acuña, 2007), including some nationally significant areas of forest (WWF, 2008). There have also been initiatives to engage indigenous people in forest management through projects such as those facilitated by the International Model Forests Program (IMFP, 2008).

However, the overall level of indigenous and community involvement in forest management in Chile remains very limited.

Mexico

Forests and forest ownership

One-third of Mexico's land area is forested; around half the forests are classified as primary, and nearly half as modified natural (FAO, 2006). The total extent of forest is estimated at 55 million ha, of which 1 million ha are plantations.

The management of Mexico's forests is internationally distinctive, with the majority of its forest under forms of communal tenure. Mexico's land tenure reflects both the country's tradition of indigenous communal organization and dramatic political changes over the last century. At the turn of the 20th century, 1 per cent of Mexican landowners controlled 97 per cent of all agricultural land in Mexico and 92 per cent of the rural population was landless (Chavez, 1995). In the decades following Mexico's revolution of 1910–1920, a national agrarian reform programme transferred both agricultural and forest land to *ejidos*, groups of individuals who hold rights to communally owned property, and recognized *comunidades*, traditional indigenous common property regimes (Antinori and Bray, 2005).

There are about 8500 *ejidos* and *comunidades* with rights over forest areas; in total, it is estimated that these tenure arrangements prevail over 50–80 per cent of Mexico's forests (Antinori and Bray, 2005; ITTO, 2006). An estimated 15 per cent of forestlands are privately owned and 5 per cent are state-owned, the latter consisting mostly of protected areas (Mery et al, 2001; ITTO, 2006). FAO Forest Resource Assessment data, however, report 59 per cent of forestlands as publicly owned and 41 per cent as 'other' ownerships (FAO, 2006). This discrepancy may be explained in part by variable definitions of 'forestland' and by different levels of tenure rights. *Ejidos*, for example, may have legal ownership of land, but lack the documentation to make this ownership fully effective; further, they may lack the legal skills to file a management plan and thus rely on private contractors (Matthews, 2009). For the purposes of consistency with other case studies in this book, we have used FAO data to classify forest policies by ownership type and have interpreted 'other' as community-based tenure.

Native forests

There are three principal eco-climatic zones in Mexico, of approximately equal area: the arid/semi-arid, the temperate/sub-tropical and the tropical. Forests occur principally in the latter two zones, in each of which are found about half of the country's approximately 55 million hectares of forest. Consistent with Mexico's status as one of the world's most biodiverse countries, many of Mexico's forest ecosystems are globally distinctive and significant. These include the extensive mountain pine/oak ecosystems, and dry, lowland and montane tropical forests.

Mexico's annual rate of deforestation between 2000 and 2005 was 0.4 per cent, with forest disturbance and consequent degradation of similar magnitude (FAO, 2006; ITTO, 2006). As elsewhere in the developing world, the causes of deforestation in Mexico are complex and multifaceted; however, it has been estimated that around 49 per cent of deforestation can be traced to cattle ranching and 14 per cent to agriculture, with 80 per cent of this loss occurring in the lesser-developed southern and central Mexico (Masera et al, 2001). Illegal logging, mostly organized by criminal gangs, is widespread and has serious impacts on both conservation values and local communities' livelihoods. It is estimated that the volume of illegally harvested wood approximates that of legally harvested industrial round-wood (ITTO, 2006; Chatham House, 2008b).

Plantations

Mexico currently has just over 1 million hectares of forest plantations, almost all (93 per cent) of which are for protection purposes. Eucalypt and pine are common species for production plantations, although other species are also sometimes planted, including Spanish cedar, mahogany and teak (FAO, 2006; ITTO, 2006). Two national development programmes promote reforestation of degraded areas and commercial plantations, respectively (ITTO, 2006). Estimates of reforestation rates are, however, imprecise. A recent ITTO report listed an annual planting rate of 35,000ha, based on the 2001 FAO Global Forest Resources Assessment (ITTO, 2006).

Forest governance

Mexico is a federal republic of 31 states and a federal district. The Mexican Constitution of 1917 provides the basis for national laws relating to natural resources and the environment, including forests. It also grants states a high degree of autonomy for establishing state-level laws; however, to date the federal government has remained the primary source of the forest practice policies covered in this book. One of the major goals of recent national environmental legislation has been to achieve better coordination between federal, state and municipal levels of government (Székely et al, 2005).

Mexico has had a series of national forest laws; the first was established in 1926 and focused on conservation (World Bank, 1995) and the seventh and most recent was in 2005. Among the intervening laws, the 1986 Forestry Law was instrumental in institutionalizing sustainable forest management plans as a framework for regulating timber harvest. The 1992 Forest Act has been characterized as an attempt to bolster forest trade through deregulation, and included the lifting of various restrictions on forest products transport (Guerrero et al, 2000; Liverman and Vilas, 2006). The most recent Act – the 2003 General Act for Sustainable Forestry Development – included a number of new provisions related to biodiversity protection and also envisaged the devolution of authority to state level agencies and the transfer of forest management, conservation and protection to forestland owners and producers (Székely et al, 2005).

The basis for current institutional arrangements was established under the 1992 Forest Act: the Secretariat for Environment, Natural Resources and Fisheries (SEMARNAP, later changed to SEMARNAT with the exclusion of fisheries) is responsible for the protection and management of environmental and natural resource assets, including forests. A number of its Directorates, particularly that for Forest and Soil Management (DGFS), have responsibility for specific activities. The Federal Office for Environmental Protection (PROFEPA) has responsibility for compliance monitoring. A National Forestry Commission, CONAFOR, was created within SEMARNAT in 2001, with responsibility for enhancing sustainable development through sustainable forest management, consistent with the objectives of the national Strategic Forestry Plan (PEF) 2025 (Patiño Valera, 2001; ITTO, 2006).

Forest governance in Mexico is strongly influenced by national land tenure and environmental policies. The former are discussed under 'Indigenous and community forests' below. In terms of the latter, the 1988 General Act for Ecological Balance and the Protection of the Environment (LGEEPA) has been described as 'the most important environmental legislation ever enacted in Mexico'. Key components of this Act include:

1   guaranteeing the right of every person to live in an environment conducive to his/her development, health and well-being;
2   defining the principles of environmental policy and supporting policy instruments;
3   providing for biodiversity conservation;
4   establishing protected areas:
5   achieving sustainable utilization of natural resources; and
6   ensuring prevention of air, water and soil pollution (Székely et al, 2005, p87).

In summary, trends in Mexican forest governance and policies mirror many evident internationally, in terms of an emphasis on decentralization, conservation and sustainable forest management, with, however, some modulation and contradiction among priorities for environment versus trade. Mexico is also internationally distinctive in the historical and contemporary role and significance of community forests under *comunidad* and *ejidos* arrangements. Conversely, some aspects of forest governance – notably those concerned with budget allocations – continue to be strongly centralized or centrally determined, meaning that little has changed in terms of implementation (Matthews, 2004). The capacity of the Mexican state to implement its laws remains weak in many respects: as Székely et al (2005, p87) note:

> The true challenge for Mexico is not the continuous enactment of laws, adminis-trative rules, norms or provisions, but the strengthening of the country's rule of law to ensure that effective enforcement and implementation of environmental legislation becomes the rule rather than the exception, as is the case now.

Forest practice systems

Forest practices are regulated under the General Act for Sustainable Forestry Development, which requires that any forest harvesting, forestation or reforestation, or transport, storage and processing of forest products be authorized. Authorizations for harvesting require the approval of a forest management programme and may require an environmental impact study (Székely et al, 2005).

For natural forests, the form of forestry management programme required is determined by the scale of harvesting: a simplified programme is required for areas of less than 20ha, an intermediate programme for areas between 20 and 250ha and an advanced programme for areas greater than 250ha. Where relevant, the programme must include an annual harvesting plan and forest owners must employ at least one forestry professional to oversee forest management. *Comunidades* and *ejidos* must also present proof of a formal resolution approving use of the site and documentation of land ownership including maps and proof of tax payments (Fonseca, 2006), much of which is difficult for many communities to produce (Matthews, 2004). SEMARNAT is required to make a decision on applications within 30 days, unless an environmental impact study is required; up to two additional 60-day periods are allowed in this case. SEMARNAT may impose conditions in approving the forestry management programme (ITTO, 2006).

For plantation forests, the requirements also vary with scale. For commercial plantations less than 800ha, written notification of harvesting is required. A forestry management programme is compulsory for commercial plantations of greater than 800ha. Forestation and reforestation for conservation and restoration and agro-silvi-cultural practices in degraded forests do not require authorization (Székely et al, 2005). It is estimated that, in 2001, about one-third of Mexico's 21 million hectares of forestland were managed under authorized management plans (Madrid and Chapela, 2003).

Silvicultural systems for Mexico's temperate and pine forests, notably the Silvicultural Development Method, are well developed. Tropical rainforest manage-ment has generally involved selective harvesting to diameter limits, of both high value species – particularly Spanish cedar (*Cedrela odorata*) and mahogany (*Swietenia macrophylla*) – and other species. The quality of forest management varies consider-ably between *comunidades* and *ejidos*, reflecting both internal and external circumstances and constraints (ITTO, 2006).

Forest production

About 38 per cent of Mexico's forestlands are suitable for commercial timber production, but only about 15 per cent are currently so managed. Half of this production occurs in Durango and Chihuahua states (Huppe, 2007) and over 86 per cent of total production is pine originating from the highland pine and pine/oak forests of the Sierra Madres Oriental and Occidental (Matthews, 2004). Temperate hardwoods are harvested in relatively small quantities and largely processed as charcoal.

Tropical species make up 6 per cent of total wood production by volume (Huppe, 2007) and one-third of total hardwood production (ITTO, 2006). Quintana Roo and Campeche are the largest tropical wood producing states. The species that have been most heavily exploited include *Cedrela odorata* (*cedro rojo*/Spanish cedar) and *Swietenia macrophylla* (*caoba*/mahogany) (Huppe, 2007); the latter is listed in CITES Appendix II. Illegal logging frequently targets these and other high value species (Chatham House, 2008b).

Despite the intention of agrarian reforms, parastatal forest companies controlled much of the management and harvesting of community-owned forests until the 1990s (Matthews, 2004). Currently, however, the majority of timber harvested legally now originates from about 2500 community forests (Charnley and Poe, 2007). Total roundwood harvest was estimated at 45.5 million m^3 in 2003, including an 'official' industrial roundwood production of 6 million m^3. The volume of illegally harvested wood that same year was estimated at 5–7 million m^3. The majority of all industrial roundwood harvested is consumed domestically (ITTO, 2006). It has been estimated that unregulated firewood and charcoal production account for roughly half of all wood production (Díaz, 2000; Matthews, 2005). More than 1000 species are used for non-wood forest products (ITTO, 2006).

Indigenous and community forests

The significance of indigenous and community forests and forestry in Mexico has been noted in preceding sections; only Papua New Guinea has a higher proportion of its forest under community ownership and/or management (Antinori and Bray, 2005). The origins of Mexico's contemporary community land and forest governance system, represented by comunidades and *ejidos*, can be traced back to ancient indigenous social systems and to colonial and postcolonial governance initiatives for indigenous people (Alcorn and Toledo, 1998). Contemporary forms of communal forestry are sanctioned by the state and exist as corporate entities for a particular land base, with specific membership and governance rules (Antinori and Bray, 2005). The eighth *ejido* census of 2001 reported 30,305 *comunidades* and *ejidos* in Mexico with ownership of a total of 105 million hectares of land with natural vegetation cover. Nearly 9000 of these own forests, totalling around 45 million ha, and one-third of these conduct commercial extraction of timber or manage their forest resources with forestry activities as a central objective. The absolute sizes of communities' forest holdings ranged from under 300ha to 450,000ha, with an average of 7000ha (Madrid and Chapela, 2003).

A series of agrarian reforms since the Mexican revolution have sought to empower indigenous and agrarian communities. However, timber production from community-owned forests continued to be controlled by parastatal and private companies until the 1970s, with little benefit to communities. Subsequently, and with significant levels of state and civil society support, the communities themselves have progressively assumed greater rights and responsibilities for forest management and forest products processing (Antinori and Bray, 2005). However, the extent of regulation of forest management by communities suggests that 'co-management' might be the most appropriate characterization of contemporary arrangements (Charnley and Poe, 2007).

BIODIVERSITY CONSERVATION

Protected areas

In Brazil, federal law (Substitute for the Law n. 2.892/91, Art. 2) establishes and defines the following different types of conservation units: national parks (federal, state and municipal), biological reserves and national forests (federal, state and municipal). Parks and biological reserves are considered conservation units that have integral protection, while national forests are designated for sustainable use (Magalhães Lopes, 2000).

More recently, in 2000, SNUC (Sistema Nacional de Unidades de Conservação da Natureza) has provided further legal guidance for conservation areas and added two additional categories, Extractive Reserves (for mostly non-timber forest products) and Sustainable Development Reserves. All forms of these conservation units together covered about 21 per cent of the Amazon in 2006 (ITTO, 2006). With regard to national-level reporting through UNEP-WCMC, as of 2008 Brazil reported a notable 26 per cent of its land area protected, including 5.6 per cent under the stricter IUCN Categories I–IV, 9.2 per cent under I–VI and the remainder as yet unclassified (Chapter 2, Figure 2.15).

In Chile, the selection of protected areas is guided by the 1984 Law 18362, 'Creating the system of government-protected wildlife reserves – SNASPE' (Pauchard and Villarroel, 2002). According to UNEP-WCMC, as of 2008 Chile had 16.7 per cent of its land in protected areas, almost all of which was in the stricter IUCN Categories I–IV (Chapter 2, Figure 2.15). Chile's National Biodiversity Strategy sets a target of 10 per cent of the area of each of the country's major ecosystems to be included in the network of protected areas (CONOMA, 2004).

Mexico's goal is to conserve representative samples of existing, intact ecosystems. The 1988 General Act for Ecological Balance and the Protection of the Environment (LGEEPA) addresses the selection and categorization of protected areas. As amended in 1996, the categories include biosphere reserves, national parks, natural monuments, areas for the protection of flora and fauna, sanctuaries and state parks and reserves. In the year 2000, a separate branch of SEMARNAT, the National Commission of Natural Areas, was established to oversee protected areas management

(SEMARNAT, 2008). As of 2008, 7.3 per cent of Mexico's lands held protected area status, including 1.3 per cent under IUCN Categories I–IV (Chapter 2, Figure 2.15).

Protection of species at risk

In 1993, Brazil established the National Council for the Amazon Region to oversee the protection of habitats in the Amazon region (1993). The following year, in 1994, Brazil launched the National Biodiversity Programme and an Inter-Ministerial Commission on Sustainable Development and Coordination for Biological Diversity. National policies aimed at habitat protection include the National Integrated Policy for the Amazon Region (enacted December 1994), the National Policy on Ecotourism (enacted September 1994), the National Policy to Control Desertification, the Brazilian National Policy and Strategy for Biodiversity (ESNABIO) and the National Biodiversity Programme. As of 2005, Brazil had also listed 59 plant species in CITES Appendix I and 610 in Appendix II (ITTO, 2006). Brazil has not, however, enacted an endangered species protection law.

Until recently, Chile had addressed biodiversity concerns on an individual species-by-species and region-by-region basis. Numerous regulations in Chile protect specific tree species and ecosystems of special concern. In 2003, however, important steps were taken to coordinate the country's efforts. At this time, the Committee of Ministers of CONAMA approved a system for protecting ecosystems, species and genetic resources and approved the 'Regulation for the Classification of Species' that calls for the creation of a national list of species at risk, based on categories defined by the IUCN (CONOMA, 2004).

There are a variety of Mexican *normas* (rules) that address species and habitat protection. Among these are NOM-059-ECOL 1994 and NOM-059-SEMARNAT-2001 that list species and subspecies endemic to Mexico, which are under threat of extinction, rare or otherwise subject to special protection, and establish specifications for their protection. The rules also outline procedures for new species listings or changes to those listings.

FOREST PRACTICE REGULATIONS: NATURAL FORESTS

Riparian zone management (Indicator: Riparian buffer zone rules)

The Brazilian Amazon has particularly stringent regulations on riparian zone protection, involving permanent reserve areas from 30 to 500 metres wide depending on the width of the river (Brazil, 1965).

In Chile, D.S. 2374 (1937) prohibits cutting within 400 metres above mountain springs, 200 metres of continuous uphill slope from a given stream and a 200-metre radius from springs on flat, non-irrigated terrain. While this rule has not been repealed, it is not enforced (Gayoso and Gayoso, 2008). Instead, CONAF has developed voluntary guidelines for no-harvest buffer zones based on forest type (CONAF, 2007).

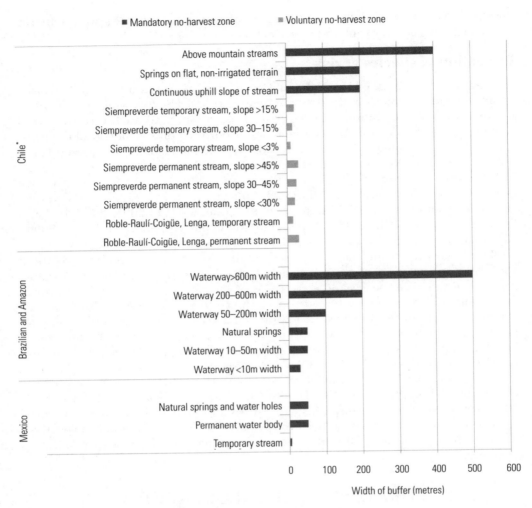

Figure 7.1 *Riparian buffer zone policies in Brazil, Chile and Mexico*

■ Mandatory no-harvest zone   ■ Voluntary no-harvest zone

Note: * Chile's mandatory no-harvest zones are reportedly not enforced (Gayoso and Gayoso, 2008)
Sources: Chile, 1937; Brazil, 1965; Mexico, 1994a, 1994b, 2001; CONAF, 2007

The regulation of riparian zone management in Mexico includes detailed buffer zone requirements. Specified buffer zone widths are 20 metres on either side of perennial streams, five metres on either side of temporary streams and 50 metres around natural springs and water holes. Management in these zones must minimize disturbance and habitat damage to riparian species of flora and fauna (Mexico, 1994a, 1994b, 2001; Pérez, 2004).

Roads (Indicators: Culvert size at stream crossings, road abandonment)

Environmental impact assessments and detailed road plans are required for road building in the Brazilian Amazon. However, no standardized specifications regarding road width, culvert size at stream crossings or road abandonment have been identified.

There are no standardized minimum culvert sizes or requirements for road decommissioning in Chile or other legal requirements for road design or location.

Mexican forest law states that the damage from building roads must be minimized. According to the 1992 Forest Law, roads must not be built through streams. Managers must develop a long-term programme for maintaining roads, including abandonment. No specific regulations on road width, culvert size at stream crossings or road decommissioning have been identified.

Cutting rules (Indicator: Clearcutting or other cutting rules)

Forest operations outside legal reserves on Amazonian property must have an approved sustainable forest management plan. The plan must include rules for harvesting (ITTO, 2006), based on a 100 per cent inventory, and nominate specific trees to either be harvested, identified and retained as seed trees, or retained for other reasons. Minimum diameter limits are specified for each species and all trees of 35–45cm diameter at breast height (dbh) must be retained. A maximum of 35m^3 per hectare can be harvested and a minimum period of 25–30 years is required before re-entry (Seneca Creek Associates and Wood Resources International, 2004).

In Chilean natural forests, silvicultural prescriptions – clearcut, seed tree, protection and selection systems – are specified for particular forest types and slope classes. Clearcutting is allowed only in two forest types (Roble – Hualo, and Roble – Raulí – Coihue) that occur between latitudes 34° and 40° south (Espinosa and Acuña, 2007). In these forests there is a 20ha size limit on clearcuts on slopes of 30–45 per cent and there must be a 100-metre buffer in between cuts. No clearcutting is allowed on slopes greater than 45 per cent (DS 259/1980). The observance of the appropriate silvicultural system is verified in the required management plans.

In Mexico, harvest design is governed by the requirement for management approval. Approved management plans must include appropriate justification for the harvesting method selected (Article 15, Decree No. 1. 282/94). Selective cutting is the norm in forests managed for timber production, although patch clearcuts and group selection are also practised in pine and pine/oak forests (Matthews, 2004). Clearcuts have also been used on an experimental basis, including strip cutting implemented by the Fiprodefo (Trust for Forest Development) Institute (Pérez, 2004). Some states have imposed minimum diameter cutting limits in tropical forests (ITTO, 2006).

Reforestation (Indicators: Requirements for reforestation, including specified time frames and stocking levels)

In Brazil, requirements for reforestation were made more stringent under the Decreto Redesenha Normas Para Elaboração De Planos De Manejo Sustentável. Forest management plans must include the assessment of existing stocking levels and obligations for reforestation. In addition, the decree outlines standardized stocking levels for harvests of fuel wood, timber and charcoal.

In natural forests in Chile, reforestation is required by DL 701 and stocking levels were prescribed as at least ten seedlings planted for each native species cut, or regeneration resulting in a match of the original mix of species. DS 259 of 1980 specifies a minimum regeneration stocking of 3000 plants per hectare (Espinosa and Acuña, 2007). Time frames for achieving adequate stocking levels are also specified.

Under Mexican forest law, plans for reforestation must be included in the required Forest Management Plans.

Annual allowable cut (AAC) (Indicator: Cut limits based on sustained yield)

In Brazilian, Chilean and Mexican natural forests, the annual allowable cut is calculated for each management unit as part of required management planning processes. Harvest must not exceed yields that can be sustained over the longer term.

FOREST PRACTICE REGULATIONS: PLANTATIONS

Plantations in Brazil are subject to some of the same forest practice requirements as native forests, in terms of the proportion of forest that must be retained on private land, the protection of riparian zones and the requirement for harvesting permits. There are also additional requirements for plantations greater than 1000ha. These include the preparation of an environmental impact assessment prior to the issuance of a licence to operate. The impact assessment must specify the actions to be taken for environmental protection and, frequently, restoration. In addition, some states have developed state-level legislation for plantation management.

In Chile, there are voluntary guidelines for no-harvest zones 25 metres wide on each side for both permanent and temporary streams. Reforestation is required for plantations established with government support and stocking levels are specified based on species planted; these requirements do not apply to plantations established without subsidy, outside the provisions of DL 701 (Morales, 2006). There are no culvert size specifications or decommissioning requirements on forest roads. Clearcutting is not regulated, nor are there required annual allowable cut limits. There are no restrictions on the use of exotic species in plantations.

Plantations in Mexico are subject to the same laws as those governing natural forests.

ENFORCEMENT

In Brazil, following the Public Forest Law (11.284/2006), the authority for enforcing forest practice regulations has been incrementally devolving from IBAMA to the state level. Auditing mechanisms include the patrolling of wood product transportation corridors, sawmills and forestry sites and the use of remote sensing imagery. Capacity, however, is limited and the costs of monitoring the frontier areas of the Amazon are enormous. Meanwhile, illegal deforestation and forest degradation continue at alarming rates. Brazil ranks the highest of our case study countries in the area of forest lost annually[3] and some estimate that the proportion of wood produced illegally (in other words, lacking the proper licensing procedures) is similarly high – an estimated 50 per cent in the Amazon in 2004 (Verissimo and Lentini, 2007).

In Chile, CONAF is responsible for enforcement and monitoring of forest practices, but has lacked the capacity and institutional status to do so adequately. The illegal harvest of tree species at risk, such as *alerce*, the illegal conversion of native forests to plantation and the lack of implementation of management plans are the principal areas of concern (Silva, 2004; Morales, 2006; Espinosa and Acuña, 2007).

The Federal Office for Environmental Protection (PROFEPA) is responsible for monitoring and enforcement of forest practices in Mexico. Only about one-third of the country's forests are estimated to be actually governed by the required sustainable forest management plans (Masera et al, 2001) with many of the laws in place governing forest management inconsistently enforced (Matthews, 2004). The scale of illegal harvesting and conversion is substantial; PROFEPA has identified 100 critical zones in which these activities are occurring (Fonseca, 2006).

Conversely, the community-based institutions of *comunidads* and *ejidos* provide a local-level basis for enforcement of forest regulations. However, their capacity is variable and it is also very difficult for them to challenge the organized illegal exploitation of forests, in which violence is a threat or reality (Fonseca, 2004).

FOREST CERTIFICATION

Brazil had some 4.9 million hectares FSC certified as of January 2008; about 43 per cent of that certified area is in plantations. The PEFC-accredited Brazilian CERFLOR certification system has certified 974,000ha.

Chile has 321,000ha FSC certified, 90 per cent of which are plantation. The PEFC-accredited Chilean CERTFOR, which applies only to plantation forests, has certified 1.7 million ha.

Mexico had about 694,000ha of FSC certified forest as of January 2008. The vast majority of this certified area consists of Community Forest Enterprises (CFE). Currently, most CFEs sell to the domestic market where demand for FSC certified products is almost non-existent and many certified CFEs are finding themselves

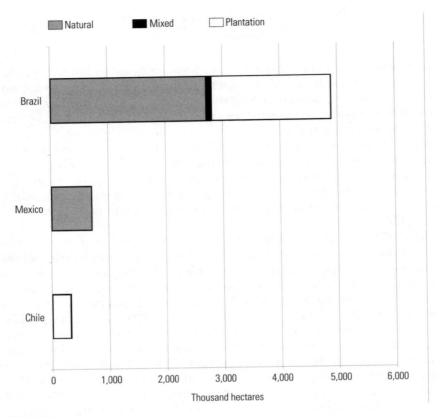

Source: FSC-AC, 2008

Figure 7.2 *Area FSC certified by forest type, January 2008*

unable to support the cost of maintaining their certified status (Segura, 2000).

For all three countries, the total percentage of forestlands certified is quite small, 2 per cent or less. However, in both Brazil and Chile, the proportion of plantations certified by either or both the FSC and PEFC-accredited national schemes is relatively high, at around 43 per cent for FSC in Brazil and in Chile, 80 per cent for FSC and 100 per cent for PEFC.

Figure 7.2 shows the area FSC certified by forest type and Figures 2.16–2.19 in Chapter 2 show figures for both FSC and PEFC certified areas.

SUMMARY

The case study countries in this chapter differ substantially in size, forest endowment and political and institutional frameworks. Nevertheless, there are certain key similarities among forest policies in all three cases. All take a highly procedural approach centred on the preparation and approval of management plans, coupled with relatively high environmental thresholds for some policy indicators.

Brazil is unique in its two-pronged approach to forest exploitation and conservation on private lands. This includes the requirement to retain an unprecedented 80 per cent of forested private property in the Amazon. Ironically, however, it is more difficult to obtain approval for forest management plans in these retained forests than to obtain legal permits for conversion of the other 20 per cent.

Changes over the last few years in Brazilian forest practice policy have been towards an increasingly prescriptive approach. This is due to recent progress in the development of the sustainable forest management planning process and regulations. Within the evolving SFM system, regulatory updates in 2006 replaced procedural regeneration requirements with quantitative prescriptions.

Some of the nominal forest practices requirements in both Brazil and Chile are strikingly stringent. For example, Brazil's riparian no-harvest zones are by far the largest of all case studies worldwide, although the difference in buffers is less acute on rivers of moderate width. Likewise, Chile's formal buffer requirements are quite stringent, but apparently are not well enforced.

In all three countries, harvest patterns are outlined in individual forest management plans. Standardized cutting rules are quite prescriptive in the Brazilian Amazon and there are well-developed silvicultural guidelines for particular forest types in Mexican pine/oak forests and in Chile. The cutting rules for tropical forests in Mexico, however, are less standardized.

Forest road development is relatively unregulated in Chile. Brazil and Mexico take a procedural approach to forest roads, requiring environmental impact assessments and/or road building plans.

Plantation expansion has been subsidized in all three countries. In Mexico, a primary objective for plantation development has been landscape conservation. In Brazil and Chile, the focus is on timber and pulp production and growth in plantation area has been remarkable. This expansion is concentrated in areas best suited for high-yield, intensive management. In Brazil, this means a concentration of plantations in the south and southeast where there is little closed natural forest and hence little issue of 'conversion'. Nevertheless, conversion is technically allowed within specified limits. In Chile, most plantations occur in the south central area, overlapping with forest types of key conservation concern. Conversion is prohibited only in some forest types. Rules governing Chilean plantations are less restrictive than those governing natural forests.

The table below provides a comparative summary of policy approaches to our five forest practice regulation indicators in each of our case study jurisdictions.

Table 7.1 *Matrix of policy approaches in natural forests to the five forest practice criteria*

Case Study	1) Riparian	2) Roads	3) Clearcuts	4) Reforestation	5) AAC
Brazilian Amazon (Private)	■	▓	■		▓
Brazilian Amazon (Public)	■	▓			▓
Chile (Private)	■	□		▓	▓
Chile (Public)	■	□		▓	▓
Mexico (Public/Communal)	■	▓	▓	▓	▓

■ Mandatory substantive
▓ Mandatory procedural
□ No policy

NOTES

1 The Amazon Basin forest extends to Bolivia, Colombia, Ecuador, French Guyana, Guyana, Peru, Surinam and Venezuela; more than 50 per cent is located in Brazil.
2 The Legal Amazon is defined by Article 2 of Law no 5,173, October 1966, and comprises about 5 million km^2, or 59 per cent of Brazil (IBGE, 2008).
3 Given the vast size of the forest resource, the *rate* of deforestation (in other words, percentage of forest lost) in Brazil is 0.4 per cent, a relatively modest rate compared to the case study countries of Indonesia and Mexico.

REFERENCES

Alcorn, J. and V. Toledo (1998) 'Resilient resource management in Mexico's forest ecosystems: The contribution of property rights', in F. Berkes and C. Folke (eds) *Linking Social and Ecological Systems*, Cambridge: Cambridge University Press

Ambiente Brasil (2006) 'Decreto redesenha normas para elaboração de planos de manejo sustentável', Ambiente Brasil, 12 December 2006, http://noticias.ambientebrasil.com.br/ noticia/?id=28189, accessed August 2009

Antinori, A. and D. B. Bray (2005) 'Community forest enterprises as entrepreneurial firms: Economic and institutional perspectives from Mexico', *World Development*, 33, pp1529–1543

ARD Inc and Grupo Darum (2003) 'Biodiversity and tropical forest conservation, protection, and management in Mexico: Assessment and Recommendations', ARD Inc and Grupo Darum

Being Indigenous (2009a) 'Mapuche Region', Chile, www.beingindigenous.org/regions/ mapuche/region_mapuche.htm, accessed April 2009

Being Indigenous (2009b) 'Regions', Chile, www.beingindigenous.org/regions/regions.htm, accessed April 2009

Brazil (1965) Código Forestal (Forest Law), Government of Brazil

Brown, Katrina and Sérgio Rosendo (2000) 'The institutional architecture of extractive reserves in Rondónia, Brazil', *The Geographical Journal*, 166, 1, pp35–48

Bryant, Dirk, Daniel Nielsen, and Laura Tangley (1997) 'The last frontier forests: Ecosytems and economies on the edge. What is the status of the world's remaining large, natural forest ecosystems?', Washington, DC: World Resources Institute, Forest Frontiers Initiative

Charnley, Susan and Melissa R. Poe (2007) 'Community forestry in theory and practice: Where are we now?', *Annual Review of Anthropology*, 36, pp301–336

Chatham House (2009a) 'Brazil', www.illegal-logging.info > Countries > Brazil, last accessed January 2009

Chatham House (2009b) 'Mexico', www.illegal-logging.info > Countries > Mexico, accessed January 2009

Chavez, Luis Felipe Valladolid (1995) 'Privatization of Mexican *ejidos*: The implications of the new Article 27', unpublished report available from www.aad.berkeley.edu/95journal/LuisChavez.html, accessed March 2004

Chile (1937) DS No 2374 de 1937 Reglamenta la explotación de bosques en hoyas hidrográficas declaradas forestales. Santiago: Government of Chile

CONAF (2007) 'Normas de Manejo', Government of Chile, Corporación Nacional Forestal (CONAF), www.conaf.cl/ > normas de manejo, accessed February 2008

CONAF (2009) 'Ley de bosque nativo', Government of Chile, Corporación Nacional Forestal www.conaf.cl > ley de bosque nativo, accessed March 2009

CONAMA (2004) Protección de la Naturaleza, Comisión Nacional Del Medio Ambiente (CONAMA), available at www.conama.cl/portal/1255/propertyvalue-10399.html (last accessed September 2007)

Del Lungo, A. and J. Ball (2006) 'Global planted forests thematic: Country responses to reporting tables for planted forests survey', in *FAO Planted Forests and Trees Working Paper FP/35a*, Rome: Food and Agriculture Organization of the United Nations (FAO)

Díaz, Jiménez R. (2000) 'Consumo de leña en el sector residencial de México: Evolución histórica de emisiones de CO_2', Mexico City: Division de Estudios de Posgrado, Facultad de Ingenieria, Universidad Nacional Autonoma de Mexico

Durigan, G., M. F. Siqueira, and G. A. D. C. Franco (2007) 'Threats to the *cerrado* remnants of the state of Sao Paulo, Brazil', *Scientia Agricola*, 64, 4, pp355–63

Espinosa, Miguel and Eduardo Acuña (2007) 'Chile: Forest resources, forest types, forest retention, current issues, and forest protection', in F. Cubbage (ed) *Forests and Forestry in the Americas*, WikiHome, Society of American Foresters, International Society of Tropical Foresters, www.encyclopediaofforestry.org/index.php?title=Chile01, accessed November 2009

FAO (2006) 'Global forest resources assessment 2005: Progress towards sustainable forest management', in *FAO Forestry Paper 147*, Rome: Food and Agricultural Organization of the United Nations

FAO (2007) *State of the World's Forests – 2007*, Rome: United Nations, Food and Agriculture Organization of the United Nations

FAOSTAT (2008) 'ForesSTAT', Rome: United Nations, Food and Agriculture Organization of the United Nations

Fonseca, S. A. (2004) 'Why is there violence in some forest areas of Mexico?', *EFTRN Newsletter*, pp43–44, www.etfrn.org/etfrn/newsletter/news4344/articles/3_7_Anta.pdf, accessed November 2009

Fonseca, S. Anta (2006) 'Forest certification in Mexico', in B. Cashore (ed) *Confronting Sustainability: Forest Certification in Developing and Transitioning Countries*, New Haven, CT: Yale School of Forestry and Environmental Studies

FSC-AC (2008) 'FSC certified forests', 10 January 2008, Forest Stewardship Council, fact sheet accessed January 2008, www.fsc.org

Gayoso, S. and Jorge Gayoso (2008) Personal communication with Sylvana Gayoso and Jorge Gayoso, Faculty of Forest Sciences, Universidad Austral de Chile, 24 January

Geist, Helmut J. and Eric F. Lambin (2002) 'Proximate causes and underlying driving forces of tropical deforestation', *BioScience*, 52, 2, pp143–150

Gregersen, H., A. Contreras-Hermosilla, A. White and L. Phillips (2004) 'Forest governance in federal systems: An overview of experiences and implications for decentralization', paper read at Interlaken Workshop on Decentralization in Forestry, 27–30 April, Interlaken, Switzerland

Guerrero, María Teresa, Cyrus Reed, Brandon Vegter and George Kourous (2000) 'The timber industry in Northern Mexico: Social, economic, and environmental impacts', *Borderline*, 8, 2, pp1–4, 15, 16

Hall, M. (2004) 'Extractive reserves: Building natural assets in the Brazilian Amazon', University of Massachussetts, www.peri.umass.edu/fileadmin/pdf/working_papers/ working_papers_51-100/WP74.pdf, accessed January 2009

Huppe, Heather (2007) 'The forests of Mexico: Sustaining Mexico's cultural, biological and economic values for the future', in F. Cubbage (ed) *Forests and Forestry in the Americas*, WikiHome, Society of American Foresters, International Society of Tropical Foresters, www.encyclopediaofforestry.org/index.php?title=Mexico01, accessed November 2009

IBGE (2008) 'IBGE makes available a database for the Legal Amazon vegetation', Instituto Brasileiro de Geografia e Estatística, www.ibge.gov.br/english/presidencia/noticias/ noticia_visualiza.php?id_noticia=1152&id_pagina=1, accessed May 2009

IMFP (2008) 'Chile', International Model Forest Program (IMFP), www.imfp.net > fact sheets, accessed April 2009

INPE (2008) 'Estimativas Annuais desde 1988 até 2006: Taxa de desmatamento anual (km²/ano)', Instituto Nacional de Pesquisas Espaciais, www.obt.inpe.br/prodes/ prodes_1988_2006.htm, accessed April 2008

ITTO (2006) 'Status of tropical forest management 2005', ITTO Technical Series No. 24, Yokohama: International Tropical Timber Organization

Liverman, Diana M. and Silvina Vilas (2006) 'Neoliberalism and the environment in Latin America', *Annual Review of Environment and Resources*, 31, 1, pp327–363

Macqueen, D. J., M. Grieg-Gran, E. Lima, J. MacGregor, F. Merry, V. Prochnik, N. Scotland, R. Smeraldi and C. E. F. Young (2003) 'Growing exports: The Brazilian tropical timber industry and international markets', in *Small and Medium Enterprise Series No.1*, London: International Institute for Environment and Development

Madrid, Sergio and Francisco Chapela (2003) 'Certification in Mexico: The cases of Durango and Oaxaca', Washington, DC: Forest Trends

Magalhães Lopes, S. R. (2000) 'Procedimentos legais para exploração das florestas naturais da bacia amazônica', Belém, Pará: Tropical Forest Foundation

Masera, Omar R., Alma Delia Cerón and Antonio Ordón (2001) 'Forestry mitigation options for Mexico: Finding synergies between national sustainable development priorities and global concerns', *Mitigation and Adaptive Strategies for Global Change*, 6, pp291–312

Matthews, Andrew Salvador (2004) 'Forestry culture: Knowledge, institutions and power in Mexican forestry 1296–2001', PhD, Forestry and Environmental Science, New Haven, CT: Yale University

Matthews, A. S. (2005) 'Power/knowledge, power/ignorance: Forest fires and the state in Mexico', *Human Ecology*, 33, 6, pp795–820

Matthews, A. S. (2009 Personal communication with Assistant Professor Andrew Matthews, Department of Anthropology, University of California, Santa Cruz, 11 May 2009

May, P. (2006) 'Forest certification in Brazil', in B. Cashore (ed) *Confronting Sustainability: Forest Certification in Developing and Transitioning Countries*, New Haven, CT: Yale School of Forestry and Environmental Studies

Medina, G., B. Pokorny and B. M. Campbell (2008) 'Favoring local development in the Amazon: Lessons from community forest management initiatives', in *CIFOR Livelihood Brief*, Bogor, Indonesia: Center for International Forestry Research

Mery, Gerardo, Sebastiao Kengen and Concepcion Lujan (2001) 'Forest-based development in Brazil, Chile and Mexico', in M. Palo, J. Uusivuori and G. Mery (eds) *World Forests, Markets and Policies*, Dordrecht: Kluwer Academic Publishers

Mexico (1994a) NORMA Oficial Mexicana NOM-060-SEMARNAT-1994, that establishes the specifications for mitigating the adverse effects caused to the soil and bodies of water from forest utilization, Mexico City: Government of Mexico

Mexico (1994b) NORMA Oficial Mexicana NOM-062-SEMARNAT-1994, that establishes the specifications for mitigating the adverse effects on biodiversity caused by changes in soil use on forest and agricultural lands and fisheries, Mexico City: Government of Mexico

Mexico (2001) NORMA Oficial Mexicana NOM-020-SEMARNAT-2001 that establishes the procedures and guidelines that should be observed for the rehabilitation, improvement and conservation of forest and pasture lands, Mexico City: Government of Mexico

Morales, E. (2006) 'Early experience of total divestment: Chile', in M. Garforth, N. Landell-Mills and J. Mayers (eds) *Changing Ownership and Management of State Forest Plantations*, London: International Institute for Environment and Development

Morton, Douglas C., Ruth S. DeFries, Yosio E. Shimabukuro, Liana O. Anderson, Egidio Arai, Fernando del Bon Espirito-Santo, Ramon Freitas and Jeff Morisette (2006) 'Cropland expansion changes deforestation dynamics in the southern Brazilian Amazon', *Proceedings of the National Academy of Sciences in the United States of America*, 103, 39, pp14637–14641

Neira, Eduardo, Hernán S. Verscheure and Carmen Revenga (2002) 'Chile's frontier forests: Conserving a global treasure', Washington, DC: Global Forest Watch

Patiño Valera, F. (2001) 'Regional update for Mexico', in *Papers for 12th Session of the FAO Panel of Experts on Forest Gene Resources*, Rome: Food and Agriculture Organization of the United Nations (FAO)

Pauchard, Aníbal, and Pablo Villarroel (2002) 'Protected areas in Chile: History, current status, and challenges', *Natural Areas Journal*, 22, pp318–30

Pérez, José Juan González (2004) Personal communication with José Juan González Pérez, La Comisión Nacional Foresta (CONAFOR), Aguascalientes, 23 February 2004

Pokorny, B. and J. Johnson (2008) 'Community forestry in the Amazon: The unsolved challenge of forests and the poor', in *ODI Natural Resources Perspectives*, London: Overseas Development Institute (ODI)

RENACE (2002) *Bosque Sustentable: Marco de Referencia y Propuestas para la Formulación de una Ley de Conservación y Manejo Sustentable del Bosque*, Santiago: Red Nacional de Acción Ecológica (RENACE)

Ruiz-Pérez, Manuel, Mauro Almeida, Sonya Dewi, Eliza Mara Lozano Costa, Marian Ciavatta Pantoja, Atie Puntodewo, Augusto de Arruda Postigo and Alexandre Goulart de Andrade (2005) 'Conservation and development in Amazonian extractive reserves: The case of Alto Juruá', *Ambio*, 34, p3

SBS (2008) 'Fatos e números do Brasil florestal, Dezembro de 2007', Sociedade Brasileira de Silvicultura, www.sbs.org.br, accessed January 2009

Segura, Gerardo (2000) 'Mexico's forest sector and policies: A general perspective', paper presented at 'Constituting the Commons: Crafting Sustainable Commons in the New

Millennium', the Eighth Conference of the International Association for the Study of Common Property, Bloomington, Indiana, US, 31 May–4 June 2000

SEMARNAT (2008) '¿Quiénes Somos?, Secretaría de Medio Ambiente y Recursos Naturales', www.semarnat.gob.mx/queessemarnat/Pages/quienes_somos.aspx, accessed June 2008

Seneca Creek Associates and Wood Resources International (2004) 'Summary: "Illegal" logging and global wood markets: The competitive impacts on the U.S. wood products industry', report prepared for the American Forest and Paper Association, www.illegal-logging.info/uploads/afandpa.pdf, accessed November 2009

Silva, Eduardo (2004) 'The political economy of forest policy in Mexico and Chile', *Singapore Journal of Tropical Geography*, 25, 3, pp261–280

Singer, B. (2006) 'Brazilian forest-related policies', Paris: Institut d'Études Politiques and Centre de coopération internationale en recherche agronomique pour le développement (CIRAD)

Siqueira, J. D. P., M. F. R. Souza, and G. P. Saraiva (2009) 'Brazil', in F. Cubbage (ed) *Forests and Forestry in the Americas: An Encyclopedia*, Bethesda, MD: Society of American Foresters and International Society of Tropical Foresters

Smeraldi, Roberto (2003) 'Expedient plunder? – The new legal context for Amazonian logging', in D. Macqueen, M. Grieg-Gran, E. Lima, J. MacGregor, F. Merry, V. Prochnik, N. Scotland, R. Smeraldis and C. Young (eds) *Growing Exports: The Brazilian Tropical Timber Industry and International Markets*, London: International Institute for Environment and Development

Survival International (2009) 'Brazilian Indians', www.survival-international.org/tribes/brazilian, accessed May 2009

Székely, A., L. O. Martínez Morales, M. J. Spalding and D. Cartron (2005) 'Mexico's legal and institutional framework for the conservation of biodiversity and ecosystems', in J.-L. E. Cartron, G. Ceballos and R. S. Felger (eds) *Biodiversity, Ecosystems and Conservation in Northern Mexico*, New York: Oxford University Press

Verissimo, A. and M. Lentini (2007) 'The Brazilian Amazon', in F. Cubbage (ed) *Forests and Forestry in the Americas*, WikiHome, Bethesda, MD: Society of American Foresters and International Society of Tropical Foresters

Wilcox, Ken (1996) *Chile's Native Forests: A Conservation Legacy*, Washington State: NW Wild Books

World Bank (1995) 'Mexico resource conservation and forest sector review', Washington, DC: The World Bank

WWF (2008) 'Southern Chile – people', www.worldwildlife.org/what/wherewework/southernchile/people.html, accessed April 2009

WWF (2009) 'Chile – problems', www.panda.org > About WWF > Where We Work > Latin America > Solutions by Country > Chile > Problems, accessed January 2009

Chapter 8

Oceania: Australia and New Zealand

INTRODUCTION

Australia and New Zealand together contain about 4 per cent of the world's forests. Although they differ biogeographically and in terms of scale, these countries share comparable forest histories. The native forests of both countries originate from the Gondwanan flora, have high levels of biodiversity and endemism and were significantly impacted by indigenous peoples – the Australian Aboriginals and Aotearoa Maori – principally through their hunting of wildlife and use of fire (see for example Flannery, 1994; Whitehead et al, 2003). Maori reduced the area of forest in New Zealand by about one-third prior to the arrival of Europeans and European settlers then cleared a further third (Roche, 1990). In Australia, both Aboriginal and subsequent European use of fire altered the landscape pattern and structure of forests and European settlers converted about one-third of Australia's forests to other, principally agricultural, land uses (Australian National Forest Inventory, 2003).

Both countries were colonized by British settlers in the late 18th century and both inherited British institutions and the Westminster system of government, which persist to the present. In the early stages of European colonization of both countries, the native forests proved rich sources of export income as well as barriers to agricultural development and it was not until the turn of the 20th century that forests began to be reserved from conversion and brought under management by fledgling forest agencies (Bonyhady, 1993; Dargavel, 1995; Roche; 1990). A century later, 77 per cent of New Zealand's remaining native forests, including all those in public ownership, and 18 per cent of Australia's remaining native forests, including 54 per cent of those in public ownership and management,[1] are formally reserved for conservation. Around a quarter of each country's native forests are privately owned and a further 46 per cent of Australia's native forests are privately managed under leasehold tenure. In both countries, a strong focus has emerged since the 1980s on restoring trees to agricultural landscapes, through both commercial and environmental plantings (Australian National Forest Inventory, 2003; Ministry of Agriculture and Forestry, 2003).

In both countries, plantation forests – initially of softwoods and more recently of eucalypts – were established on large scales during the 20th century. These now total about 1.8 million ha in each country and now supply the majority – in the case of New Zealand, virtually all – of industrial wood. The plantation-based forest industries are among New Zealand's most important, representing the nation's third largest export earner and responsible for about 4 per cent of GDP (MAF, 2003); the Australian forestry sector, based on both plantations and native forests, is Australia's second largest manufacturing industry and represents about 1 per cent of GDP (Australian National Forest Inventory, 2007). In both countries, the majority public ownership of plantation forests which prevailed until the 1990s has been transformed, through both privatization of standing plantations and policies encouraging private investment in plantation expansion, to majority private ownership.

In both countries, social conflicts over the extent to which native forests should be managed for wood production have led to 'forest agreements' of various forms, principally the New Zealand Forest Accord (Ministry of Agriculture and Forestry, 2002b) and Australia's Regional Forest Agreements (Australian National Forest Inventory, 2003). Thus, wood production from New Zealand's public native forests ceased in 2002. There is no harvesting in public native forests in two Australian states,[2] the Australian Capital Territory and South Australia, and it is being phased out in southeast Queensland. While wood production from native forests remains significant in five of the eight Australian states, it now comprises only one-third of harvest volume nationally (ABARE, 2006) and its absolute and relative magnitudes are diminishing as native forests are reserved from harvesting and plantation forests and production expand. There have also been social tensions associated with the rapid expansion of plantations onto farmland in some regions of both countries (Schirmer and Kanowski, 2005; Schirmer and Roche, 2005) and over the conversion of native forest to plantations in the two Australian states in which it is still permitted, the Northern Territory and Tasmania[3] (WWF Australia, 2004; Environment Centre Northern Territory, 2007).

Both Australia and New Zealand are active participants in the international Montreal Process for sustainable forest management[4] and in other multilateral forums focused on forests. The criteria and indicators developed under the Montreal Process provide the basis for both countries' reporting of progress in SFM and for Australia's State of the Forest reporting (MAF, no date; Montreal Process Implementation Group, 2008).

Australia

An overview of forests and forest ownership

Australia's forest extent is 164 million ha, 19 per cent of the continent's land area; 1.8 million ha of these are plantation forests (FAO, 2007). Australia's native forests are globally distinctive; they are dominated by the genus *Eucalyptus* and its close relatives *Corymbia* and *Angophora* and a high proportion of Australian forest flora and fauna are endemic. Eucalypts and other native species have historically provided the majority of wood and fibre, but plantations of exotic softwoods have been important

Table 8.1 *Australian forest areas by tenure and jurisdiction (1000ha)*

State	Native forest							Plantation forest				
	Public lease-hold	Public nature conser-vation	Public produc-tion	Other public	Private*	Un-resolved	Total	Public	Private	Joint	Un-known	Total
ACT	8	108	0	7	0	0	123	10	0	0	0	10
NSW	9,891	5,148	1,980	943	8,076	170	26,208	245	82	5	13	345
NT	13,920	16	0	674	16,317	83	31,010	0	16	0	10	26
Qld	34,304	4,576	1,991	1,598	8,908	1,204	52,518	191	32	3	7	233
SA	3,083	4,029	0	277	1,399	67	8,855	87	80	0	5	172
Tas	0	1,121	1,026	85	855	0	3,116	27	142	58	21	248
Vic	35	3,505	3,163	109	1,025	0	7,837	3	374	7	12	396
WA	3,891	3,868	1,248	7,169	1,489	0	17,665	77	271	29	12	389
Total	65,132	22,369	9,410	10,862	38,099	1,524	147,397	641	998	101	80	1,818
%	44	15	6	7	26	1	100	35	55	6	4	100

Notes: * Includes forests to which Aboriginal Australians have formal title.
Forest and plantation area totals from Montreal Process Implementation Group, 2008, from 2007 data; 'public, private & joint' areas from National Plantations Inventory, 2006, from 2005 data; 'unknown' reflects increase 2005–2007. Total forest area figures differ from the FAO 'State of the World's Forests 2007' report cited elsewhere in this chapter and book.
Plantations 'joint' area includes 4000ha 'unknown' for Victoria
Sources: National Plantations Inventory, 2006; Montreal Process Implementation Group, 2008

for more than a century and now provide the majority of the country's domestic solid wood production (Montreal Process Implementation Group, 2008).

The largest single tenure category of native forest (44 per cent; 65 million ha) is that held by individuals or companies under long-term leasehold arrangements from the state designed to facilitate pastoral enterprises; the terms of the lease typically allow the lessee rights to graze animals and manage understorey vegetation, but preclude forest management or commercial harvesting of forest products without specific permission. Twenty-six per cent (38 million ha) is privately owned and managed, about half by Aboriginal Australians; 15 per cent (22 million ha) is formal conservation reserve; 6 per cent (9 million ha) is public land available for wood production, and 7 per cent (11 million ha) is in various other public tenures, the majority of which preclude wood production (see Table 8.1; Montreal Process Implementation Group, 2008).

For plantations, state governments, usually through various forms of state trading enterprise, are the largest single category of owner (37 per cent), managed investment schemes[5] own 23 per cent, forestry companies 15 per cent and super-annuation funds and smaller private owners around 13 per cent each; the distribution of plantation ownership varies greatly between states.

Forest areas by tenure and jurisdiction are summarized in Table 8.1.

Native forests

About 80 per cent of the Australian continent is too arid for tree growth. Within the remaining 20 per cent, there are eight broad forest types. Eucalypt woodland is the dominant of these, comprising two-thirds of native forest area; open forests, dominated by but not confined to eucalypts, comprise 30 per cent of native forest area; closed forests, including rainforests and tall wet eucalypt forests, comprise only 3 per cent (Montreal Process Implementation Group, 2008).

Native forest ownership varies between forest types: around three-quarters of woodlands are under leasehold or private management; the various open forest types are approximately evenly distributed between public and private owners in most states; most closed forest is under public ownership. As noted in the introduction, management objectives for public native forests have progressively – and completely in a number of states – shifted to conservation; nationally, 18 per cent of Australia's native forests are now formally protected (Montreal Process Implementation Group, 2008). Private native forests, although important for wood production in NSW, Queensland and Tasmania, have only recently been the focus of policy attention in the former two states.

Plantation forests

Australia's plantation forest estate comprises 1 million ha of softwoods, grown principally for sawnwood, and 0.8 million ha of eucalypts, almost all grown for pulpwood (Montreal Process Implementation Group, 2008). Plantation forests are significant in six states (see Table 8.1) and are the only forests harvested for wood production in two states: the ACT and South Australia. The longer-rotation softwood plantation estate is relatively mature and generally static in area; the shorter-rotation hardwood plantation estate is immature and has been expanding at an average rate of 70,000 ha annually over the past decade, due almost completely to managed investment schemes (Montreal Process Implementation Group, 2008).

Forest governance and policy

Constitutionally, Australia is a federation of eight states and territories,[6] which retained their rights and responsibilities for native resource management at the time of national federation in 1901. These arrangements continue to be reflected in a state-level basis for forest governance and management. Nevertheless, the Australian government has played a progressively increasing role in forest policy. The national government facilitated the substantial expansion of Australia's state-owned plantations in the 1960s and 1970s. It later concluded, with all state governments, the *National Forest Policy Statement* (1995), which established foundations for the principal policy initiatives focused on native and plantation forests: respectively; the Regional Forest Agreement process, conducted between 1996 and 2001, and the *Plantations for Australia – The 2020 Vision*, implemented from 1997, as well as initiatives focused on each of farm and indigenous forestry (DAFF, 2007a). The Environmental Protection and Biodiversity Conservation Act, which came into effect in 2000, is the principal national conservation legislation (Department of Environment, Water, Heritage ad the Arts, 2007a).

The Regional Forest Agreements, which focused on native forests, and the Plantations for Australia initiative have been significant in expanding the areas of forest in conservation reserves and plantations, respectively, over the past decade. The Regional Forest Agreements also strengthened sustainable forest management regimes in public forests that remained available for wood production and strengthened the recognition of cultural heritage and other forest values.

National and state governments also collaborate in delivering on Australia's international commitments, such as those under the Montreal Process[7] and in other policy arenas such as the development and adoption of the Australian Forest Certification System for forest certification and product labelling (Australian Forestry Standard (AFS), 2007a). The third tier of government, local councils, generally has only limited planning and consent authority over forestry activities, although this is nevertheless important in some states.

Forest practices systems

Notwithstanding these national initiatives and collaborations, Australian forest practices policies and systems remain strongly state-based. The forest-rich state of Tasmania, in which forestry is also of greatest relative economic importance amongst the Australian states, was the first to implement a Forest Practices Code, in 1985, and apply it to all forest types and tenures; other states have followed, to varying degrees. While forests on leasehold lands represent the majority of forest by area, commercial timber harvesting is usually precluded by the terms of the lease. Where harvesting occurs, such as for sandalwood in Western Australia, forest practices are regulated by codes applying to public land. Forests owned by Aboriginal communities are considered as private land.

All Australian states in which commercial harvesting of native forests is permitted have codes of forest practice applying to public and private tenures, with the exception of New South Wales and Western Australian private forests;[8] the extent of harvesting in the latter is trivial (see Table 8.2). All states other than Queensland also have codes applying to plantation forests, although those in South and Western Australia are voluntary. Tasmania and Victoria have codes that are common across both public and private tenures, and for native and plantation forests. The Northern Territory's, and the South and Western Australian, plantation codes are common to both public and private tenures, as is the Northern Territory code for native forests. New South Wales has a different code for each forest type and tenure category, as does Queensland for native forests.[9]

Formal operational or management plans for forest harvesting are required by forest practices systems in most Australian states. Most plans are required to specify proposed forest operations in sufficient detail to demonstrate their compliance with the relevant code. However, in the case of private native forests in Queensland, the landowner has only to provide formal notice of their intention to conduct 'forest practices' (Department of Natural Resources and Water, 2005). For South Australian plantations, the voluntary guidelines suggest only a checklist, although management plans are likely to be prepared by the larger-scale forest owners who collectively manage almost all of South Australia's plantation forests.

Sandalwood harvesting, which usually involves removing the tree's roots as well as its stem, is important in three Australian states. There is a separate code in

Western Australia, where sandalwood is harvested from public land, including that leased by pastoralists. In Queensland, sandalwood harvesting is governed specifically by the public native forest code, and non-specifically by the private native forest code. There is no specific regulation of its harvest in the Northern Territory.

Non-commercial harvesting on private land is generally exempt from the provisions of codes, although it may be subject to other native vegetation management regulations. Customary harvest by Aboriginal people from any native forests in the Northern Territory is exempt from the provisions of that Territory's code (Department of Infrastructure, Planning and Environment, 2003).

Forest production and trade

Industrial wood is the dominant product, by volume and value, harvested from Australia's forests. Total wood harvest (2005–2006) approximated 27 million m^3, the majority of which was from plantations (about 14 million m^3 from softwoods and 4 million m^3 from eucalypts), with a gross value of AUD$1.7 billion (Australian National Forest Inventory, 2007). Firewood harvesting, for both individual use and commercial sale, is locally important, but there is no consistent national data. A diversity of non-wood forest products – including apiary products, eucalyptus and sandalwood oil, native plants and cut flowers and bark and wood for Indigenous art products – are commercially important, but data on quantities harvested and value are limited. The estimated value of the three largest of these is AUD$49 million for apiary products, AUD$20 million for native plants and cut flowers and at least AUD$4 million for bark and wood for Indigenous art products (Montreal Process Implementation Group, 2008). Environmental services from forests are important but largely unpriced. In total, the forestry sector accounts for 1 per cent of Australia's GDP (Montreal Process Implementation Group, 2008).

Australian forest practice policies focus on industrial wood production, for reasons outlined above. Harvest volumes by state are summarized in Table 8.2. Data are available for both public and private native forest tenures, but are not segregated on this basis for plantations (Montreal Process Implementation Group, 2008). Harvesting of public native forests for wood production is significant in five states – NSW, Queensland, Tasmania, Victoria and Western Australia – and from private native forests in three of these: NSW, Queensland and Tasmania. Native forest harvesting does not take place in the ACT, is limited to firewood from dead trees on private land in South Australia and largely to low volumes of speciality products – including those for Indigenous art products – in the Northern Territory. Commercial wood production from native forests is focused on the taller open and closed eucalypt forest types and almost completely excluded from rainforests.

Plantation production is significant in all states but the ACT and Northern Territory (Table 8.2), although the latter have locally important plantation resources. New South Wales and Victoria are the largest producers, reflecting the scale of their established softwood plantations; short-rotation eucalypt plantations have been expanding in most Australian states, principally in South Australia, Tasmania, Victoria and Western Australia (Montreal Process Implementation Group, 2008).

Table 8.2 *Australian and New Zealand roundwood production volumes by tenure and jurisdiction (1000m³)*

	Native		Plantation	Total
	Public	Private	Public & Private	
ACT	No harvest	No harvest	1	1
%	–	–	100	0
NSW	1044	323	4377	5744
%	18	6	76	21
NT	Not significant	Not significant	Not significant	0
%	–	–	–	0
Qld	304	243	2139	2686
%	11	9	80	10
SA	No harvest	No harvest	2361	2361
%	–	–	100	9
Tas	2532	2108	1387	6026
%	42	35	23	22
Vic	2152	Not significant	4393	6545
%	33	–	67	24
WA	506	Not significant	3018	3614
%	14	–	86	13
Australia	6642	2724	18,011	26,977
	24	10	66	100
New Zealand	No harvest	26	18,728	18,754
	–	1	99	100

Sources:

- Australian roundwood totals: ABARE, 2006, Table 51 (rounded up for ACT); July 2005–June 2006;
- Australian native forests: three-year average 1998/99–2000/01, Table 48, National Forest Inventory 2003; except for NSW private, which is estimated from the most recent available three years;
- Australian plantations: estimated as difference between total and total native forest removals;
- New Zealand MAF, 2007a (data for July 2005–June 2006)

Nationally, plantation forests provide two-thirds, public native forests one-quarter and private native forests one-tenth, of Australia's industrial wood production (Table 8.2).

Australia exports around AUD$2.4 billion, and imports around AUD$4.3 billion, of forest products annually. A forest products trade deficit of AUD$1–2 billion annually has typified the Australian forest products sector for 20 years. The majority of exports by volume and value are woodchips, principally to Japan and China, and the majority of imports by value are paper and panel products, principally from China, Indonesia and New Zealand (Bhati, 2003; Montreal Process Implementation Group, 2008).

Indigenous and community forestry

Aboriginal Australians have formal title to 21 million ha (13 per cent) of forest nationally, most of which is woodland or open forest in the northern tropics and 98 per cent of which is in northern and western Australia, in the states of Queensland, the Northern Territory and Western Australia. Aboriginal-owned forests comprise nearly 50 per cent of forests in the Northern Territory; around 6 per cent in each of Queensland, South Australia and Western Australia and are of negligible extent elsewhere (Montreal Process Implementation Group, 2008), reflecting the history of European colonization of Australia and the nature of judicial and political decisions about Aboriginal land rights.

There is little industrial wood production from forests formally owned by Aboriginal Australians, reflecting – variously – those forests' limited value for industrial wood production, their inaccessibility and the choices of their owners. A notable exception is the plantation forests established on the Tiwi Islands, near Darwin, in partnership with forestry companies. The 2005 National Indigenous Forestry Strategy (DAFF, 2007a) seeks to expand Aboriginal Australians' engagement with the forestry sector, but there are a number of constraints to realizing this goal (Feary, 2007).

There has been only one initiative to introduce community forestry, in its internationally understood sense, to Australia, in Victoria's Wombat State Forest (Petheram et al, 2004). Opinions on the success of that initiative vary (e.g. Poynter, 2005). However, there are many initiatives nationally which involve members of the community in forest and landscape restoration and rehabilitation on a voluntary basis (e.g. Kanowski, 2006; Greening Australia, 2009).

New Zealand

An overview of forests and forest ownership

The flora and fauna of New Zealand's native forests are also globally distinctive and highly endemic, reflecting the country's long isolation from other land masses. The dominant indigenous tree species in these forests are the conifers kauri (*Agathis australis*) and rimu (*Dacridium cupressinum*) and the broadleaves beech (*Nothofagus* spp.), taraire (*Beilschmedia taraire*) and tawa (*Beilschmedia tawa*) (MAF, 2002b, 2007c).

New Zealand's remaining 6.5 million ha of native forests occupy around one-quarter of its land area (FAO, 2007). The entire public native forest estate, comprising 77 per cent of these forests, is now in conservation reserves. New Zealand's indigenous Maori people own one-third, and non-Maori private owners the balance, of native forests. More than 90 per cent of private native forests are managed primarily for conservation and other values, rather than for wood production (MAF, 2007b; New Zealand Government, 2007a).

Exotic plantations, principally of *Pinus radiata*, have been important in New Zealand for more than a century. They now total 1.8 million ha, corresponding to 7 per cent of New Zealand's land area (NZFOA, 2007). More than 90 per cent of New Zealand's plantations are now owned by businesses, both large and small scale, following privatization of almost all former state-owned plantations in the 1990s. Five large companies own about 40 per cent of the forests; the smaller-scale businesses include some Maori companies and around one-third of the resource is the property of small private owners. National and local governments own around 8 per cent (MAF, 2003; NZFOA, 2007).

Native forests

New Zealand's unique native forests are highly valued for their biodiversity and for non-consumptive values, as well as for their unique timbers. Many native fauna species are endangered or threatened (MAF, 2007b). All public native forests are formally managed for conservation, as are – informally – the majority of private native forests (New Zealand Government, 2007a). Sustainable Forest Management Plans and Permits, which allow wood production, have been issued for about 112,000ha (around 10 per cent) of private native forest, representing nearly 600 owners (MAF, 2007c), although few of these are for Maori-owned forests (Hammond, 2001). The native forest wood harvest of about 26,000m^3 represents only about 1 per cent of New Zealand's annual wood production (see Table 8.2).

Plantation forests

Ninety per cent of New Zealand's plantations are *Pinus radiata*. The plantation estate is distributed throughout the country, with the largest single concentration in central North Island. The majority of the plantation estate is less than 15 years of age, reflecting the substantial expansion of the 1990s (NZFOA, 2007). The land use is dynamic and other primary industries are strong; for example, around one-third of the 39,000ha of forests harvested in 2006 was then converted to agricultural use, compared to a historic average of less than 5 per cent (MAF, 2007a).

Forest governance and policy

New Zealand forest governance and policy are determined by the national government. Policies for the management of New Zealand's native forests are guided by the New Zealand Forest Accord (1991) between forest industries and environmental non-government organizations, which began the process of ending wood production from public native forests and restricting sites on which plantations could be established to protect remnant native forest (Schirmer and Roche, 2005; New Zealand Government, 2007b).

The primary responsibility for environmental policy and planning in New Zealand is devolved to the second and thirds tiers of government, regional and district councils, under the Resource Management Act 1991. The Act provides an integrated, outcomes-based framework for assessing and managing all resource-related activities, including forestry (New Zealand Ministry of Environment, 2006).

Forest practices system

Forest management activities in native forests must comply with the Standards and Guidelines for Sustainable Management of Indigenous Forests (MAF, 2002b); the only exception is 24,000ha of native forest vested in Maori owners under the 1906 South Island Landless Natives Act (SILNA) (Hammond, 2001; Shepherd, 2004). With this exception, the Forests Act requires the Ministry of Agriculture and Forestry to approve Forest Management Plans or Permits for all native production forests and to approve and monitor Annual Logging Plans. The Ministry is required to consult with the Department of Conservation and, where appropriate, the

Ministry of Māori Development (Te Puni Kokiri), prior to approving a Forest Management Plan or Permit.

Depending on the district plans established by district councils, a 'resource consent' under the Resource Management Act 1991 may be required for native or plantation forestry activities (MAF, 2002b; Schirmer and Roche, 2005). The New Zealand Forest Owners Association launched, in mid-2007 with the support of all forest industry actors, a voluntary New Zealand Environmental Code of Practice for Plantation Forestry (NZFOA, 2008).

Forest production and trade

New Zealand's annual wood harvest is around 19 million m^3 (see Table 8.2) and forest sector exports of NZ$3.1 billion represent 10 per cent of the total value of New Zealand exports (NZFOA, 2007). New Zealand's forest industries are now based almost entirely on plantation forests of exotic species. Around one-third of production is exported in log form, the majority to China, India, Japan and Korea. The major export destinations by value are Australia and Japan, followed by the USA, Korea and China. New Zealand imports of forest products are around half the value of exports, at NZ$1.5 billion, and originate principally from Australia and China (MAF, 2009).

Indigenous and community forestry

Maori people own around 400,000ha (6 per cent) of New Zealand's native forests and 238,000ha (13 per cent) of its plantation forests. The employment and income derived from these forests are important to the Maori, who are expected to assume ownership of further areas of forest – perhaps up to 40 per cent of New Zealand's plantation forests – under anticipated Treaty settlements (Miller et al, 2007).

The situation of community forestry in New Zealand is very similar to that described for Australia. There is no 'community forestry', in the internationally understood sense, but there are many initiatives which involve members of the community on a voluntary basis in forest restoration and rehabilitation (see, for example, New Zealand Landcare Trust, 2009).

BIODIVERSITY CONSERVATION

In Australia and New Zealand, biodiversity conservation is now a dominant theme in native forest policy and practices. Commitments to conserve biodiversity have led to major policy decisions and processes in both countries. In New Zealand's case, the most significant of these was the exclusion of wood production from all public native forest, representing 77 per cent of its remaining native forest. Australia has established and subsequently expanded a 'comprehensive, adequate and representative' conservation reserve system in all forested regions included in the Regional Forest Agreement process, resulting in a national total of 18 per cent of remaining native forests conserved in formal reserves. In New Zealand and most Australian states, there has been a cessation of conversion of native forests to plantations. Both countries have also increased regulation of forest practices on private property. In the Australian case, each of these measures is part of the biodiversity conservation strategy envisaged in the National Forest Policy Statement, which advocated ecologically sustainable forest management of a permanent native forest estate comprising protected areas, public multiple-use forests and private forests (Commonwealth of Australia, 1995).

The analytical framework adopted in this study reviews biodiversity conservation measures in terms of those directed at protection of species at risk, including those focused on their habitat, and the existence of protected areas reserved from wood production. Both Australia and New Zealand have also adopted a range of other measures for biodiversity conservation, outlined in their respective national strategies (Australia: Department of the Environment, Sport and Territories, 1996; New Zealand: New Zealand Government, 2000). In Australia, all states have also developed and implemented formal state biodiversity strategies, or the equivalents, or are in the process (Northern Territory and Western Australia) of doing so.

Protected areas

Protected forest areas are fundamental to both Australia's and New Zealand's approaches to biodiversity conservation. Both countries have made substantial additions to their protected area systems since 1990; in Australia, the extent of forest in the protected areas has doubled since 1990, as a result of the Regional Forest Agreement and precursor, and successor, processes; in New Zealand, the increase was associated principally with the decisions made under and subsequent to the New Zealand Forest Accord 1991 (New Zealand Government, 2007b), to transfer all public production forests to conservation reserve status.

The Australian Regional Forest Agreement process had an explicit goal of establishing a 'comprehensive, adequate and representative' forest protected area system, in which representative samples of 15 per cent of the estimated pre-1750 extent of each forested ecosystem would be conserved in each bioregion (DAFF, 2007b). The Regional Forest Agreement process focused on public land; subsequent state-based processes, such as Tasmania's Private Property Conservation Program (Department of Primary Industries and Water, 2007), have sought to address the resultant gaps by establishing formal conservation agreements with private forest owners. Progress towards these goals is summarized in Australia's State of Environment (Department of Environment, Water, Heritage and the Arts, 2007d) and State of the Forests (Montreal Process Implementation Group, 2008) reports. New Zealand's Biodiversity Strategy (New Zealand Government, 2000) identified the need to increase both the extent and representativeness of New Zealand's protected forest areas, noting that – given the distribution of land and forest ownership – much of this would need to be realized with 'the active assistance of willing landowners'. It noted, for example, that although most (at the time; now all) public forests were conservation reserves, only 16 per cent of protected forest areas were of lowland forest types, despite these comprising 50 per cent of New Zealand's original forests.

Protected area representation

According to UNEP-WCMC data, in 2008 Australia had designated 9.4 per cent of its total land area as protected areas, including 6.6 per cent under the stricter IUCN Categories I–IV (Chapter 2, Figure 2.15). Forestlands are particularly well represented within this larger protected areas network.

The extent of Australian forest protected areas, by IUCN category and state, is summarized in Table 8.3;[10] summary data are also available by forest type (Australian National Forest Inventory, 2008, Table 14). The proportion of major forest ecosystem type conserved nationally varies from 5 per cent, for *Acacia* forests, to 55 per cent, for rainforest. The proportion of forest in each state in protected areas ranges from 9 per cent, for Queensland, to 91 per cent for the Australian Capital Territory.

In New Zealand, 2008 UNEP-WCMC data suggests that 26.2 per cent of the total land area is in protected status, with 20.6 per cent under the stricter IUCN Categories I–IV (Chapter 2, Figure 2.15). These figures are notably high relative to the other case study countries, but somewhat lower than those reported by the Ministry of the Environment New Zealand (2007a), of 32 per cent (8.6 million ha) under legal protection, including 0.2 million ha of private land. According to this latter source, 37 per cent of 'broadleaved native hardwoods' and 80 per cent of 'native forest' (in other words, that dominated by coniferous taxa) are protected (Ministry of the Environment, 2007a, Table 12.4). The 4.8 million ha of public native forests in protected areas represent 77 per cent of remaining native forest and 18 per cent of New Zealand's land area. Although about 90 per cent of New Zealand's 1.4 million ha of private native forests is not used for wood production (New Zealand Government, 2007b), little of this area is managed under formal conservation agreements (New Zealand Government, 2007a).

Protection of species at risk

Australia's Environment Protection and Biodiversity Conservation Act 1999 and corresponding state legislation (Department of Environment, Water, Heritage and the Arts, 2007b), and a suite of New Zealand legislation, including the Conservation Act 1987, Native Plants Protection Act 1934, Wildlife Act 1953 and Resource Management Act 1991 (Ministry of Environment, 2007b), oversee the protection of species at risk in each country. In each country, species and ecosystems meeting nominated criteria (variously defined as, for example, 'threatened' or 'at risk') are required to be identified and protected and species- or ecosystem-specific conservation strategies (variously described as, for example, 'threat abatement plans' or 'recovery plans') developed and implemented. Historically, these measures have been most advanced on public land, but achieving biodiversity conservation on private land is now also an issue of policy and programme focus in both countries (Department of Environment, Water, Heritage and the Arts, 2007c; New Zealand Government, 2007a).

Public native forests

In those regions of Australia for which Regional Forest Agreements were concluded, the species and habitat protection provisions agreed as part of the Regional Forest

Table 8.3 *Extent of Australia's native forest in IUCN protected area categories by jurisdiction (units in 1000ha)*

	Ia	Ib	II	III	IV	V	VI*	Forest in IUCN categories I–IV	Forest in IUCN categories I–VI	Total native forest	% of forest in IUCN categories I–VI
ACT	–	28	84	–	–	–	–	112	112	123	91
NSW	672	1,636	2,184	–	16	7	215	4,506	4,730	26,208	18
NT	13	–	377	1	–	148	896	392	4,536	30,927	15
Q'land	36	–	4,114	45	9	–	657	4,204	4,861	52,582	9
SA	1,194	1,306	340	111	103	41	1,059	3,054	4,155	8,855	47
Tasmania	14	–	604	12	23	46	292	653	991	3,116	32
Victoria	387	781	2,187	49	30	30	66	3,434	3,531	7,838	45
WA	1,913	–	1,636	–	1	2	46	3,550	3,598	17,664	20
Total	4,229	3,752	14,626	218	182	274	3,232	23,005	26,514	147,311	18
IUCN areas as % of total forest area	3	3	10	0.1	0.1	0.2	0.2	16	18		

Note: * Multiple-use public forest could be classified under IUCN category VI; however the Collaborative Australian Protected Areas Database, which provides estimates of forest areas in IUCN categories, does not do so if the multiple-use public forest is not principally managed for the conservation of biodiversity (see Dudley and Phillips, 2006). Areas of forest in IUCN categories calculated using the Collaborative Protected Areas Database for IUCN data, except for Tasmania and Victoria, where state-supplied data were used.
Source: Reproduced from Montreal Process Implementation Group, 2008, Table 13

Agreement have been deemed by the Australian government to satisfy the requirements of the Environment Protection and Biodiversity Conservation Act 1999 and forestry operations consistent with the Regional Forest Agreement do not require specific approval under the Act (Montreal Process Implementation Group, 2008). The specific implementation of species and habitat protection arrangements varies from state to state: for example, in New South Wales, the National Parks and Wildlife Service is required to assess forestry operations that may impact adversely on a threatened 'species, population or ecological community', determine whether the forestry operations should proceed, and – if so – issue specific licence conditions governing those operations (Government of NSW, 1999); in Tasmania, a management prescription for 'threatened species or inadequately reserved plant communities' must be agreed between the Forest Practices Authority and the state's conservation agency and incorporated into the Forest Practices Plan (Forest Practices Board, 2000) after consultation with relevant experts. Equivalent arrangements apply in Victoria and Western Australia.

Although the prerequisite assessments were completed, no Regional Forest Agreement was ultimately concluded for Queensland; instead, the South-East Queensland Forest Agreement was agreed by the state government and forest sector stakeholders (EPA (Queensland), 2007). Consequently, the provisions of the Queensland Nature Conservation Act 1992, as well as those of the Australian EPBC Act

1999, apply throughout Queensland; these are effectively identical to those implemented in RFA regions. Wood production from the Northern Territory's native forests was not of sufficient magnitude to require development of an RFA; here, in addition to the provisions of the Australian EPBC Act 1999, 'no harvesting will be permitted in areas where threatened species are known to occur unless there is evidence that harvesting will promote the development of suitable conditions for the threatened species' (Department of Infrastructure, Planning and Environment, 2003).

No harvesting of forest products is permitted in the public native forests of the Australian Capital Territory, South Australia or New Zealand.

Private native forests
In the Australian states, the Environment Protection and Biodiversity Conservation Act 1999 and the state Acts relevant to species and habitats at risk apply to forestry operations on private land. In all states other than the Northern Territory and Tasmania, there are no specific additional forest practice requirements relating to species and habitats at risk; in the Northern Territory and Tasmania, these are specifically addressed, in terms identical to the approach taken on public land, as discussed above.

In New Zealand, specific measures for the protection of threatened flora and fauna, based on advice from the Department of Conservation, are required to be incorporated in the Sustainable Forest Management Plan which must be prepared and approved for each private production forest (MAF, 2002b).

FOREST PRACTICE POLICIES: NATIVE FORESTS

Riparian zone management

Riparian zone management is a strong element of codes of practice for native forest management in all Australasian jurisdictions in which forest practices occur.[11] In all Australian states, no-harvest zones (usually described as buffer zones) of specified minimum width are required adjacent to streams or other water bodies. The specified width varies with the size of the stream, usually assessed by the nature of the watercourse or water body (e.g. perennial or intermittent stream; drainage line or gully; wetland; drinking water supply area) and, for streams, by stream order.[12] Additional special management zones, usually described as filter zones, are specified in Queensland, Tasmania and Victoria. In the Northern Territory, Queensland, Tasmania and Victoria, soil properties such as erodability or permeability are also used to determine the minimum widths of buffer and/or filter zones.

In New Zealand's native forests, 'adequate riparian protection zones in keeping with the terrain, soil stability and proposed management systems' and 'consistent with rules in Regional Plans' are required (MAF, 2002b), but their specific dimensions are not prescribed.

In addition to precluding the harvest of trees in no-harvest (buffer) zones, all Australasian codes but the Northern Territory's specifically preclude machinery entering these zones, except at designated stream crossings, and generally place conditions on the felling of trees into the zones. The sole exception is the case of sandalwood harvesting on public land in Queensland, in which the harvesting of sandalwood stems is permitted throughout the buffer zone provided certain conditions relating to erosion and residual trees are met. Sandalwood roots may also be harvested in buffer zones, but not within two metres of the defining stream bank (EPA (Queensland), 2002). Where additional special management (filter) zones are required (in the Northern Territory, Queensland, Tasmania and Victoria) a number of conditions apply:

- machinery access is usually precluded or heavily restricted;
- harvesting is restricted – either in terms of the proportion of canopy removal allowed (limited to 20 per cent in the Northern Territory and 30 per cent in Tasmania), and/or there are restrictions on how harvesting is conducted: for example, trees can only be felled so they do not fall into no-harvest zones in Victoria, or in Queensland if this can be done 'with minimal disturbance' to the zone (EPA (Queensland), 2002); and
- logging debris is required to be removed or dispersed from no-harvest and special management zones, to minimize fire risk.

No-harvest and special management zone dimensions

Figures 8.1 and 8.2 compare the specified no-harvest (buffer) and special management (filter) widths for all Australasian jurisdictions. The widest buffers, 200m wide either side of the stream, are those required for Western Australian rivers.[13] In eastern Australia, combined buffer and filter widths of around 50m are typical for major streams. Riparian protection zone widths and management restrictions are specifically increased for drinking water supply areas in Tasmania and Western Australia.

Among those six jurisdictions where native forest harvesting occurs on both public and private land, riparian protection measures are identical across tenures in the Northern Territory, Tasmania and Victoria; are differently specified but largely equivalent in Queensland; and – at the census date – did not apply to private land in NSW.[14]

Road stream crossings and road decommissioning

Australia is geologically stable and its landscape is ancient, heavily weathered and relatively flat. It is the world's driest inhabited continent, with the most variable rainfall and streamflow of any continent. Due to this highly variable rainfall, many Australian streams are ephemeral (Wasson et al, 1996). Whilst Australia's forested catchments support the majority of the continent's perennial streams, they also include many watercourses that flow only intermittently, as a consequence of either sustained rainfall or storm events. Forest practices relating to stream crossings have

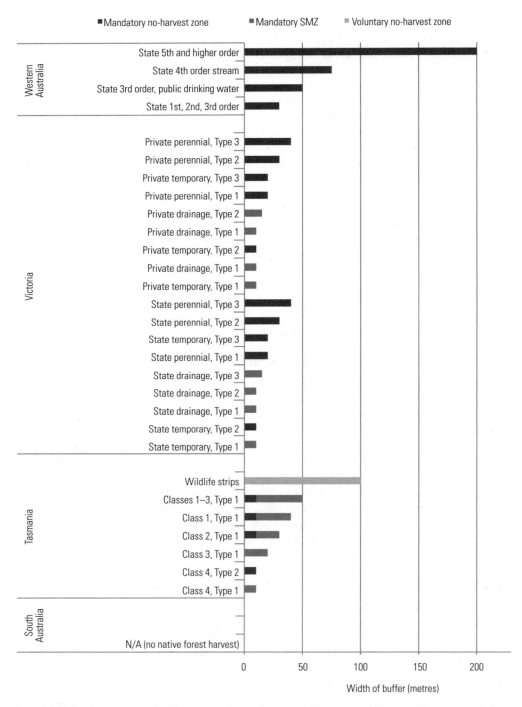

Figure 8.1 *Riparian buffer zone policies for Australia (SA, Tasmania, Victoria, WA)*

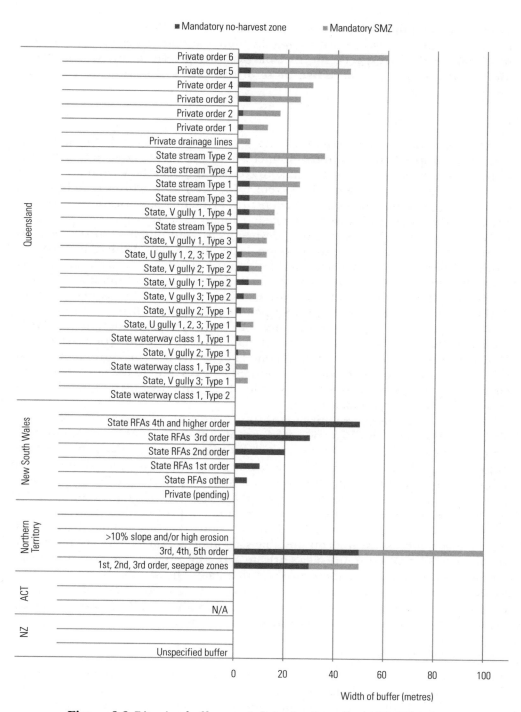

Figure 8.2 *Riparian buffer zone policies for Australia (ACT, NT, NSW,*
Queensland) and New Zealand

to accommodate these characteristics. In contrast, New Zealand is geologically a young and active landscape, with generally higher and more regular rainfall than Australia. In these respects, the hydrological characteristics of New Zealand's forested landscapes are more like those of the Pacific Northwest of the North American continent.

Road stream crossings

Codes of practice for public native forests in NSW, Queensland, Tasmania and Victoria each specify minimum requirements for road stream crossings. Structures in NSW and Queensland must be able to contain the peak flow from a 1:5 storm event and withstand that from a 1:10 event.[15] In Victoria, structures must be able to contain the peak flow from a 1:10 event. In Tasmania, specifications vary with stream classification, with structures able to contain 1:50, 1:20 and 1:10 peak flows for class 1, 2, and 3 or 4 streams, respectively. Additional requirements are common; for example, there are additional specifications for bridges in NSW and Tasmania, and for particular slope and erosion risk classes in Tasmania. In Western Australia, the *Harvesting Contractors' Manual* (Forest Products Commission, 2004) notes that 'Road drains, culverts and soaks are designed to maintain natural drainage patterns and serve to prevent ponding of water, scouring of surfaces and increasing turbidity loading in streams', but does not specify design standards. Similarly, the Northern Territory Management Program (Department of Infrastructure, Planning and Environment, 2003) states that 'measures to minimize soil erosion will be required both in areas used to access and to harvest timber. This covers such things as proper construction of stream crossings ...', but does not nominate specific requirements.

Road stream crossing requirements in private native forests in the Northern Territory, Tasmania and Victoria are identical to those for public forests in those states. At the census date, there were neither code nor road crossing specifications applying to NSW private native forests. The requirements specified in the code subsequently introduced mirror those in the state's public native forests. Those in Queensland are less specific than their public forest counterpart, requiring that 'streams, drainage lines and adjacent vegetation are protected ... [that] crossings over streams... must be at least 200m from any other crossing ... [and that] a crossing ... does not cross through an unstable section of stream'. There are no specific requirements for Western Australia's private native forests, but this is consistent with the lack of production from these forests in that state. The Standards and Guidelines for New Zealand's private native forests (MAF, 2002b) require 'adequate riparian protection zones in keeping with the terrain, soil stability and proposed management systems ... consistent with rules in Regional Plans'.

Road decommissioning

All Australian public native forest codes, other than the Northern Territory's, specifically require decommissioning of roads no longer required for forest operations. The specificity with which decommissioning requirements are stated varies: for example, NSW regulations state simply, 'A road or fire trail must be closed, and the relevant land rehabilitated, as soon as practicable after it is no longer required for the carrying out of forestry operations', although a specific 'road and fire trail management plan' is also required (Government of NSW, 1999). The Tasmanian requirements (Forest Practices Board, 2000) are typical of those which are more specific: 'Roads of no further use will be outsloped, water barred, or otherwise left in a condition to minimize erosion, with clean drains and blocked to vehicular traffic. In some situations it may be desirable to recover existing gravel pavements and rehabilitate the road by ripping and sowing a suitable local native species seed mix'.

The only Australian state in which decommissioning of roads following private native forest operations was required on the census date is Tasmania, where the specification is identical to that for public forests. The Victorian code stated that roads no longer required should be decommissioned. There were no decommissioning requirements for private native forests in other Australian states at the census date. The codes subsequently introduced (1 August 2007) for NSW and Victorian private native forests require road decommissioning (Department of Environment and Climate Change, 2007; Department of Sustainability and Environment, 2007, respectively). Road decommissioning is not specifically required for New Zealand private native forests.

Clearcut size limits and cutting rules

Clearcutting has been the preferred silvicultural system in the tall, wet eucalypt forests of Tasmania, Victoria and Western Australia since foresters observed, in the first half of the 20th century, that the dominant eucalypt species in these forests did not regenerate well under more selective harvesting regimes (Florence, 1996). For similar ecological reasons, clearcutting was the preferred system for harvesting and regenerating light-demanding species in New Zealand's indigenous forests, such as beech (*Nothofagus* spp.) and rewarewa (*Knightia excelsa*). When large-scale harvesting of forests for pulpwood began in some southern Australian forests in the 1970s, clearcutting was often used as the preferred silvicultural system because it maximized the harvested volume and created good conditions for subsequent regeneration.

As awareness of the adverse impacts of clearcutting on other forest values, particularly hollow-dependent fauna and species associated with old forests, and associated public opposition to the practice grew, limits to clearcut coupe sizes were introduced and alternative silvicultural regimes were explored and implemented to varying degrees (see for example Forestry Tasmania, 2005; Lindenmayer and Franklin, 2005). Other measures have also been introduced to mitigate impacts. Tasmania requires the retention of wildlife habitat strips of 100m width every 3–5km, based on streamside reserves but also linking up slopes and across ridges (FPA, 2000). 'Clearcutting' systems which retain seed and habitat trees within coupes are already operating in Western Australian karri forests (Department of Conservation and Land Management, 2005) and hollow-bearing habitat trees are generally retained in Victorian coupes (Blackburn, 2008).

In Victoria and Western Australia, clearcutting is restricted to specific 'wet eucalypt' forest types, namely the 'mountain ash' and 'mixed species' forests of Victoria (Department of Sustainability and Environment, 2005) and the karri-dominated forests of Western Australia (Department of Conservation and Land Management, 2005). In Tasmania, clearfelling is used in a wider range of forest types, depending on existing and intended stand characteristics (FPA, 2000, Table 9).

The ecological characteristics of other commercially important forests in Australia and New Zealand – principally the mixed species and drier eucalypt forests of Australia and the kauri and podocarp forests of New Zealand – meant that clearcutting was never the preferred silvicultural system in these forests, although it was occasionally tried experimentally on a small scale.

Coupe size

Maximum clearcut block (usually referred to as 'coupe' in Australasia) sizes are presented in Figure 8.3. The largest, of 100ha on slopes less than 20°, is in Tasmania and is twice that allowed on steeper slopes in that state. In Victoria, and in mature karri (*Eucalyptus diversicolor*) forests of Western Australia, the maximum is 40ha. However, forest policy in Western Australia now limits harvesting to regrowth karri forests (Conservation Commission of Western Australia, 2004), with a maximum coupe size of 20ha. Twenty hectares is also the maximum allowed, in exceptional cases, in New Zealand's indigenous beech forests, where the maximum coupe size is usually restricted to 0.5ha (MAF, 2002b).

Cutting rules

The Victorian code allows a maximum aggregation of coupes 'up to 120ha over five years'. The Tasmanian code requires that, 'where practicable', adjacent coupes are not harvested until regenerating coupes are satisfactorily stocked and the new stand has reached a dominant height of five metres. The Western Australian silvicultural practice guidelines (Department of Conservation and Land Management, 2005) require a separation between neighbouring coupes of either 'uncut forest, such as a stream reserve', or that the neighbouring coupe 'has at least been regenerated (i.e. a separation of about 3 years)'. In New Zealand private native forests, 'regeneration on the harvested coupe must have reached a predominant mean height of 4 metres and have reached a stocking of the harvested species equal to or greater than the forest before any further harvesting can take place within a coupe's width of the harvested coupe' (MAF, 2002b).

Other silvicultural systems

Clearcutting as a forest management practice is not allowed in the other Australian states where native forests are harvested – NSW, the Northern Territory, and Queensland – or in forest types other than those specified in Victoria and Western Australia. At the census date for this study, this policy was reflected in the content of codes of forest practice and supporting documents other than for NSW private

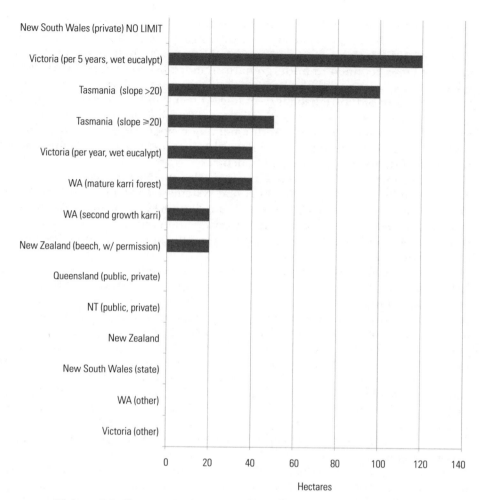

Figure 8.3 *Clearcut size limits in New Zealand and Australia*

forests, where it was effected through regulations governing native vegetation management in general (Said, 2004). In each of these codes, there are specific requirements for silvicultural systems, most of which are based on either small group or single tree selection and designed to maintain the uneven-aged structure of these forests as well as other forest values. Thus, requirements for habitat tree retention, and for aesthetics, apply in addition to silvicultural system specifications.

The retention of habitat trees is a specific requirement in all Australasian jurisdictions[16] because of their importance to native fauna. Typically, requirements are specified in terms of a minimum number of retained trees with particular characteristics per unit area. Specifications range from simple (e.g. for the Northern Territory a minimum of five trees per hectare which have hollows with an entrance dimension of greater than 5cm width, with no harvesting allowed if there are fewer than six such trees per hectare (Department of Infrastructure, Planning and Environment, 2003); for New Zealand, 'A proportion of old trees with high habitat values shall be identified and retained to undergo natural mortality processes' (MAF, 2002b)) to specifications which vary with forest type and structure and which include provisions for recruitment of future habitat trees (e.g. Queensland: EPA, 2002, Department of Natural Resources and Water, 2005; Western Australia: Department of Conservation and Land Management, 2004). Additional requirements apply in most jurisdictions for trees identified as feed trees for arboreal mammals, as active nest sites and for the retention of stand composition and structure.

Small group and single tree selection is also the basis of harvesting allowed in New Zealand's native forests. Shelterwood systems are used in some southern New South Wales, Tasmanian and Western Australian forests (Forests NSW 2008; Forest Practices Board, 2000; Department of Conservation and Land Management, 2004, respectively) and have also been used on a small scale in some New Zealand beech forests (MAF, 2002b).

In NSW public forests, other than those in the Eden region, these requirements are specified in terms of either the maximum gap size (ranging from 0.13 to 0.79ha) or the maximum basal area removed (40–45 per cent in northern regions) or retained ($10m^2$ per hectare in the southern region) and the proportion of a coupe that can be harvested (no more than 22.5 per cent). In the Eden region in the southeast of NSW, where both fire history and previous forest management have created more even-aged stands, a modified shelterwood system – which typically retains 10–25 per cent of the canopy – is practised (Forests NSW, 2008) and adjacent coupes cannot be harvested at intervals of less than five years (Forests NSW, 2007a). The private native forestry codes introduced in NSW on 1 August 2007 (Department of Environment and Climate Change NSW, 2007) also specify minimum basal areas to be retained, and – with the exception of western NSW forests – limit the extent of canopy opening to no more than 20 per cent of the net harvestable area.

In the Northern Territory, a minimum diameter limit applies to harvesting on both public and private land (25cm for species in the genera *Corymbia*, *Erythrophleum* and *Eucalyptus*; 20cm for other species) and returning to a stand that has been harvested is not normally permitted for at least 20 years (Department of Infrastructure, Planning and Environment, 2003).

In Queensland's public forests outside the area defined by the SouthEast Queensland Forest Agreement (EPA (Queensland), 2007), single tree selection is preferred, but more intensive regimes are permitted, provided that biodiversity and habitat requirements are met and if they are 'based on sound ecological principles ... and maintain or improve the long-term productive capacity and area of the forest' (EPA (Queensland), 2002). In the public forests governed by the SouthEast Queensland Forest Agreement, a once-only harvest of merchantable trees greater than 40cm, on a single tree selection basis, generally removing about 50 per cent of basal area or canopy cover, is permitted prior to the transfer of these public production forests to conservation reserve status (EPA (Queensland), 2002). In Queensland's private native forests, group selection creating gaps of less than 0.5ha is permitted in wet or moist coastal forests; only single tree selection is permitted in other forest types (Department of Natural Resources and Water, 2005).

For the two general forest types subject only to selection harvesting in Victoria, 'box-ironbark' and 'riverine', only single tree selection is permitted in the former; group selection around retained seed trees over areas of 2–5ha is permitted in the latter, where the existing stand is even-aged (Department of Sustainability and Environment, 2005).

In Western Australia's jarrah (*Eucalyptus marginata*) forests and mixed jarrah and marri (*Corymbia calophylla*) forests, a variety of selective harvesting regimes are employed, depending on stand history, characteristics and location (Department of Conservation and Land Management, 2004, Appendices 1 and 2). For each of these, specific silvicultural regimes are defined by various combinations of gap size, basal area retention and diameter limit.

In New Zealand, for all but specified light-demanding species, harvesting 'is confined to single trees or small groups of three to five trees' and 'stand composition and structure shall as far as possible be maintained consistent with unmanaged forest' (MAF, 2002b).

Reforestation

The single tree and small group selection silvicultural systems practised in most Australasian native forests (see above) rely on natural regeneration as the reforestation method. Consequently, reforestation requirements for these systems are commonly expressed in terms of restoring the stand to either or both the pre-harvest or an ecologically 'natural' condition. In New South Wales public forests, for example, the forest management agency 'must ensure' that regeneration of the overstorey delivers species post-harvest that are 'the same' as, and in relative proportions 'similar' to, those present prior to harvesting (Government of NSW, 1999). In Queensland's public native forests, 'timber harvesting is conditional upon effective regeneration ... in terms of species mix, stocking and site productivity... Regeneration is considered adequate when species, species mix and density approximate the characteristics of a healthy growing stand' (EPA (Queensland)), 2002). In Queensland's private native forests, 'floristic composition and stand structure will be restored within 20 years' (Department of Natural Resources and Water, 2005). In New Zealand's private native forests, the forest management standard requires that 'forest modified by logging or other practices shall be managed so as to enable forest composition and structure to return to a near natural state over time' (MAF, 2002b).

In some cases, such as for some New South Wales public forests or Western Australian jarrah forests, or New Zealand's private native forests, planting or artificial seeding is required if natural regeneration is determined to be inadequate. The clearcut-based systems in all jurisdictions typically rely on artificial reseeding rather than on replanting, although natural seedfall from retained seed trees is preferred in some cases, such as in some Western Australian karri forests or in New Zealand's beech forests. In all clearcut-based systems, minimum stocking levels are specified: for example, in New Zealand beech forests, it is 'no less than about 500 sph [stems per hectare]'[17] after five years (MAF, 2002b); in Tasmania and Victoria, different stocking standards are specified for different forest types and silvicultural systems (Forest Practices Board, 2000; Department of Sustainability and Environment, 2005); in Western Australian karri forests, a stocking rate of 1666 species per hectare is required on 75 per cent of areas within 18 months and on 100 per cent of areas within 30 months of harvesting (Conservation Commission of Western Australia, 2004).

Codes of forest practice in most Australasian jurisdictions also specify the period within which regeneration standards must be achieved. The exceptions were for Queensland's public native forests, where no specific time period was nominated, and for New South Wales and Western Australian private forests, where no code existed at the census date. New South Wales' new private native forests codes, introduced on 1 August 2007, specify both stocking levels and time periods for assessment of regeneration (Department of Environment and Climate Change, 2007). As discussed above, there is almost no commercial wood harvested from Western Australia's private forests.

Annual allowable cut

At the census date for this study, the volume that could be harvested from public native forests available for wood production was regulated in all Australasian jurisdictions, typically on a regional or statewide basis. Harvest from private native forests was also regulated at stand level in some jurisdictions. As the wood production objectives of Australian public native forest management focus on solid wood (in other words, pulpwood is seen as a byproduct), annual allowable cut regulations typically apply only to sawlogs or equivalent products (Montreal Process Implementation Group, 2008). In those Australian states in which wood production from public native forests is significant (see Table 8.2), annual allowable cuts have generally diminished over the past decade as a result of the Regional Forest Agreement process and, in some states, subsequent forest allocation and planning processes and may diminish further in subsequent allocation periods. The area of public forest available for wood production decreased by around one-third as a result of the RFA process (Bureau of Rural Sciences, 2008) and greater restrictions were placed on the harvesting of many remaining production forests.

The most common form of annual allowable cut regulation for Australian public native forests is that applied in New South Wales, Queensland, Tasmania, Victoria and Western Australia. In these states, a maximum allowable cut is specified for a given region and/or species group over a nominated period, typically 5–15 years. This allowable cut is based on the predicted sustained yield of various categories of wood product for forests in the particular region, estimated over a long-term planning period (e.g. 90 years, Forestry Tasmania, 2007c) and making allowances for various risk and uncertainty factors, such as fire. In some cases (e.g. the Tumut sub-region of southern NSW; Forests NSW, 2007b), the annual allowable cut for the nominated period is that which is predicted to be available over the long term and so is not expected to decline in subsequent allocation periods. In others, the harvest level for a particular allocation period may be set above or below that predicted for the long term, as a consequence of strategic planning decisions (e.g. Tasmania's Integrated Forestry Strategy; Forestry Tasmania, 2007c) or improved inventory information (Australian National Forest Inventory, 2003). Typically, a level of fluctuation of the harvest level is permitted over a specified period (e.g. by 25 per cent in any year but no more than 5 per cent over a five-year period for NSW, or over a three-year period for Western Australia; Australian National Forest Inventory, 2003; Montreal Process Implementation Group, 2008), so long as allowable cut is not exceeded on average over the specified period.

There are three exceptions to this generality. The first is the forests of the Southeast Queensland Forest Agreement region, which – as described above – are subject to a once-only harvest prior to transfer to conservation reserve status. The second exception is Tasmania, where the Forestry Act 1920 stipulates that Tasmania's public forests must be managed to provide an annual minimum of $300,000m^3$ of high-quality eucalypt sawlogs. Following the 1997 Regional Forest Agreement, annual production was raised to $350,000m^3$ for a ten-year period from 2001, to 'make suitable land available to enable plantation establishment as part of the Forestry Growth Plan' (Forestry Tasmania, 2002). Projections were reviewed in 2007 to take account of subsequent policy changes with respect to old-growth forest harvesting and natural forest conversion and to confirm that the statutory annual minimum sawlog supply could be sustained (Forestry Tasmania, 2007c). The third exception is the Northern Territory, which has very low levels of harvesting from public native forests and for which the knowledge base necessary to implement sustained yield principles is limited (Department of Infrastructure, Planning and Environment, 2003). Consequently, on both public and private land in the Northern Territory, sustained yield principles are currently implemented through the imposition of the selective harvesting rules described above and the requirement for a minimum 20-year return interval to any previously harvested stand. No annual allowable cut is specified.

Given their regional or statewide basis, annual allowable cut specifications of the form typical of public forests are, by definition, not relevant to small-scale private forest owners such as those who manage most of Australasia's private native forests. In the case of these smaller-scale forests, the level of harvest in relation to the productive capacity of forests at the property level is the closest approximation of annual allowable cut. This concept is embodied in regulations governing private native forest management in a number of Australasian jurisdictions. In the Northern Territory, it is as described above. In Queensland it is required that 'there is a demonstrated high probability that within 20 years the area [harvested] ... will be capable of being mapped as remnant vegetation'[18] (Department of Natural Resources and Water, 2005). In New Zealand it is required that 'harvests do not exceed rates of species/stand replacement' and this requirement is implemented – over the ten-year life of any particular Sustainable Forest Management Permit – by specifying the maximum volume (either $250m^3$ or $500m^3$ in total, according to species groups) and the maximum proportion of standing timber (10 per cent) that can be harvested, and by not allowing any subsequent harvesting until the regulatory authority is satisfied that 'the quantity of indigenous timber standing in the area to which the permit shall apply is at least equivalent to the quantity standing in the area at the date of the grant of the previous permit' (MAF, 2002b).

FOREST PRACTICE POLICIES: PLANTATIONS

As discussed above, plantations currently supply 66 per cent of wood harvested in Australia and 99 per cent of that harvested in New Zealand. Plantations are the only forests harvested commercially in two Australian states, the ACT and South Australia, and provide the majority of wood in four others – New South Wales, Queensland, Victoria and Western Australia (see Table 8.2). Whereas the plantation area in New Zealand is now relatively stable, after significant expansion during the 1990s (MAF, 2003), that in Australia is continuing to increase as a result of a national policy and related taxation arrangements, promoting plantation expansion (Montreal Process Implementation Group, 2008). For these reasons, forest practices associated with plantations are significant in both countries.

At the 1 January 2007 census date for this study, codes governing plantation forestry practices were established in all Australian states other than Queensland. Those in South and Western Australia were voluntary. A voluntary code has subsequently been introduced in New Zealand. The codes that govern plantation practices are different from those governing native forest practices in all Australasian jurisdictions other than Tasmania and Victoria.

Australasian codes of forest practice for plantation forests generally require forest management unit level measures for biodiversity conservation, in addition to those relating to species and habitats at risk. Other than in exceptional cases discussed below, these measures generally require retention of native vegetation in the development of plantation estates and preclude the clearing of remnant native vegetation for plantation establishment, except on a very small scale (e.g. less than 1ha, NSW, Government of NSW, 2001; less than 5ha, New Zealand, NZFOA, 2001) or where explicitly approved (e.g. Western Australia, Forest Industries Federation, 2006).

In Tasmania (FPA, 2000), riparian zone protection in plantations is equivalent to the corresponding requirements for native forests, with some modifications for harvesting and remediation in pre-existing plantations. Road drainage and decommissioning requirements are identical. Although dispersed harvesting is encouraged, clearcut coupe sizes are not formally limited other than for slopes greater than 20°, where they are restricted to the same 50ha maximum as native forest coupes. Tasmania's general reforestation standards apply, but are expressed as guidelines rather than requirements. Finally, annual allowable cut requirements do not apply, other than to the extent that Tasmania's publicly owned plantations[19] are included in the calculations of resources contributing to that state's legislated high-quality sawlog production requirements (discussed above; Forestry Tasmania,

2007c). Requirements in Victoria (Department of Natural Resources and Environment, 1996) were generally more liberal: harvesting and replanting were allowed in riparian buffer zones; road drainage and decommissioning requirements were identical to those for native forests; there were no restrictions on coupe sizes; there were no reforestation requirements; and annual allowable cut requirements did not apply.

In New South Wales and the Northern Territory, the same requirements apply to plantations on both public and private land. In the Australian Capital Territory, only public land is available for plantation activity. In New South Wales, the requirements of the Plantations and Reafforestation Code have been further developed and elaborated for public plantations in a Forest Practices Code for Timber Harvesting in Plantations (Forests NSW, 2005). In South Australia, Western Australia and New Zealand, the voluntary code or guidelines are intended to apply to both public and private tenures. There is no code for either public or private plantations in Queensland. The state plantation forestry agency imposes voluntary standards on its operations (DPI Forestry, 2005) and some local authorities impose restrictions (e.g. minimum riparian buffer widths) on new plantation development.

In comparison to the corresponding codes for native forest practices, in states where both are in effect but are different (New South Wales, Western Australia), riparian protection policies for plantations generally require no-harvest and special management zones of lesser width. This reflects a number of factors: the prior history of management of land on which plantations are being established or harvested, in which riparian native vegetation was typically cleared, largely or partly; that plantations are not being established on sites converted directly from native forest; and that riparian zone restoration may be assisted by removing plantation trees already established in riparian zones and allowing regeneration of native vegetation. In Western Australian drinking water catchments, minimum buffer widths apply regardless of land use (Department of Environment, 2005), although their interpretation of pre-existing plantations is handled on a case-by-case basis (Forest Industries Federation, 2006). Road stream crossing and decommissioning requirements are equivalent.

Australasian plantation codes do not generally specify reforestation requirements, beyond recommending – for example – achievement of 'optimal stocking determined by the management system used' (Forest Practices Board, 2000). Associated operational plans typically specify stocking levels and assessment procedures.

Conversion of native forests to plantations

The conversion of native forest to plantation in Australia accounted for less than 10 per cent of the 1 million ha of woody vegetation clearance nationally over the period 2000–2004 (Montreal Process Implementation Group, 2008). At the census date, Tasmania was the only Australian state that allowed conversion of public native forest to plantation; it did so under a process agreed with the Australian government under the Tasmanian Regional Forest Agreement, to offset the loss of wood production from native forests as a result of the Regional Forest Agreement. The extent of conversion was restricted by Tasmania's 'Permanent Native Forest Estate Policy' to a maximum of 5 per cent of the 1996 area of native forest in the state and specified the minimum area and proportion of each forest community which had to be retained in each bioregion (Department of Infrastructure, Energy and Resources, 2007). Subsequently, the conversion of public native forest, and of private native forest by the principal Tasmanian private forestry company, Gunns Ltd, ceased on 31 December 2007 (Forestry Tasmania, 2007a; WWF Australia, 2007).

Conversion of private forestland to plantation is permitted in the Northern Territory, under provisions which require 'no net loss of biodiversity' through compensatory activities elsewhere on the estate (Department of Business, Industry and Resource Development, 2003). Conversion of native vegetation to plantation is not permitted in other Australian states, other than on the very minor scale noted earlier.

In New Zealand, conversion of native vegetation to other land uses is regulated by the Resource Management Act 1991, which does not preclude such conversion, but regulates it at a district council level (Shepherd, 2004). Under the New Zealand Forest Accord (1991) and the associated Principles for Commercial Plantation Forest Management in New Zealand (1995), all members of the New Zealand Forest Owners Association agreed that – with the exception of very small areas that were not of conservation significance – native vegetation would 'remain undisturbed' in any plantation establishment operations (NZFOA, 2001).

ENFORCEMENT AND COMPLIANCE POLICIES

Third-party oversight of forest practices, in other words, oversight involving entities separate from those governing operations, is now almost universal for the native forests of Australia and New Zealand and common for plantation forests.

Tasmania is the only Australasian jurisdiction with an agency specifically dedicated to oversight of forest practices; its Forest Practices Authority has responsibility across all tenures and forms of forestry, reflecting Tasmania's forest practices system and code. In other Australian states, and in New Zealand, different government agencies are responsible for oversight of different tenures and/or forms of forest. In general, a state's environmental protection agency or equivalent is responsible for the oversight of public native forests and, in some states, public plantations. An environment, forestry or natural resources agency is typically responsible for oversight of private native forests, although local authorities have this role in Victoria. A variety of arrangements, including voluntary self-regulation in four of the nine Australasian jurisdictions, apply to either or both public and private plantation forests.

The basis for enforcement and compliance in almost all cases is some form of forest management plan, prepared by the managing agency or forest owner proposing forest operations. These are required for all native forest operations on both public and private tenures, although their form varies greatly. They are identical on public and private tenures in the Northern Territory and Tasmania, but different in other jurisdictions. In the latter case, those prepared for public forests are typically more comprehensive, although the requirements of private forest owners have increased substantially where codes have been recently introduced or revised. Requirements for private native forests are least substantial in Queensland, which requires simply a notice of intent to conduct forest practices. They are most comprehensive in Tasmania and New Zealand.

Management plans for plantation operations are required in the ACT, New South Wales, the Northern Territory, Tasmania and Victoria, and – on a voluntary basis – in New Zealand and Western Australia. Although public agency and corporate plantation growers in Queensland and South Australia typically prepare such plans, they are not required.

The Tasmanian forest practices system requires that Forest Practices Plans be prepared and supervised by a qualified Forest Practices Officer registered with the Forest Practices Authority (FPA, 2007a). Elsewhere, responsibility for the preparation and supervision of the management plan may be vested in a nominated individual (e.g. a 'Supervising Forestry Officer' for NSW public plantations (Forests NSW, 2005)), but in most cases responsibility is simply vested in the managing agency or landowner.

The form of oversight also varies across jurisdictions. All jurisdictions with forest practices codes have provisions for investigation of individual cases; most codes require the forest manager to inform the oversight agency of any forest practices breaches and also allow other parties to report alleged breaches. Various forms of penalty, typically fines, may be imposed for breaches.

In addition, most jurisdictions require or allow some level of compliance auditing, on either a random or targeted basis. For example, the Tasmanian Forest Practices Authority conducts an audit of a 'representative sample' of 15 per cent of Forest Practices Plans annually, across both public and private tenures, and native and plantation forests. This model is also followed for native forests in the Northern Territory, which audits 10 per cent of timber harvesting permits, and in Victoria, where 8 per cent of public native forest operations are audited annually. Independent audits of a sample of forest operations are conducted in the public native forests of New South Wales, Queensland and Western Australia, and in New South Wales' public plantations. In New Zealand's private native forests, the Ministry of Agriculture and Forestry will typically audit previous forestry operations before approving applications to conduct further operations. The results of forest practices audits are publicly reported annually in Tasmania, for all forests, and in Victoria for public native forests (EPA (Victoria), 2007; FPA, 2007). Public reporting is also the intention for Queensland public forests (EPA (Queensland), 2005). In other Australasian jurisdictions, results may be reported in agency annual reports, and – for public forests – in self-reporting by the management agency (e.g. Forests NSW, 2007b).

As discussed in the next section, most Australian state forestry agencies, and many corporate forest growers, have also pursued forest certification, which represents a non-state form of compliance monitoring.

FOREST CERTIFICATION

Forest certification in both Australia and New Zealand began with promotion of the FSC, following its launch internationally in 1993. As elsewhere (discussed in Chapter 1), certification in Australasia was contentious from the outset. This conflict centred on two areas: the view of many in the Australasian forestry sector that the FSC scheme was too strongly influenced by environmental interests (e.g. Lang, 1999) and the insistence of many in the Australian and New Zealand environmental movement that certification not be used to legitimize native forest harvesting (e.g. Cadman, 2002). The latter issue was particularly strong in Australia, where the inter-governmental RFA process had, as one of its principal objectives, enhancing the security of supply from public native forests which remained available for wood production (Kanowski, 1997). In contrast, in New Zealand, the dialogue between forest sector stakeholders that culminated in the New Zealand Forest Accord largely ended harvesting from public native forests (MAF, 2007a). As a result, Australian NGOs championed the FSC process for plantation forests, but not for native forests, and in doing so sought to divide the forest sector between those businesses based solely on FSC-eligible plantation forests (in other words, excluding those established on sites converted from native forest after November 1994) and those which drew wholly or partly from native forests (Cadman, 2002). In New Zealand, a national process to develop FSC-endorsed National Standards for each of native and planta-tion forests faltered in 2004 because of lack of consensus among participants. In the case of plantation forests, the only issue preventing agreement was that of conserva-tion reserve criteria (New Zealand Forest Certification, 2004), which had been agreed by all parties other than one major non-government organization (Dyck, 2008).

The growing international pressure for certification, coupled with the political dynamic described above and the motivation of Australasian forest growers and processors to demonstrate their sustainable forest management credentials, led to the development of nationally based certification initiatives. In the Australian case, this resulted in the Australian Forestry Certification Scheme (AFS, 2007b), which is now the dominant scheme operating in Australia. In the New Zealand case, the New Zealand Forest Industry Council developed a Verification of Environmental Performance scheme (Kanowski et al, 1999), but it did not become established. More recently, the New Zealand Forest Owners Association has stated that it 'hopes to introduce a national certification scheme which is third-party certified and inter-nationally accredited' (NZ FOA, 2006). The FSC has operated in both Australia and New Zealand using interim standards. While the New Zealand national standards initiative has not advanced since 2004 (New Zealand Forest Certification, 2004), a national standards development process commenced in Australia in 2007 (FSC Australia, 2007a).

Australia

Some 9.2 million ha of forest are certified in Australia under two schemes, the PEFC-accredited Australian Forestry Certification Scheme and the FSC. The former has certified nearly 8.7 million ha of both native and plantation forest in six states and the latter 550,000ha of plantation forest in four states (AFS, 2007a; FSC Australia, 2007b). In total, 77 per cent of public native production forests, around 0.5 per cent of private native forests and 91 per cent of plantations are certified.

The Australian Forest Certification Scheme standards were developed through the formal 'Australian Standard' process overseen by Standards Australia (Standards Australia, 2006), which accredited AFS Ltd as a standard development organization and oversaw and reviewed the standards development process. A number of Australian environmental NGOs were reluctant to participate in the AFS development process and subsequently withdrew from it, criticizing it for legitimizing existing practices, including conversion of native forest to plantation (WWF Australia, 2002; The Wilderness Society et al, 2005). However, other environmental representatives remained involved in the process's Technical Reference Committee and subsequently supported the revised standard (e.g. Peacock, 2007). Proponents of the AFS have argued strongly in its defence (Forest and Wood Products Research & Development Corporation, 2005; NAFI, 2005; FS, 2006). The Australian Forestry Certification Scheme was formally endorsed by the PEFC in 2004 (PEFC, 2007).

The state forest management agencies in NSW, Queensland, Tasmania and Victoria, and one company – Gunns Ltd in Tasmania – have been awarded AFCS certification for the native forests they manage. These certifications, totalling about 7.2 million ha and 150,000ha of public and private forests, respectively, represent around 77 per cent of public native forests managed for production and 0.5 per cent of private forests. State plantation growers in four states – NSW, Queensland, South Australia and Tasmania – and ten plantation companies, representing operations in all states but the ACT and Northern Territory, have been awarded AFCS certification for their plantation forests. One of these, Hancock Victorian Plantations, has dual certification under the AFCS and FSC (AFS, 2007a). In addition to the 240,000ha with dual certification, a further 308,000ha of private plantations in four states are FSC certified, bringing the total Australian plantation area certified under AFCS or FSC to almost 1.7 million ha, or 91 per cent of all Australian plantations. No Australian public native forest manager has yet sought FSC certification; two small private native forest managers have received FSC certification (FSC, 2008).

New Zealand

Some 760,000ha of New Zealand's plantation forests, and 12,000ha of native forests, are FSC certified (NZFOA, 2007; FSC, 2008). This represents 42 per cent of the area of New Zealand's plantation forests, across 15 owners and 33 per cent of the plantation harvest volume, and 10 per cent of New Zealand's native forests managed for wood production.

Table 8.4 *Matrix of policy approaches in Australasian natural forests*

Case Study	1) Riparian	2) Roads	3) Clearcuts	4) Reforestation	5) AAC
ACT* (Public)	N/A	N/A	N/A	N/A	N/A
Queensland (Private)					
Queensland (Public)					
New South Wales (Public)					
New South Wales (Private)					
Northern Territory (Public)					
Northern Territory (Private)					
Tasmania (Public)					
Tasmania (Private)					
Victoria (Public)					
Western Australia (Public)					
South Australia* (Public)	N/A	N/A	N/A	N/A	N/A
New Zealand (Public)*	N/A	N/A	N/A	N/A	N/A
New Zealand (Private)					

Legend:
- Mandatory substantive
- Mandatory procedural
- Mandatory mixed
- No policy

Note: * No harvest is allowed in public natural forests in ACT, South Australia, and New Zealand

SUMMARY

Native forests remain important for wood production in most Australian states and plantations are of primary importance in most Australian states and New Zealand. The relative importance of plantation forests for wood production has increased substantially since 1992. Plantations now provide 99 per cent of New Zealand and 66 per cent of Australian wood production. Forest practices systems in Australia and New Zealand have developed significantly over the past decade, and now (with changes after the study's census date) encompass all native forests managed for wood production in both countries. They also encompass plantation forests in most jurisdictions, but only on a voluntary basis in about half the jurisdictions.

NOTES

1   In other words, conservation reserve, multiple-use forest, and other Crown land categories, as discussed above (Montreal Process Implementation Group, 2008).
2   Formally, states and territories; for simplicity, only the former term is used.

3   Most native forest conversion to plantations ceased in Tasmania at the end of 2007 (Forestry Tasmania, 2007a; WWF Australia, 2007).

4   www.daff.gov.au/forestry/international/fora/Montreal; www.maf.govt.nz/forestry/montreal-process/information-summary.htm

5   Managed investment schemes are investment vehicles which allow Australian taxpayers to offset the costs of purchasing shares in new woodlots against personal taxation liabilities (see Treefarm Investment Managers Australia, 2008).

6   Subsequently abbreviated to 'states'.

7   Including five-yearly State of the Forests reporting (e.g. DAFF, 2008).

8   Codes for NSW private forests in each of 4 regions were introduced on 1 August 2007 (Department of Environment and Climate Change, NSW, 2007); a new code for all Victorian forests was introduced on the same date (Department of Sustainability and Environment, 2007)

9   The codes of forest practice, or equivalent, applying at 1 January 2007 are those cited as: Department of Business, Industry and Resource Development (Northern Territory) (2003); Department of Infrastructure, Planning and Environment (Northern Territory) (2003); Department of Natural Resources and Environment (Victoria) (1996); Department of Conservation and Land Management (Western Australia) (1999); Department of Natural Resources and Water (Queensland) (2005); Environment ACT (2005); EPA (Queensland) (2002); Forest Industries Federation (Western Australia) (2006); Forest Practices Board (Tasmania) (2000); Forests NSW (1998); Forests NSW (2005); Forestry SA (1997); Government of NSW (2001).

10 Public native forests managed for wood production are not included in IUCN Category VI data for Australia (Australian National Forest Inventory, 2008).

11 Other than NSW private forests at the census date for this study; the subsequently introduced code conforms to the more general arrangements.

12 Stream order describes 'the relative size and frequency of well defined watercourses' (Department of Infrastructure, Planning and Environment, 2004) and provides an index of catchment characteristics; the classification system varies between states, although Strahler's Order is commonly used, see, for example, Department of Infrastructure, Planning and Environment (2004).

13 And Northern Territory wetlands; wetlands are not included in the international comparison.

14 The NSW Code of Practice for Private Native Forests subsequently introduced (1 August 2007) specifies riparian protection zones of identical overall width to those for public forests, but allows conditional harvesting within a specified filter zone component of the riparian protection zone.

15 1:X storm events are defined as those that occur, on average, once every X years.

16 Including NSW private forests from 1 August 2007.

17 Stems per hectare.

18 The following methodology is used when determining if vegetation is remnant:
- *50 per cent of the predominant canopy cover that would exist if the vegetation community were undisturbed;*
- *70 per cent of the height of the predominant canopy that would exist if the vegetation community were undisturbed;*
- *composed of the same floristic species that would exist if the vegetation community were undisturbed'* (Department of Natural Resources and Water (Queensland), 2007a).

19 Formally a public–private joint venture between Forestry Tasmania and GMO Renewable Resources (Forestry Tasmania, 2007b).

REFERENCES

ABARE (2006) 'Forest products statistics – March and June Quarters 2006', p72, www.abareconomics.com, accessed December 2007

AFS (2006) 'A response to incredible claims by the Wilderness Society', www.forestrystandard.org.au/6news.asp, accessed December 2007

AFS (2007a) 'Australian forest certification scheme', AFS, Canberra, www.forestrystandard.org.au > A quick introduction to AFCS, accessed December 2008

AFS (2007b) 'AFS certification register', www.forestrystandard.org.au, accessed December 2008

Australian National Forest Inventory (2003) 'Australia's state of the forests report 2003', Department of Agriculture, Fisheries and Forestry, Canberra, www.daff.gov.au/brs/forest-veg/publications, accessed December 2008

Australian National Forest Inventory (2007) 'Australia's forests at a glance 2007', Department of Agriculture, Fisheries and Forestry, Canberra, www.daff.gov.au/brs/forest-veg/publications, accessed January 2009

Bhati, U.N. (2003) 'Exports of Australian forest products increasing fast', ANU Forestry Market Report 23, http://fennerschool-associated.anu.edu.au/marketreport/, accessed January 2009

Blackburn, W. (2008) Personal communication, Department of Sustainability and Environment (Australia)

Bonyhady, T. (1993) *Places Worth Keeping: Conservationists, Politics and Law*, Sydney: Allen & Unwin, p192

Bureau of Rural Sciences (2008) 'The changing face of Australia's forests', www.daff.gov.au/brs/forest-veg > Publications, accessed December 2008

Cadman, T. (2002) 'FSC Standards development in Australia 2002', www.certifiedforests.org.au/corr/fsc.html, accessed September 2004

Commonwealth of Australia (1995) 'National forest policy statement', www.daff.gov.au/forestry/policies/statement, accessed December 2008

Conservation Commission of Western Australia (2004) 'Forest management plan 2004–2013', www.conservation.wa.gov.au/downloads.htm?docCatID=3&TLCN=Forest+Management+Plan, accessed December 2008

DAFF (Australia) (2007a) 'Forestry', www.daff.gov.au/forestry, accessed December 2007

DAFF (Australia) (2007b) 'Regional forest agreements home', www.daff.gov.au/rfa, accessed December 2007

DAFF (Australia) (2008) 'Australia's State of the Forest Report 2008', Australian Government, Department of Agriculture, Fisheries and Forestry, Canberra

Dargavel, J. (1995) *Fashioning Australia's Forests*, Melbourne: Oxford University Press, p312

Department of Business, Industry and Resource Development (NT) (2003) 'Northern Territory codes of practice for plantation forestry', p3, www.planningplantations.com.au/assets/content/plantation_management/regulation_planning/nt4.html, accessed December 2007

Department of Conservation and Land Management (Western Australia) (1999) 'Code of practice for timber harvesting in Western Australia', Perth, p30

Department of Conservation and Land Management (Western Australia) (2004) 'Silvicultural practice in the jarrah forest', Sustainable Forest Management Series, SFM Guideline No. 1, Perth, p46

Department of Conservation and Land Management (Western Australia) (2005) 'Silvicultural practice in the karri forest', Sustainable Forest Management Series, SFM Guideline No. 3, Perth, p31

Department of Environment (Western Australia) (2005) 'Water quality protection note', Perth, p14, http://portal.environment.wa.gov.au/pls/portal/docs/PAGE/DOE_ADMIN/ GUIDELINE_REPOSITORY/WQ6.PDF, accessed December 2007

Department of Environment and Climate Change NSW (2007) 'Private native forestry', www.environment.nsw.gov.au/pnf/index.htm, accessed December 2007

Department of Environment, Water, Heritage and the Arts (2007a) 'Environment Protection and Biodiversity Conservation Act 1999', www.environment.gov.au/epbc, accessed December 2007

Department of Environment, Water, Heritage and the Arts (2007b) 'Threatened species and ecological communities', www.environment.gov.au/biodiversity/threatened/index.html, accessed December 2008

Department of Environment, Water, Heritage and the Arts (2007c) 'Conservation incentives', www.environment.gov.au/biodiversity/incentives/index.html, accessed December 2008

Department of Environment, Water, Heritage and the Arts (2007d) 'State of the environment 2006', www.environment.gov.au/soe/2006/index.html, accessed December 2007

Department of Infrastructure, Energy and Resources (Tasmania) (2007) 'Permanent native forest estate policy', www.dier.tas.gov.au/forests/permanent_native_forest_estate_policy, accessed December 2007

Department of Infrastructure, Planning and Environment (NT) (2003) 'A management program for the commercial harvesting of timber from native vegetation in the Northern Territory of Australia', Draft 2004–2009, Darwin, p33

Department of Infrastructure, Planning and Environment (NT) (2004) 'Stream orders and setbacks in the Northern Territory', www.nt.gov.au/nreta/natres/natveg/brochures/ index.html, accessed January 2009

Department of Natural Resources and Environment (Victoria) (1996) 'Code of forest practices for timber production 1996', Revision No. 2, Melbourne

Department of Natural Resources and Water (Queensland) (2005) 'Code applying to a native forest practice on freehold land', www.nrw.qld.gov.au/vegetation/clearing/ forestpractice.html, accessed December 2007

Department of Natural Resources and Water (Queensland) (2007a) 'Vegetation communities', www.nrw.qld.gov.au/vegetation/bioregions.html, accessed December 2007

Department of Natural Resources and Water (Queensland) (2007b) 'Regional vegetation management codes', www.nrw.qld.gov.au/vegetation/regional_codes.html, accessed December 2007

Department of Primary Industries and Water (Tasmania) (2007) 'Private forests reserve program', www.dpiw.tas.gov.au/inter.nsf/WebPages/LBUN-6JD735?open, accessed December 2007

Department of Sustainability and Environment (Victoria) (2005) 'Management procedures for timber harvesting and associated activities in state forests in Victoria', www.dse.vic.gov.au > Forests > Publications > Code of Practice [updated 2007], accessed December 2007

Department of Sustainability and Environment (Victoria) (2007) 'Code of practice for timber production 2007', www.dse.vic.gov.au > Forests > Publications > Code of Practice, accessed December 2007

Department of the Environment, Sport and Territories (1996) 'National strategy for the conservation of Australia's biological diversity', www.environment.gov.au/biodiversity/ publications/strategy/index.html, accessed December 2007

DPI Forestry (Queensland) (2005) 'Sustainable forest management', Version 1, p29, www.fpq.qld.gov.au/asp/index.asp?sid=5&page=publications, accessed December 2007

Dudley, N. and A. Phillips (2006) *Forests and Protected Areas: Guidance on the Use of IUCN Protected Area Management Categories*, Gland and Cambridge: IUCN, p58

Dyck, W. (2008) Personal communication, Former Interim National Coordinator, Forest Certification NZ

Environment ACT (2005) 'ACT code of forest practice', Version 1, Department of Territory and Municipal Services, Canberra, p60

Environment Centre Northern Territory (2007) 'Landclearing – Tiwi Islands', www.ecnt.org/html/cur_land_tiwi.html, accessed December 2007

EPA (Queensland) (2002) 'Code of practice for native forest timber production', p84, www.epa.qld.gov.au/publications/p00069aa.pdf/, accessed December 2007

EPA (Queensland) (2005) 'Annual environmental audit report 2003–2004', www.epa.qld.gov.au/publications?id=1552, accessed December 2007

EPA (Queensland) (2007) 'Forest transfer processes in Queensland', www.epa.qld.gov.au/parks_and_forests/managing_parks_and_forests/forest_transfer_processes_in_queensland, accessed December 2007

EPA (Victoria) (2007) 'Environmental auditing of forestry in Victoria', www.epa.vic.gov.au/envaudit/forestry.asp, accessed December 2007

FAO (2007) 'State of the World's Forests – 2007', United Nations, Food and Agriculture Organization, Rome

Feary, S. (2007) 'Forests and forestry: An overview of indigenous involvement in forest management in Australia', in S. Feary (ed) 'Forestry for indigenous peoples', ANU *Fenner School Occasional Paper No. 1*, http://fennerschool.anu.edu.au/publications/occasional, accessed January 2009

Flannery, T. (1994) *The Future Eaters*, Melbourne: Reed Books, p423

Florence, R. G. (1996) *Ecology and Silviculture of Eucalypt Forests*, Melbourne: CSIRO Publishing, p413

Forest and Wood Products Research and Development Corporation (2005) 'Forests for tomorrow', www.fwpa.com.au/content/pdfs/PR05.5029.pdf, accessed December 2007

Forest Industries Federation (WA) (2006) 'Code of practice for timber plantations in Western Australia', www.fpc.wa.gov.au/content/environment/codes_of_practice.asp, accessed December 2008

Forest Practices Authority (Tasmania) (2007a) 'The forest practices authority', www.fpa.tas.gov.au/, accessed December 2007

Forest Practices Authority (Tasmania) (2007b) 'Annual report 2006–2007', www.fpa.gov.au > Publications > Annual Reports, accessed December 2007

Forest Practices Board (Tasmania) (2000) 'Forest practices code 2000', www.fpa.tas.gov.au > Publications, accessed April 2009

Forest Products Commission (Western Australia) (2004) 'Contractor's timber harvesting manual for south west native forests in Western Australia', Perth, p87

Forestry Tasmania (2002) 'Sustainable high quality eucalypt sawlog supply from Tasmanian state forest – Review No. 2', www.forestrytas.com.au/forestrytas/pdf_files/hq_euc_sawlog_supply_review2.pdf, accessed December 2007

Forestry Tasmania (2005) 'New silviculture for oldgrowth forests', www.forestrytas.com.au/forest-management/old-growth-forests, accessed December 2007

Forestry Tasmania (2007a) 'End of conversion of native forests to plantations', www.forestrytas.com.au/news/2007/06, accessed December 2007

Forestry Tasmania (2007b) 'Stewards of the forest 2007 Annual Report',
    www.forestrytas.com.au/publications/annual-reports, accessed December 2007
Forestry Tasmania (2007c) 'Sustainable high quality eucalypt sawlog supply from Tasmanian
    state forest', www.forestrytas.com.au/sfm/sustainable-high-quality-eucalypt-sawlog-
    supply-from-tasmanian-state-forest, accessed January 2008
Forestry (SA) (1997) 'Environmental management guidelines for plantation forestry in South
    Australia', p36, www.forestry.sa.gov.au/enviro.stm, accessed December 2007
Forests NSW (1998) 'Forest practices code for native timber harvesting', www.forest.nsw.gov.au/
    publication/forest_facts/harvesting/default.asp, accessed December 2007
Forests NSW (2005) 'Timber harvesting in forests NSW plantations. Forest Practices Code –
    Part 1', p46, www.forest.nsw.gov.au/publication/forest_facts/plantations/code/default.asp,
    accessed December 2007
Forests NSW (2007a) Personal communication with I. Barnes, Southern Region Planning
    Manager, 12 December 2007
Forests NSW (2007b) 'ESFM Plans', www.forest.nsw.gov.au/esfm/, accessed December 2007
Forests NSW (2008) Native Forest Silviculture Manual, Sydney: Forests NSW
FSC (2008) 'FSC certificate search', www.fsc-info.org/, accessed December 2008
FSC Australia (2007a) 'Update August 2007', www.fscaustralia.org/, accessed December 2007
FSC Australia (2007b) 'Certified companies in Australia', www.fscaustralia.org/
    fsc-in-australia/certified-companies-in-austral/, accessed December 2007
Government of NSW (1999) 'Integrated forestry operations approval [for each of 3 RFA
    regions]', www.forest.nsw.gov.au/ifoa/, accessed December 2007
Government of NSW (2001) 'Plantations and reafforestation (code) regulations 2001', p42,
    www.austlii.edu.au/au/legis/nsw/consol_reg/parr2001438/, accessed December 2007
Greening Australia (2009) 'Volunteer with Greening Australia',
    www.greeningaustralia.org.au/get-involved/volunteer, accessed January 2009
Hammond, D. (2001) 'Development of Maori owned indigenous forests', MAF Technical
    Paper No. 2003/4, p80, www.maf.govt.nz/mafnet/publications/, accessed December 2007
Kanowski, P. J. (1997) 'Regional Forest Agreements and future forest management', in
    Proceedings of the National Agricultural and Resources Outlook Conference, Canberra, 4–6
    February, 1, pp 225–235
Kanowski, P. J. (2006) 'Communities and participatory forest management', in A. G. Brown
    (ed) Forests, Wood and Livelihoods, Melbourne: Crawford Fund pp63–66
Kanowski, P. J., D. Sinclair and B. Freeman (1999) 'International approaches to forest
    management certification: A review', Canberra: Agriculture Fisheries and Forestry
    Australia, p47
Lang, W. (1999) 'Certification and labelling in relation to environmental policies and sustain-
    able forest management', in Proceedings, Institute of Foresters of Australia Conference,
    Practising forestry today, Hobart, 3–8 October
Lindenmayer, D. B. and J. F. Franklin (eds) (2005) Towards Forest Sustainability, Melbourne:
    CSIRO Publishing,
MAF (New Zealand) (no date) 'The Montreal Process', www.maf.govt.nz/forestry/montreal-
    process/#mprs, accessed April 2009
MAF (New Zealand) (2002a) 'Relevant legislation and industry agreements',
    www.maf.govt.nz/mafnet/sectors/forestry/forind/forind07.htm, accessed December 2007
MAF (New Zealand) (2002b) 'Standards and guidelines for sustainable management of
    indigenous forests', www.maf.govt.nz/forestry/indigenous-forestry/index.htm, accessed
    December 2007

MAF (New Zealand) (2002c) 'Indigenous forestry: Sustainable management – A guide to plans and permits', www.maf.govt.nz/mafnet/sectors/forestry/indig/, accessed December 2007

MAF (New Zealand) (2003) 'The New Zealand forestry industry', www.maf.govt.nz/mafnet/rural-nz/overview/nzoverview015.htm, accessed December 2007

MAF (New Zealand) (2007a) 'A national exotic forestry description, 1 April 2006, Part 2 – Overview', www.maf.govt.nz/forestry/publications, accessed December 2007

MAF (New Zealand) (2007b) 'New Zealand's forest resources', www.maf.govt.nz/forestry/resources/, accessed December 2007

MAF (New Zealand) (2007c) 'MAF Indigenous Forestry Unit', www.maf.govt.nz/forestry/indigenous-forestry/, accessed December 2007

MAF (New Zealand) (2009) 'Annual forestry import statistics', www.maf.govt.nz/statistics/forestry/annual/annualimports.htm, accessed April 2009

Miller, R., Y. Dickinson and A. Reid (2007) 'Maori connections to forestry in New Zealand', in S. Feary (ed) 'Forestry for indigenous peoples', ANU *Fenner School Occasional Paper No. 1*, http://fennerschool.anu.edu.au/publications/occasional/, accessed January 2009

Ministry of Environment (New Zealand) (2006) 'Your guide to the Resource Management Act', www.mfe.govt.nz/publications/rma/rma-guide-aug06/html/index.html, accessed December 2007

Ministry of the Environment (New Zealand) (2007a) 'Environment New Zealand 2007', www.mfe.govt.nz/publications/ser/enz07-dec07/index.htm, accessed January 2009

Ministry of the Environment (New Zealand) (2007b) 'Laws and treaties', www.mfe.govt.nz/laws/, accessed December 2007

Montreal Process Implementation Group (Australia) (2008) 'Australia's State of the Forests Report 2008', http://adl.brs.gov.au/forestsaustralia/publications/sofr2008.html, accessed January 2009

NAFI (Australia) (2005) 'Nobody does it better – AFS is the best', www.nafi.com.au/media/, accessed December 2007

National Plantations Inventory (Australia) (2006) 'Australia's plantations 2006', www.brs.gov.au/plantations > Reports, accessed December 2007

New Zealand Forest Certification (2004) 'News and updates', www.nzcertification.com/news_developments.asp?pageid=215, accessed January 2008

New Zealand Government (2000) 'The New Zealand biodiversity strategy', www.biodiversity.govt.nz, accessed December 2007

New Zealand Government (2007a) 'Guidance about biodiversity on private land', www.biodiversity.govt.nz/land/guidance/index.html, accessed December 2007

New Zealand Government (2007b) 'Third national report [to CBD]', www.cbd.int/countries/?country=nz, accessed December 2007

New Zealand Landcare Trust (2009) 'Think', www.landcare.org.nz/

NZFOA (2001) 'The New Zealand Forest Accord – A unique undertaking', www.nzfoa.org.nz/file_libraries_resources/agreements_accords, accessed December 2007

NZFOA (2006) 'Sustainable plantations', www.nzfoa.org.nz/index.php?/new_zealand_plantation_forestry/sustainable_plantations, accessed December 2007

NZFOA (2007) 'New Zealand forest industry facts and figures 2006–2007', www.nzfoa.org.nz/file_libraries_resources/facts_figures, accessed December 2007

NZFOA (2008) 'New Zealand environmental code of practice for plantation forestry', www.nzfoa.org.nz/index.php?/File_libraries_resources/Standards_guidelines

Peacock, R. (2007) 'Forest Standard', *Bulletin of the Ecological Society of Australia*, 37, 2, 12–15 June, www.ecolsoc.org.au/bulletin.html, accessed January 2008

PEFC (2007) 'Members and schemes', www.pefc.org > PEFC members & schemes > Australian Forestry Standard Limited, accessed December 2007

Petheram, R. J., P. Stephen and D. Gilmour (2004) 'Collaborative forest management: A review', *Australian Forestry*, 67, pp137–146

Poynter, M. (2005) 'Collaborative forest management in Victoria's Wombat State Forest – Will it serve the interests of the wider community?', *Australian Forestry*, 68, pp192–201

Roche, M. (1990) *History of Forestry*, Wellington: NZ Forestry Corporation

Said, A. (2004) 'Private native forestry in Australia: Status, issues and possible directions for policy', paper read at Private Forestry Consultative Committee, Department of Agriculture Fisheries and Forestry, Canberra, p22

Schirmer, J. and P. J. Kanowski (2005) 'A mixed economy commonwealth of states – Australia', in M. Garforth, N. Landell-Mills and J. Mayers (eds) *Changing Ownership and Management of State Forest Plantations*, London: IIED, pp101–125

Schirmer, J. and M. Roche (2005) 'Corporatization, commercialisation and privatisation – New Zealand', in M. Garforth, N. Landell-Mills and J. Mayers (eds) *Changing Ownership and Management of State Forest Plantations*, London: IIED, pp200–222

Shepherd, I. (2004) The long white cloud over New Zealand's forests', *Ecos*, 122, pp28–29

Standards Australia (2006) 'Preparing standards: Standards Australia', https://committees.standards.org.au/POLICY/SG- 001/STANDARDIZATION GUIDE-SG-001.HTM, accessed November 2006

The Wilderness Society, Australian Conservation Foundation, Friends of the Earth and Greenpeace Australia (2005) 'Call to reject "worst forestry practice" standard', www.wilderness.org.au/campaigns/forests/afs/, accessed December 2007

Treefarm Investment Managers Australia (2008) 'About TIMA', www.afg.asn.au/tima/about_tima.html, accessed January 2008

Wasson, B., B. Banens, P. Davies, W. Maher, S. Robinson, R. Volker, D. Tait and S. Watson-Brown (1996) 'Inland waters', in 'Australia's State of Environment 1996', www.environment.gov.au/soe/1996/publications/report/index.html, accessed December 2007

Whitehead, P. J., D. J. M. S. Bowman, N. Preece, F. Fraser and P. Cooke (2003) 'Customary use of fire by indigenous people in northern Australia: Its contemporary role in savannah management', *International Journal of Wildland Fire*, 12, pp415–425

WWF Australia (2002) 'Australian Forestry Standard lacks credibility in the marketplace', www.nativeforest.net/pressrel/alert28.html, accessed December 2007

WWF Australia (2004) 'A blueprint for the forest industry and vegetation management in Tasmania', https://secure.wwf.org.au/publications/tasmaniablueprint2004/, accessed December 2007

WWF Australia (2007) 'Tasmanian government must end landclearing and forest conversion', www.panda.org/about_wwf/where_we_work/oceania/index.cfm?uNewsID= 105961, accessed December 2007

Chapter 9

Africa: The Democratic Republic of Congo and South Africa

INTRODUCTION

Our two African case study countries, the Democratic Republic of Congo (DRC) and South Africa, have about 4 per cent of the world's forests. There is stark contrast between these countries, both generally and in the forestry context. The Congo Basin forests, of which the DRC has more than half, contain the second largest area of dense tropical rainforest in the world (CARPE, 2007). As a result, the DRC is tremendously forest-rich, but has suffered from a long, ongoing civil war that has – amongst other things – greatly damaged its capacity and infrastructure for sustainable forest management and kept its people poor. While the DRC is ranked seventh globally in forest area, it is the lowest ranked amongst our case study countries in terms of economic indicators such as per capita GDP and the value of forest products trade, and in human development.

South Africa, in contrast, is poor in its natural forest resources, but has developed a globally competitive plantation industry. It has enjoyed relative political stability since the end of the apartheid regime in 1994 and its economic and social indicators – while mid-ranked globally – are among the highest in Africa. With a forest sector based on intensively managed planted forests, there has been considerable public debate over some of the environmental issues associated with intensive plantation forestry and the best means by which the plantation forestry sector can contribute to the livelihoods of the rural poor (Kanowski and Murray, 2008).

Despite these differences, there are also commonalities. Both countries face serious problems of rural poverty, which policy development has sought to address. In the DRC, poverty has been fuelled by civil and transnational war, while in South Africa, the legacy of apartheid and the impact of AIDS have been major contributing factors. The potential of forest-based economic development has led to the issuance of large-scale forest concessions in the DRC and continuing investment in the plantation sector in South Africa. The DRC has recognized the need for communities to benefit from commercial exploitation of their forests, although policies to operationalize community rights have been slow to develop. In South Africa, there has been a strong emphasis on the welfare of plantation workers and on the development of partner-ship arrangements, such as outgrower schemes, in which smallholders can participate.

Forest policies in the DRC and South Africa are generated principally at the national level. Hence, our policy analyses in this chapter are also national in focus.

The Democratic Republic of Congo (DRC)

An overview of forests and forest ownership

The forests of the Democratic Republic of Congo (DRC) cover an estimated 134 million ha (FAO, 2007). According to Debroux et al (2007), this forest area includes the 'second largest block of tropical forest in the world',

Topographically, the DRC is divided between north and south plateaus, roughly separated by the Congo River. The northern plateau is typically 600–800 metres elevation and the southern plateau 1000–2000 metres. The northern plateau is fringed by mountains along the eastern border with Rwanda and Uganda, with the tallest peaks up to 5000 metres. Tropical moist evergreen, swamp and deciduous forests grow in the plateau, while sub-montane and montane forests are found on the mountain slopes. Moist evergreen forests comprise around one-third (35–40 million ha) of the DRC's forests, and swamp forests around half that extent (ITTO, 2006). Savanna woodlands, representing nearly 40 per cent of the DRC's forest, cover the southern part of the country (Debroux et al, 2007).

The DRC is ranked third among our case study countries in area of tropical frontier or intact forests (Bryant et al, 1997). Human pressures on the forest, however, are increasing. Perhaps 40 million of the DRC's 62 million inhabitants are forest-dependent to some extent (World Bank, 2009); they include the indigenous Pygmy and Bantu peoples, as well as more recent immigrants (ITTO, 2006; CARPE, 2007). Population growth rates are high and impacts on forests have been consider-able in some parts of the DRC, such as the more accessible coastal forests or those of the eastern Albertine Rift (ITTO, 2006; CARPE, 2007).

Deforestation and forest degradation have occurred at much lower rates in the DRC than in Indonesia and Brazil, our other forest-rich tropical case studies. FAO data indicates forest loss of 319,000ha between 2000 and 2005, amounting to a deforestation rate of 0.2 per cent (FAO, 2007). Most of this loss has occurred in the more populated areas of the coastal forests, eastern DRC, and the forest–savanna interface (CARPE, 2007).

Rates of forest 'degradation' are harder to measure. On the basis of defining degradation as the 'transition from dense forest to degraded forest through small canopy openings', the DRC experienced an average degradation rate of 0.15 per cent per year between 1990 and 2000 (CARPE, 2007). This definition, however, does not allow for variation in the environmental impacts of canopy loss. Furthermore, if degradation is defined in these terms, the DRC's relative rate of forest degradation may be less than those of many developed countries with forests under active forest management.

The most widespread, direct causes of degradation in DRC have been swidden farming, escaped fires, fuel wood harvest and the bushmeat trade. The greatest impact from commercial logging is indirect, in the financing and development of roads that increase general forest access. Poor logging practices play a direct role in degradation in a few areas (Debroux et al, 2007).

The Forest Law of 1973, and the subsequent 2002 Forest Code, asserted state ownership over all land. Three categories of forests are recognized, each with different use and management objectives and restrictions. 'Classified' forests are those managed for conservation, in which customary rights are restricted; 'permanent production forests' are managed for wood production on an ongoing basis, through the letting of concessions, and customary rights are discontinued; 'protected' forests are those in which customary rights are most strongly recognized and in which both traditional uses and commercial harvesting, by communities as concessionaires, may be practised (CARPE, 2007). However, implementation of these arrangements has been slow (ITTO, 2006).

Native forests

Seven of the fourteen terrestrial ecoregions identified for the Congo Basin occur in the DRC, as do 11 of the 16 aquatic ecoregions to which the forests are intimately linked (CARPE, 2007). Farming and the bushmeat trade have exerted significant impacts near roads and settlements but, as noted above, most of the moist forests remain little modified by people.

The relative intactness of the majority of the DRC's forests is very significant for conservation and environmental services. While the total number of species in the Congo Basin forests is less than those in the tropical forests of the Americas and Asia, there are very high levels of endemism in both flora and fauna – including iconic species such as both the lowland and mountain gorilla (CARPE, 2007). Tree species number more than 700, some 200 of which are currently of commercial value (ITTO, 2006).

Planted forests and plantations

The FAO's 2005 Forest Resources Assessment reports no area of planted forest or plantation in the DRC (FAO, 2007). However, a recent ITTO report lists a government estimate of 55,500 hectares of tree plantations, with *Terminalia superba* (limba) a common species planted (ITTO, 2006).

Forest governance and policy

The DRC has a long history of political strife including a decade of civil war. Although the war ended officially in 2002 fighting has continued in the east of the country. This continued conflict has favoured lawlessness over the rule of law for much of the DRC's recent past. Amongst the many adverse consequences have been severe social and economic hardships and the loss of governance and administrative capacity (ITTO, 2006).

The Ministry of the Environment, Nature Conservation, Water Resources and Forests is the primary ministry in charge of forests. Its Directorate of Forest Management is responsible for overseeing forest management, through the letting of concessions for timber harvest. As noted above, all forests are considered state property and all commercial forest harvesting – in either the 'protected' or 'permanent production' categories of forest – is regulated through issuance of concessions by the state (CARPE, 2007).

The 2002 Forest Code places express emphasis on forest protection, as well as on the recognition of the rights of forest-dwelling peoples. It also envisages a hierarchy of forest planning from the national level to the individual forest management unit. However, much of the supporting legislation needed to further operationalize the Code has yet to be completed, including the establishment of mechanisms to allocate forest rights to communities.

The mechanisms for allocating large-scale concessions, however, have advanced relatively rapidly. Law 11/2002 establishes a system of 25-year, renewable concessions, of a maximum area of 500,000ha. Local consultation is required prior to awarding concession areas.

Concessionaires must then follow a four-stage process to secure harvesting rights, each of which requires a separate authorization. The first authorization, valid for a year, is to conduct a forest inventory of the areas proposed for harvesting. No harvesting is allowed during this phase. The second, valid for three years, requires that the concessionaire invest a minimum of 50 per cent of their planned investment, particularly in processing facilities, in which they will ultimately have to process 70 per cent of their production. Some harvesting may be permitted during this period. The third authorization guarantees supply, formally establishes the concession and allows timber harvesting. The fourth stage is the issuance of cutting permits, each of which is limited to 1000ha and includes a harvesting map and any protection measures (ITTO, 2006; CARPE, 2007).

With assistance from the World Bank, the DRC cancelled illegal or questionable logging contracts over the period 2002–2009. This led to a reduction in the area legally available for timber harvesting from 43 million ha to around 10 million ha (World Bank, 2009).

Forest practices system

The DRC's forest practices system is based on the timber harvesting rights process outlined above. In principle, cutting permits are required to address issues such as social considerations and protection measures. However, ITTO (2006) notes that 'none of these steps (required for the issuance of harvesting rights) are administered in a transparent way'; that none of the concessions operating in 2005 had a management plan and that, for the reasons discussed above, the capacity of the relevant authorities to oversee forest operations is very low. There have been many criticisms of the means by which concessions have been awarded and operated (e.g. Chatham House, 2009).

As a result, the government of the DRC has issued a moratorium on the issuing of new concessions and is currently recruiting an independent observer to oversee forest operations (World Bank, 2009).

Forest production and trade

Estimates of forest production and trade are poor because of the DRC's recent history of social conflict. Official estimates in 2003 were of about 77 million m^3 of roundwood harvest (ITTO, 2006). Ninety-five per cent of wood harvested in DRC is consumed as fuel wood (FAO, 2007). The long period of war has had a substantial impact on the forest sector, with road and port infrastructure falling into disrepair and legal production and export of logs dropping dramatically. Conversely, illegal logging, hunting and other forest-based activities are rife (ITTO, 2006). Much wood production is currently conducted by the informal sector for domestic consumption (CARPE, 2007).

Of the small percentage of wood harvested legally for commercial purposes, perhaps 500,000 m^3 (Chatham House, 2009), almost all is exported as raw logs (FAO, 2007). There are estimated to be about a dozen logging concessions in formal operation in the DRC covering about 10 million ha (Debroux et al, 2007). European firms are the largest concession holders, while other concessions are held by firms based in Africa, the Middle East, India and China (Karsenty, 2007). Illegal production is thought to be more than double the volume of legally harvested commercial wood (Chatham House, 2009).

Indigenous and community forestry

The large number of indigenous and forest-dependent people in the DRC, around one-third of its population, emphasizes the importance of indigenous and community forestry. There are an estimated 200 different ethnic groups in the DRC, speaking 700 different local languages and dialects. Pygmy people, whose numbers are estimated to be between 400,000 and 600,000, are a particular focus of concern, given their almost complete reliance on forests (Debroux et al, 2007; World Bank, 2009).

The recognition accorded indigenous peoples and communities in the 2002 Forest Code has been slow to translate into practice. Priority actions were to include participatory mapping of indigenous and local peoples' territories and use rights and the development of locally appropriate community forest regimes that meet local needs and minimize the potential for capture by external interests (Debroux et al, 2007). These actions may prove critical to protect indigenous and local peoples' rights and interests as the government of the DRC seeks to expand both the extent of forests set aside for conservation and those available for logging and other extractive uses. Future plans include awarding community-based concessions in the 'protected' category of forest.

South Africa

An overview of forests and forest ownership

The extent of South Africa's native forest is around 9.2 million ha, corresponding to around 8 per cent of the country's land area (FAO, 2007).[1] Commercial timber production from native forests is minimal, although the coincidence of their distribution with that of around 40 per cent of South Africa's rural poor means that levels of fuel wood and non-wood forest products harvest are likely to be high (INR, 2005). Commercial plantations comprise around 1.4 million ha and are the basis of South Africa's internationally competitive and domestically significant forest products sector.

Since the end of the apartheid era, public policy has emphasized the economic and social advancement of black South Africans, including through foci on poverty reduction and land reform. In the forestry sector, advancement is now being realized through the 2007 Forest Sector Charter as a means of giving effect to the broad-based Black Economic Empowerment Act 2003. The Charter represents a partnership between the forest industries, labour and government to 'ensure that the opportunities and benefits from the forest sector are extended to black South Africans previously excluded from meaningful participation in the sector ... The Forest Sector ... confirms its role as a high growth potential sector for the economy, which should contribute substantially to the Accelerated and Shared Growth Initiative for South Africa (ASGI-SA)' (DWAF, 2007). The forestry sector currently contributes around 1.5 per cent of GDP, provides employment for 62,000 people and access to essential wood and non-wood products to many of South Africa's poorest people (INR, 2005).

DWAF (2009) data on forest ownership is more restrictive than that of FAO, due to different definitions of 'forest'. The former limits 'forest' to closed formations with more than 75 per cent crown cover. Under that classification, just over half (54 per cent; 0.29 million ha) of native forests are in public ownership and just under half (46 per cent; 0.25 million ha) in private or communal ownership. Of the 29 million ha of woodlands (defined by DWAF as 5–75 per cent crown cover), just under one-quarter (24 per cent; 7 million ha) are publicly owned and more than three-quarters (76 per cent; 22 million ha) are in private or communal ownership. More than 70 per cent (1 million ha) of plantations are privately owned; the state owned only 300,000ha in the census year of 2005 and the extent of public plantations has been decreasing since (DWAF, 2009).

Native forests

South Africa's extensive savanna woodlands comprise a diversity of formations, ranging from wooded grasslands, with tree cover of between 5–10 per cent and thickets with 75 per cent crown cover. Twelve major types are recognized, ranging from 0.5–20 metres in height and occurring across the range of altitudinal and rainfall zones (DWAF, 2009). These woodlands are important to local people as sources of fuel wood and non-wood products and to the local and the national economies as home to the wildlife on which a significant tourist industry is based.

The denser and typically moister native 'forests' of South Africa occur discontinuously, and typically in small patches of less than 50ha, both along the southeast and east coasts and some hundreds of kilometres inland, and in the mountains of its northern provinces (INR, 2005; DWAF, 2009). All of South Africa's native forests are classified as 'modified semi-natural' (FAO, 2007).

Planted forests and plantations

Over the past century, South Africa has developed a substantial and commercially successful plantation forestry sector, concentrated in the south and east of the country where rainfall exceeds 800mm annually. Plantations were initiated by the public sector, but private sector investment expanded from the 1960s and most state plantations were privatized in 2000 (Dlomo and Pitcher, 2005). Around 20 per cent of the state-owned plantation resource was transferred to other land uses, principally conservation and land reform, at the time of privatization (Dlomo and Pitcher, 2005). The government retained a minority stake in SAFCOL, a state-owned for-profit company; it also retained land ownership and leased the majority of the forest rights to private companies (Ham, 2006). Plantation ownership now occurs on three scales: the two large-scale companies, Mondi and SAPPI, each own more than 400,000ha of plantations; some 1800 smaller-scale growers, mostly white South Africans, together own around 360,000ha of plantations, and some 19,000 small-scale growers, mostly black South Africans, manage an average of around 2ha and a total estate of around 42,000ha of plantation. The latter typically involve some form of outgrower arrangement for pulpwood production (Mayers et al, 2001; Ham, 2006).

The plantation area in South Africa is now approximately 1.4 million hectares (FAO, 2007). Around one-third (35 per cent) of current plantations were established primarily for sawnwood, around half (55 per cent) primarily for pulp and the balance (10 per cent) primarily for mining timbers and poles (INR, 2005). The dominant species are pine (*Pinus patula, elliotti*; 54 per cent of area), eucalyptus (*Eucalyptus grandis*, hybrids; 37 per cent of area) and acacia (*Acacia mearnsii*; 8 per cent of area) (Nyoka, 2003; Forestry South Africa, 2007). Two-thirds of new forestation since 2000 has been of eucalypts (Forestry South Africa, 2007).

The impact of plantations on water yield is a central issue in water-scarce South Africa. Plantations have been allocated water use permits since 1972 and all new forestation requires a water use licence issued by the Department of Water Affairs and Forestry (Dlomo and Pitcher, 2005; INR, 2005). Planting rates have declined as a result (FAO, 2006). There have also been voluntary initiatives by forestry companies to withdraw plantations from riparian zones and national programmes to remove invasive plantation species – notably *Acacia mearnsii* and *Pinus patula* and *elliotti* – from riparian zones (Dlomo and Pitcher, 2005).

Forest governance and policy

The 1996 Forest Policy of the Government of National Unity and the 1998 National Forests Act constitute the two main pieces of forest policy currently guiding all forest management in South Africa. Other key policy measures relevant to forestry include the 1996 National Forest Strategy, the 1998 National Environmental Management Act, the 1998 National Water Act and the 2001 Amendment to the 1983 Conservation of Agricultural Resources Act (which specifies riparian zone requirements) (Dlomo and Pitcher, 2005).

The Forest Policy assigns the national government primary responsibility for overseeing forest management. The Department of Water Affairs and Forestry (DWAF) is the responsible agency.

In addition to formal legislation, industry organizations have collaborated to develop voluntary best management practice guidelines for forestry operations, including the Guidelines for Forest Engineering Practices in South Africa (formerly the South African Harvesting Code of Practice) and the Environmental Guidelines for Plantation Forestry in South Africa (FIEC, 2002).

Current policy emphases reflect both the environmental and social dimensions of forestry. Both are reflected in the Principles, Criteria and Indicators described below and the latter in the Forest Sector Charter launched in 2008. The Charter seeks to enhance the involvement of black South Africans in the forestry sector, through an increased share of ownership of existing forestry businesses, the entry of new small medium enterprises into the sector and greater participation in tree growing (DWAF, 2007).

Forest practices system
South Africa's forest practices system is now based principally on Principles, Criteria and Indicators (PCI) for Sustainable Forest Management (DWAF, 2008). These were adopted in 2005, following several years of development. Separate sets of PCI have been developed for the national and forest management unit levels and the latter is accompanied by a set of 'measures' for evaluating performance. Forest managers in both the public and private sectors are required to report on their performance according to these PCI, which now form the basis of the country's auditing system. Specific mechanisms for the implementation of this approach are still being developed (Ham, 2006; DWAF, 2008).

One of the objectives of the PCI process is to facilitate the further development of forest certification standards (DWAF, 2008). As discussed below, more than 80 per cent of South Africa's plantation forests are certified (Ham, 2006); consequently, certification standards constitute additional policies that apply to most forest practices (although, as non-governmental policies, they are not included in this book's standardized policy comparison).

Forest production and trade

Fuel wood and industrial wood harvests are of roughly comparable magnitude and each is valuable – the former in supporting livelihoods of the rural poor; the latter in supporting South Africa's wood export sector, which ranks second to mining in total value of exports (INR, 2005; Forestry South Africa, 2007). 80 per cent of rural South African households use wood as a primary energy source; fuel wood harvest – from native forests and woodlands and plantation offcuts – is estimated at 13 million m^3 annually.[2] The gross national value of this fuel wood use was estimated at 3 million Rand (INR, 2005).

South Africa produces the largest volume of industrial roundwood of any African country, almost entirely from plantations. Commercial production from native forests is limited to very small quantities for high value uses such as furniture or artefacts (DWAF, 2009). The processing sector is dominated by a small number of large, export-oriented, pulp and paper companies. In addition, there are around a dozen medium-size, mostly sawmilling, companies (Mayers et al, 2001).

Roundwood removals in 2005 were estimated at 17.5 million m³ (FAO, 2007). Intake from plantations into processing plants in 2005/6, which included imported wood and wood products, totalled 23 million m³, of which 74 per cent was to pulp, paper and board mills, and 20 per cent to sawmills. The sector is strongly export-oriented; of total sales of 15.7 million Rand, exports totalled almost 10 million Rand, comprising 37 per cent paper and around 30 per cent for each of pulp and solid wood. Imports of 8 billion Rand comprised 66 per cent paper and nearly 30 per cent solid wood (Forestry South Africa, 2007).

FOREST USE AND BIODIVERSITY CONSERVATION

Protected areas

In the DRC, the Congolese Institute for the Protection of Nature (ICCN), formerly Zaire National Parks Institute (IZCN), was established in 1975 to oversee the national park system. The DRC currently has 4.8 per cent of its land area under IUCN Categories I–IV, most of which is under Category II, national parks. Across all six IUCN protected area categories, the DRC has officially designated 12.5 per cent of its land area as protected (see Chapter 2, Figure 2.15). With regard to protected forest area, the 2002 Forest Code states a goal of 15 per cent in protected status (in other words, designated as 'classified forest') (CARPE, 2007).

The DRC's large tracts of remote primary forest, including flooded swamp forests, and its lack of transportation infrastructure play a role in limiting human impacts within its protected areas system. At the same time, civil war and severely limited resources greatly undermine enforcement capacity, while rural populations living in frontier areas remain highly dependent on the harvesting of fuel wood and bushmeat. Reflecting a complex relationship between environmental and social goals in the DRC, the ICCN is also mandated to 'ensure the socio-economic development of the communities in the protected areas in the interests of equity and security' (Baliruhya, 2003). One of the central issues in the planned expansion of the DRC's protected area system is accommodation of the interests of indigenous and local peoples (World Bank, 2009).

South Africa's protected areas comprise 6.9 per cent of its land area (see Chapter 2, Figure 2.15) and include around 11 per cent of its forests and woodlands (DWAF, 2009). About 3.3 per cent of the total land area falls under IUCN Categories I–IV. The size of many individual protected areas in South Africa is large in comparison to those in most other countries, reflecting the needs of large game animals (Hannah et al, 2002).

The relationship between protected areas and local people in South Africa has been a topic of heated debate. During the apartheid era in the 1960s and 1970s, large numbers of people were forcibly removed from newly declared protected areas. This relocation contributed to their impoverishment and generated a negative attitude towards conservation agendas, thereby exacerbating problems with poaching and resource degradation. More recently, there is some evidence that ongoing land restitution processes have led to positive developments in relation to conservation. In many cases, local people have chosen to remain in resettlement areas and negotiate various methods of compensation and benefit-sharing. These include the leasing of their land rights within parks to state conservation agencies, the development of tourism operations and – in some cases – limited hunting rights (Fabricius and de Wet, 2002).

Protection of species at risk

According to the DRC Forest Code of 2002, the forest minister is responsible for maintaining a list of protected species. As discussed above, agencies' capacity to fulfil these responsibilities is very limited.

In South Africa, species are protected on a case-by-case basis through the South Africa Natural Heritage Programme and the Sites of Conservation Significance Programme.

FOREST PRACTICE REGULATIONS: NATURAL FORESTS

Riparian zone management (Indicator: Riparian buffer zone rules)

The 2002 Forest Code in the DRC prohibits all logging within a 50-metre buffer on either side of water courses (stream size not specified) and within a radius of 100 metres around a water source.

In natural forests in South Africa, the Department of Water and Forestry (DWAF) classifies sensitive areas, including riparian zones, as 'effective', in other words harvestable, areas, and 'non-effective', reserve areas, based on degree of environmental sensitivity. Selective logging is allowed in effective areas. DWAF uses its own internal guidelines for riparian zone management which are not publicly distributed (Ackerman, 2004).

Roads (Indicators: Culvert size at stream crossings, road abandonment)

In the DRC, according to the 2002 Forest Code, concessionaires are responsible for financing the road infrastructure necessary for forest harvest. No regulations were identified that addressed technical specifications for culvert size at stream crossings or policies on road decommissioning.

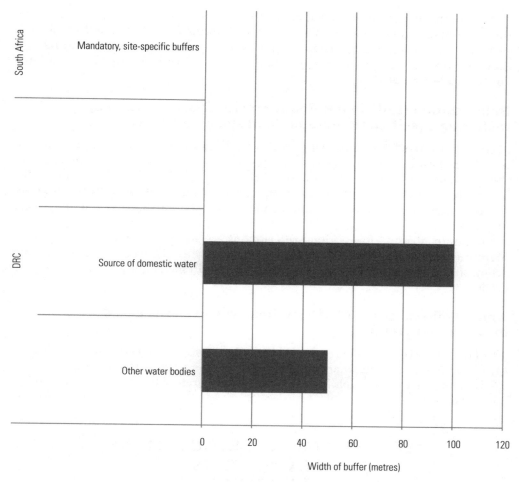

Source: DRC, 2002

Figure 9.1 *Riparian (no-harvest) buffer zone policies for the DRC and South Africa*

In South Africa, the DWAF is responsible for the environmental impacts of forest roads (Ackerman, 2004). The *South African Forest Road Handbook* (Upfold, 2006) outlines recommended technical specifications, including minimum culvert sizing. No policies on road decommissioning were identified.

Cutting rules (Indicator: Clearcutting or cutting rules relevant to tropical forestry)

Forest extraction in the DRC requires an approved management plan, which includes prescriptions on harvest methods (DRC, 2002).

In South Africa, only selective harvest is practised in natural forests, involving the removal of commercially valuable species (Ackerman, 2004). On both public and private lands, a harvesting licence is necessary to obtain permission to harvest indigenous trees within an area classified as native forest (South Africa National Forests Act No. 84 of 1998).

Reforestation (Indicators: Requirements for reforestation, including specified time frames and stocking levels)

According to Article 78 of the DRC Forest Code of 2002, whoever is 'physically or morally responsible' for the harvest of forest plants and seeds is also responsible for the regeneration of those harvested products. Article 77 of the Code states that natural regeneration and planting programmes must be established by the administrations in charge of forest management. No standardized guidelines were identified.

In South Africa, reforestation prescriptions are a required part of management planning for harvest in natural forests. Harvest methods involve single tree and group selection and there are no standardized requirements for post-harvest regeneration.

Annual allowable cut (AAC) (Indicator: Cut limits based on sustained yield)

The DRC 2002 Forest Code requires that cut levels do not exceed sustained yield, as calculated in individual management plans according to government guidelines.

In South Africa, cut levels in natural forests are prescribed through unit management plans.

FOREST PRACTICE REGULATIONS: PLANTATIONS

There are no separate rules for plantations in the DRC.

In South Africa, riparian buffer zones are required for plantations, reflecting the significance of minimizing plantation impacts on water yield. Required widths are 30 metres from the 1 in 50-year flood mark for rivers and 50 metres for wetlands. No forestry activities are allowed in these zones (2001 Amendment to the 1983 Conservation of Agricultural Resources Act).

The voluntary Guidelines for Forest Engineering Practices and the Environmental Guidelines for Plantations outline recommendations for stream crossings. The government-endorsed and now mandatory South African Principles, Criteria and Indicators (PCI) for Sustainable Forest Management incorporate best management practices, thereby requiring that road building adhere to the specific requirements of the Code.

With regard to cutting rules, reforestation and cut levels, the South African PCI are based on the concept of sustained yield. Presumably this applies to plantation forests as well, although such distinctions are not always clear in the wording of the PCI. A number of PCI expressly refer to native forest management, while others refer more generally to 'forests'. Plantations are specifically mentioned in the context of controlling pests and disease, reporting plantation area and meeting census requirements. Regardless of legal requirements, however, global competition and certification requirements provide strong incentives to maximize and sustain plantation production.

ENFORCEMENT

In the DRC, for many years civil and transnational warfare precluded the enforcement of environmental regulations. The 2002 Forest Code introduced substantial reforms, but the government has been slow to develop accompanying regulations that would help to clarify and strengthen Code mandates. Meanwhile, forest enforcement has been highly inadequate and public consultation has been minimal. While the Code requires community input in the granting of concessions, many forest concessions were allocated before the Code's enactment and involved extensive favouritism.

As discussed previously, moratoria on issuance of new concessions, and the withdrawal of existing concessions, have been instituted since 2002 in response to these issues (Debroux et al, 2007). However, a World Bank investigation panel, initiated upon request from the Indigenous Pygmy Organizations and Pygmy Support Organizations in DRC, questioned the impact of the government's concession cancellations:

> *The Panel was informed that ... substantial areas in these concessions were not covered by forests, but had been previously logged and/or were agricultural lands, swamp lands and even villages. Moreover, a substantial portion of the concession areas cancelled in 2002 that did have forest cover appears to have re-emerged as concession areas under consideration for validation ...* (World Bank Inspection Panel, 2007)

Clearly, the challenges facing enforcement in DRC are complex and highly systemic. As will be discussed further in the next subsection, lucrative foreign markets for Congolese wood, together with international scrutiny, have motivated some private concessionaires to set themselves apart as responsible managers (McDermott et al, 2008). The intention of the DRC's government to appoint an independent observer to oversee the forest sector (World Bank, 2009) represents an important initiative to strengthen forest governance.

The South African DWAF provides guidelines for first-, second- and third-party auditing according to the PCI for Sustainable Forest Management. Private companies are required to report according to those criteria, while state enterprises must use them to plan, manage and audit their natural forest and plantation operations. State-run operations must conduct annual first party audits; annual second party audits for a sample of at least one-third of all estates and involving district managers or other officers from outside the district in question, and third-party audits for an unquantified 'sample' of estates involving auditors from outside of DWAF and organized by the Deputy Director General, Regional Coordination and Support (DWAF, 2005).

FOREST CERTIFICATION

As of January 2008, no FSC certificates had yet been issued within DRC. However, a German company, the Danzer Group, with the second largest concession area in the country, has recently committed itself to working towards the FSC certification of its full supply chain. Given general trends across the Congo Basin as a whole, it can be expected that other companies selling to European markets may follow suit. However, China is also a major buyer of raw logs from the Congo and Chinese demand for certified products has been very limited (McDermott et al, 2008).

The large companies that dominate South Africa's heavily export-dependent plantation industry have almost unanimously adopted certification. The FSC reports a total of 1.5 million hectares of South African plantations certified as of January 2008.[3] Not all South African plantations are certified, however. In particular, many small-scale forest plantations and outgrower schemes face difficulties in meeting the costs of certification. This has led to concerns about the impacts of certification on social equity. At the same time, certification has reportedly led to improvements in public participation and worker benefits among those companies that do achieve certification (Ham, 2006).

Unlike in many countries, the impetus for FSC forest certification in South Africa came largely from the industrial sector. Many environmentalists are highly critical of the certification of large-scale plantations of non-native species. Perhaps reflective of this, stakeholders in South Africa have yet to agree upon a national FSC standard. Instead, FSC-accredited certifiers have developed their own generic checklists using the FSC international principles and criteria as a baseline (Ham, 2006).

The South African government has also shown strong interest in certification as a market-based complement to its PCI for sustainable management. Forest certification is required of all private companies operating on state-owned lands and some of the state-owned operations have also been FSC certified (Ham, 2006).

Table 9.1 *Matrix of policy approaches in natural forests*

Case Study	1) Riparian	2) Roads	3) Clearcuts	4) Reforestation	5) AAC
DRC (Public)	■		▨		▨
South Africa (Private)	▨	▨	▨	▨	▦
South Africa (Public)	▨	▨	▨	▨	▦

- ■ Mandatory substantive
- ▨ Mandatory procedural
- ▦ Mandatory mixed
- ☐ No policy

SUMMARY

The forest sector of the DRC embodies the many challenges facing forest-rich nations in a post-conflict situation, where there are high levels of local dependency on the forests. The value of the DRC's forests – for globally important biodiversity and environmental services, for indigenous peoples and local communities and as a vehicle for economic development – are reflected in very high levels of international interest and partnership in seeking to work with the government of the DRC to realize the goals of the 2002 Forest Code. At the same time, chronic lack of capacity in both public and private sectors, legacies of corruption and conflict, the imperative to enable economic and social development and the very scale of the country present formidable obstacles to realizing forest conservation and sustainable forest management objectives.

South Africa, a country which is not naturally forest-rich, but in which both native forest products and services and the plantation forestry sector are important for the rural poor and the national economy, illustrates a different suite of challenges and opportunities for sustainable forest management. In the South African case, forest practices systems are well established and functional, facilitated by the widespread adoption of forest certification. The principal challenges facing the South African forestry sector are to amplify the benefits it delivers to poor South Africans, as well as to address challenges associated with the environmental and social impacts of industrial tree plantations.

NOTES

1 Note that FAO data reflect a different classification of forest and woodland to that used by DWAF (2007). The latter assigns most 'forest' to the woodland category and reports only 0.53 million ha of 'forest'.
2 Note a major discrepancy from FAO (2007) data, which reported only 0.25 million m³ of fuel wood.

3   Note this figure exceeds the 1.4 million hectares of total plantation area recorded in the
    FAO 2005 inventory, presumably reflecting both data differences and some flux in planta-
    tion area.

REFERENCES

Ackerman, P. (2004) Personal communication with Professor Pierre Ackerman, Department
    of Forest Science, University of Stellenbosch, 24 February 2004
Baliruhya, Eulalie Bashige (2003) 'Conservation amid conflict', *Our Planet*, 14, 2
Bryant, Dirk, Daniel Nielsen and Laura Tangley (1997) 'The last frontier forests: Ecosytems
    and economies on the edge. What is the status of the world's remaining large, natural
    forest ecosystems?', Washington, DC: World Resources Institute, Forest Frontiers
    Initiative
CARPE (2007) 'The forests of the Congo Basin: State of the forest 2006', Central African
    Regional Program for the Environment, http://carpe.umd.edu/resources/Documents/
    THE_FORESTS_OF_THE_CONGO_BASIN_State_of_the_Forest_2006.pdf/view,
    accessed November 2009)
Chatham House (2009) 'Democratic Republic of Congo', available at
    www.illegal-logging.info/sub_approach.php?subApproach_id=70, accessed April 2009)
Debroux, L., T. Hart, D. Kamowitz, D. Karsenty and G. Topa (2007) 'Forests in Post-Conflict
    Democratic Republic of Congo: Analysis of a priority agenda', Bogor: Center for
    International Forestry Research (CIFOR)
Dlomo, M. and M. Pitcher (2005) 'Juggling social and economic goals: South Africa', in
    M. Garforth and J. Mayers (eds) *Plantations, Privatization, Poverty and Power*, London:
    Earthscan
DRC (2002) Code Forestier (Forest Code), LOI No. 011/2002, du 29 Aout 2002, Kinshasa:
    Republique Democratique du Congo (DRC)
DWAF (2005) 'Module 5: Principles, Criteria, Indicators and Standards', South Africa
    Department of Water Affairs and Forestry, available at http://www2.dwaf.gov.za/dwaf/
    cmsdocs/3104___SFM%20Training%20Course%20Module%205%202005.pdf, accessed
    November 2009
DWAF (2007) 'Forest sector transformation charter', available at http://www.dwaf.gov.za/
    Documents/ForestBEECharter.asp, accessed February 2009
DWAF (2008) 'Principles Criteria Indicators (PCI)', Department of Water Affairs and
    Forestry, http://www2.dwaf.gov.za/webapp/Documents/SF/PCI_National2008.pdf,
    accessed November 2009
DWAF (2009) 'Forests', www2.dwaf.gov.za/webapp/ForestsOverview.aspx, accessed February
    2009
Fabricius, Christo and Chris de Wet (2002) 'The influence of forced removals and land
    restitution on conservation in South Africa', in D. Chatty and M. Colchester (eds)
    *Conservation and Mobile Indigenous Peoples: Displacement, Forced Settlement and Sustainable
    Development*, Oxford: Berghahn Books
FAO (2006) 'Global planted forests thematic study: Results and analysis', A. Del Lungo, J. Ball
    and J. Carle (eds) *Responsible Management of Planted Forests: Voluntary Guidelines*, Rome:
    United Nations Food and Agriculture Organization
FAO (2007) *State of the World's Forests – 2007*, Rome: United Nations, Food and Agriculture
    Organization

FIEC (2002) 'Environmental guidelines for commercial forestry plantations in South Africa', 2nd edition, Rivonia: Forestry Industry Environmental Committee (FIEC)

Forestry South Africa (2007) 'Abstract of South African forestry facts, 2005/6', www2.dwaf.gov.za/dwaf/cmsdocs/4318___facts2006.pdf, accessed February 2009

Ham, Cori (2006) 'Forest certification in South Africa', in B. Cashore, F. Gale, E. Meidinger and D. Newsom (eds) *Confronting Sustainability: Forest Certification in Developing and Transitioning Societies*, New Haven, CT: Yale School of Forestry and Environmental Studies Publication Series

Hannah, L., G. F. Midgley, T. Lovejoy, W. J. Bond, M. Bush, J. C. Lovett, D. Scott and F. I. Woodward (2002) 'Conservation of biodiversity in a changing climate', *Conservation Biology*, 16, 1, pp264–268

INR (2005) 'Pilot state of the forest report', *Investigational Report Number 253*, Scottsville, South Africa: Institute of Natural Resources

ITTO (2006) 'Status of tropical forest management 2005', Yokohama: International Tropical Timber Organization

Kanowski, P. J. and H. Murray (2008) 'TFD review: Intensively-managed planted forests', http://research.yale.edu/gisf/tfd/ifm.html > TFD IMPF Review, accessed November 2009

Karsenty, Alain (2007) 'Overview of industrial forest concessions and concession-based industry in Central and Western Africa', Centre National de Agronomique pour le Développement

Mayers, J., J. Evans and T. Foy (2001) *Raising the Stakes*, London: International Institute for Environment and Development

McDermott, Constance, Lloyd Irland, Camille Rebelo and Benjamin Cashore (2008) 'Congo ecoregional report', New Haven, CT: Yale Program on Forest Policy and Governance

Nyoka, B. I. (2003) 'Biosecurity in forestry: A case study of the status of invasive forest species in Southern Africa', Rome: United Nations Food and Agriculture Organization, Forestry Department

Upfold, S. (ed) (2006) *South African Forest Road Handbook*, Scottsville, South Africa: Institute for Commercial Forestry Research (ICFR), www.icfr.ukzn.ac.za/icfrfiles/publication/roadv2.pdf, accessed November 2009

World Bank (2009) 'Frequently asked questions – Forests in the Democratic Republic of Congo', web.worldbank.org/WBSITE/EXTERNAL/COUNTRIES/AFRICAEXT/CONGODEMOCRATICEXTN/0,,contentMDK:20779255~menuPK:2114031~pagePK:141137~piPK:141127~theSitePK:349466,00.html#11, accessed May 2009

World Bank Inspection Panel (2007) 'Investigation report: Democratic Republic of Congo: Transitional support for economic recovery grant (TSERO) (IDA Grant No. H 1920-DRC) and emergency economic and social reunification support project (EESRSP) (Credit No. 3824-DRC and Grant No. H 064-DRC)', Report 40746-ZR, World Bank, The Inspection Panel, http://siteresources.worldbank.org/EXTFORINAFR/Resources/IP_2007_DRC_Forestry_2.pdf, accessed November 2009

PART III

Summary and Conclusions

Chapter 10

Summary of Findings

INTRODUCTION

Chapters 3 to 9 revealed a broad diversity of environmental, social and economic contexts that shape forestry across 20 case study countries and seven world regions. Variability among extent and condition of forest cover, the distribution of forest ownership, the structure of forest governance, the principal modes of forest production and forest trade, as well as the key threats facing forest conservation, have made scholars of environmental regulation in general, and forest policy in particular, reticent in undertaking global-scale comparisons of environmental performance requirements. Our approach has been the converse: instead of viewing such differences as rendering comparisons meaningless or difficult, we view standardized comparison as all the more important *because* of contextual diversity, in that it provides a means to uncover themes and patterns as to how and why forest governance arrangements develop and evolve within complex, 'real life' settings. This chapter sets the stage for such explanatory analysis through a summary of our core contextual and policy findings. Chapter 11 builds on this effort and sets the stage for future empirically grounded research, by theorizing about how these contextual factors may both explain policy variability and influence policy effectiveness.

Following this introduction, our summary is undertaken in six sections. The first section reviews landscape-scale biodiversity measures. The purpose of this section is to provide a broad overview of land use and landscape-scale protection measures important for understanding and assessing our regulatory findings. It is highly relevant, for example, to know whether intensive plantation management in one jurisdiction sits alongside relatively large protected areas, such as in the Australian state of Tasmania, or whether forestry impacts are distributed more broadly across the landscape. To address such landscape-scale issues, we focus on the extent of protected areas as well as the distribution of protection versus production forests. The challenge this book and other efforts face is that, while much improved in recent years, global-scale data on national and sub-national forest land use and protection is constrained by inconsistent classification and reporting among countries. Nevertheless we include this data to highlight the importance of considering the distribution of forest practice impacts across the landscape as a whole.

The second section presents the results of our standardized analysis of five forest practice policy criteria as they apply to commercial practices within natural forests. This includes a criterion-by-criterion review of findings followed by an overall scoring and ranking of jurisdictions in terms of their relative policy 'prescriptiveness', in other words the extent to which policies include mandatory substantive requirements prescribing specific forest practices. Overall scores[1] are then examined in light of a sampling of contextual variables, to identify possible relationships between contextual factors and policy prescriptiveness and/or threshold requirements. The theoretical implications of these emerging patterns are explored in Chapter 11.

The third section provides a brief analysis of plantation policies in countries with extensive plantation development. The fourth section then follows with a review of our findings on policy enforcement, in other words the formal systems that have been established to ensure that forest practice policies are actually implemented as written. This is then followed by a fifth section that overviews forest certification as a market-based, 'private authority' mechanism for improving forest policies and policy implementation. The sixth and final section of this chapter provides an overview of our key findings.

LANDSCAPE-LEVEL BIODIVERSITY CONSERVATION MEASURES

As discussed in Chapter 1, landscape-level biodiversity conservation is a core component of forest conservation and sustainable forest management. Our study assessed the extent of protected areas and the protection of endangered species and their habitat as key biodiversity policy measures.

The diversity of protected area arrangements discussed in Chapters 2 to 9 highlights the challenges of standardized global comparisons. Nevertheless, demand for such comparisons continues to grow, accompanied by increasing international collaboration aimed at promoting comprehensive and representative protected area networks. Among these, and particularly relevant to the summary analysis in this chapter, is the recent global protected areas target set by the United Nations Convention on Biodiversity (CBD). As mentioned in Chapter 1, this target calls for 'at least 10% of each of the world's ecological regions effectively conserved' by 2010 (Decision VII/30, CBD, 2004). All of our case study countries are signatories to the CBD and all except the United States have joined as Parties. Many Parties, furthermore, have also adopted a 10 per cent protected areas target at the national level.

The proportion of total land area formally protected varies widely in the case study countries, from 41 per cent protected in Germany to 5 per cent protected in India. Twelve of our twenty case study countries report that at least 10 per cent of their land area is protected (Chapter 2, Figure 2.15), thus meeting the CBD target;

Australia, Canada, Finland, India, Mexico, Portugal, Russia and South Africa have yet to achieve this target. However, the degree of protection also varies considerably. Only four countries – New Zealand, Chile, Latvia and the United States – have at least 10 per cent protected under IUCN Categories I–IV, in other words those areas that generally do not allow extraction of natural resources. The proportion of land area protected differed little between developed and developing case study countries: the former have on average 7 per cent protected under IUCN I–IV and 17 per cent overall, and the latter 6 per cent and 16 per cent under these categories, respectively. However, some developing countries, such as Brazil and Chile, are among the top-ranked in terms of percentage of land area protected, while Brazil is second only to the US in total area protected.

Where sub-national-level data were available, our regional chapters also revealed substantial levels of variation within countries, with some of the most biodiverse areas the least well represented. For example, a sub-national analysis of Canada and the US highlights major differences among provinces and states, with protected areas covering 40 per cent of the land area in Alaska, 10–26 per cent of all other western case study states and provinces plus Ontario, but only 1–5 per cent in the southeastern US states (see Chapter 3, Figure 3.3).

In recognition of major gaps in protected area representation, a number of case study countries have developed public policy initiatives to improve the representativeness of different ecosystem and habitat types. Examples include Australia's Regional Forest Agreement process (Montreal Process Implementation Group, 2008), which had a target of protecting 15 per cent of each forest ecosystem in Australia's most-forested regions. In Canada and the US, national efforts such as the Canadian Council on Ecological Areas (CCEA) and the USGS Gap Analysis Program (GAP) serve to facilitate systematic and standardized assessment and protection across sub-national jurisdictions. Likewise, multilateral initiatives such as the European Habitat Directive promote cross-boundary coordination on habitat protection.

At the global scale, the CBD has sought to further promote representativeness, by elaborating (CBD Decisions XIII/15 (2006) and IX/5 (2008)) the interpretation of its original 10 per cent target as applying to each of six thematic areas, including marine and coastal, inland waters, forest, mountain, dry and sub-humid and island ecosystems. In this process, the CBD has specified that the 10 per cent level of protection should apply to 'each of the world's forest types' (see Decision XIII/15, Annex IV, CBD, 2006).

A recent gap analysis by Schmitt et al (2008) was expressly designed to assess progress towards the CBD forest target. This study found that a total of 13.5 per cent of forests worldwide are protected within IUCN categories I–IV; 7.7 per cent are within IUCN categories I–IV (in other words areas that largely prohibit extractive use) and a further 5.8 per cent in Categories V–VI. Table 10.1 presents these data by world region.

According to Table 10.1, Eurasia and Africa have the smallest percentage of forests protected, while the Neotropics and Australasia have the largest. These latter regions also stand apart has having at least 10 per cent of their forests protected under the stricter IUCN Categories I–IV.

Table 10.1 *Percentage of forest cover protected, by world realm*[*]

Realm	% Protected (IUCN I–IV)	% Protected (IUCN I–VI)
Palearctic (Bulk of Eurasia and N Africa)	5.5%	8.8%
Afrotropics (Sub-Saharan Africa)	6.4%	9.2%
Neartic (most of North America)	6.6%	15.2%
Oceania (Polynesia, Fiji and Micronesia)	7.5%	8.2%
Indo-Malay (South Asian subcontinent and Southeast Asia)	9.9%	13.6%
Neotropics (South America and the Caribbean)	10.6%	21.3%
Australasia	13.4%	14.8%
Total forest cover	7.7%	13.5%

Notes: * WWF recognizes eight distinct biogeographic 'realms', seven of which include forests. The eighth, the Antarctic, has no forest cover and hence is excluded from the above table. The shading highlights percentages at or above 10 per cent, the target established in the United Nations Convention on Biodiversity (CBD).
Source: Schmitt et al, 2008

The study of Schmitt et al (2008) also developed a working definition of 'forest types' to provide further detail on protected area distribution. They identified 20 major forest types and found that 65 per cent of the forests whose types were identifiable in the dataset[2] had less than 10 per cent of their total area protected under IUCN Categories I–IV. They also considered high priority conservation areas as defined by WWF and Conservation International and found the mean and median level of IUCN Categories I–IV protection for these ranged from 8.4 per cent and 12.1 per cent. The authors point out, however, that gaps in information on forest types, and major variability in distribution among and within regions, are such that much more detailed data would be necessary to adequately assess the representativeness of forest protection worldwide (Schmitt et al, 2009). At present, only a few countries (e.g. Australia, Montreal Process Implementation Group, 2008) report protected area representation specifically by forest type.

Formally designated 'protected areas' are just one type of policy measure to address the distribution of human impacts across the landscape. Considered in isolation, such data are insufficient for placing commercial forest harvest policies within broader land use context. For example, given the wide range of methodologies used in the designation of protected areas, it is possible that what one country reports as a 'protected area' may in another qualify only as a forest management zoning decision, although both have similar degrees of both protection and permanence. Likewise, by focusing only on a single type of land use designation, protected areas do not address the relative intensity of management across the remaining parts of the landscape.

To probe further on these issues, Table 10.2 below draws on FAO data for more comprehensive forestland designations. For the purposes of measuring impact distribution, we have aggregated the six FAO designations (see Chapter 2) to distinguish between those lands designated primarily for protection or conservation and those for production. The 'other' column aggregates designations that do not specify

Table 10.2 *FAO forest designations by primary objective*

	Conservation*	Protection	Production	Other**
New Zealand	78%	0%	22%	0%
India	22%	15%	21%	42%
USA	20%	-	12%	68%
Germany	19%	22%	-	59%
Indonesia	19%	28%	54%	0%
Portugal	16%	6%	78%	0%
Chile	14%	30%	45%	11%
Latvia	14%	6%	n.s.	81%
Australia	13%	-	8%	79%
Sweden	12%	0%	73%	15%
South Africa	10%	-	16%	74%
Brazil	8%	18%	6%	69%
Finland	7%	0%	91%	2%
Mexico	7%	2%	0%	92%
Canada	5%	-	1%	94%
Poland	5%	21%	64%	11%
China	3%	31%	58%	8%
Russia	2%	9%	77%	12%
DRC	1%	no data	no data	no data
Japan	0%	0%	0%	100%

Notes: *Includes, but is not limited to, formally designated 'protected areas'.
** Other = multiple purpose, social services, none/unknown.
Source: FAO, 2006a

the degree of protection entailed. Within each category, shaded cells indicate the five countries with the largest percentage of their forest area devoted to a particular forest use.

The range of forestland designations is striking, ranging from 78 per cent for conservation of biodiversity in New Zealand to zero per cent in Japan. Likewise, China and Chile lead in the area designated primarily for protection for soil and water. The 'other' category, however, dominates in Japan, Canada, Mexico, Latvia and Australia. This latter category encompasses such a broad range of possible levels of protection as to highlight the inconclusiveness of available data on the intended distribution of human impacts across the landscape. Looking at 'conservation' and 'protection' from the perspective of their converse, 'production', does provide clues as to which countries have distributed forest harvest quite widely across their total forest area. The top five countries in these terms are Finland, Russia, Sweden, Portugal and Poland, four of which have designated more than two-thirds of their total forest area for production.

As discussed in Chapter 1, forestland designation by itself tells us nothing about the actual past, current or future condition of the forests. The efforts of the WDPA Consortium in effectiveness monitoring (WDPA Consortium, 2009) and the remote sensing technologies adopted by the FAO (Ridder, 2007) and others promise to contribute significantly to future research examining the management effectiveness of the full range of land use designations.

In addition to land use classification, our analysis has also looked at the protection of species at risk as another important policy instrument for landscape-level biodiversity conservation. All case study countries have enacted national legislation to protect species at risk and almost all formally identify habitat protection as a means to this end. In a few cases, such as India, national legislation to protect species at risk does not specifically mention habitat protection, but the importance of protecting the habitat of endangered species is clearly understood in the implementation of protection measures. As with protected areas, however, much more in-depth study is needed to assess fully the relative adequacy of existing legislation to protect threatened species and the conflicts and synergies between such legislation and policies aimed at economic development or other land use objectives.

CASE STUDY POLICY APPROACHES TO FIVE KEY POLICY CRITERIA IN NATURAL FORESTS

The five criteria

The following analysis covers 43 different national and sub-national jurisdictions. The standardized comparisons are based on forest policies applicable to the largest land ownership type for each jurisdiction. They exclude the Australian National Territory (ACT) and South Australia, where no harvest is allowed in natural forests. In two cases, New Brunswick and the US Pacific Northwest, two land ownership categories are assessed. In the case of New Brunswick, private and public forest ownerships each cover an equal area and thus both ownerships are assessed. In the case of the US Pacific Northwest, both US Forest Service and private lands are addressed because of the large area covered by national forests and the large wood volumes produced from private forests. This leads to a total of 45 leading land ownership types spread across 43 different jurisdictions.

Riparian zone management (Indicator: Riparian buffer zone rules)

Riparian buffer zone rules are the most consistently prescriptive of all of the standardized policy criteria we assess. They vary tremendously, however, across a number of parameters. Firstly, every jurisdiction has developed its own unique system for classifying the stream types to which the policies apply. Secondly, both the classification systems and accompanying riparian policies vary in their relative complexity and the type and extent of restrictions they place on management activities.

In order to provide a comprehensible, global-scale overview of policy variation, the following analyses compares policies for a subset of specified riparian buffer parameters: the maximum buffer width specified in each jurisdiction and those required for moderate (30-metre width) and small (one-metre width) streams. This analysis is intended to be illustrative rather than exhaustive.

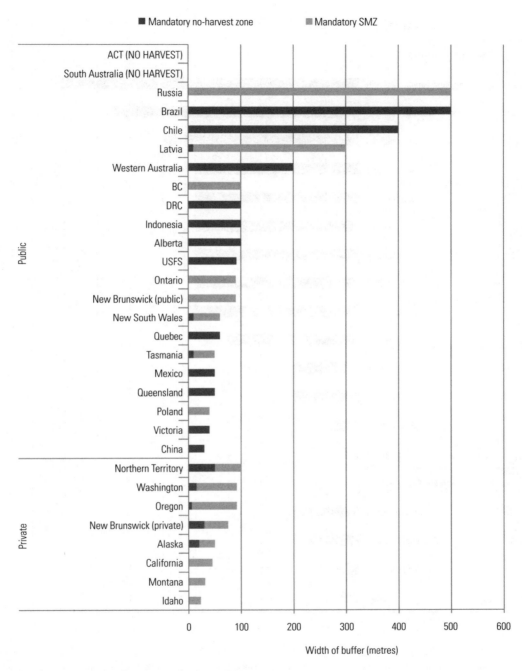

Figure 10.1 *Riparian buffer zone rules: Largest required widths*

Note: This chart includes only those cases with mandatory buffer zone requirements.
Source: See corresponding regional chapters for source information

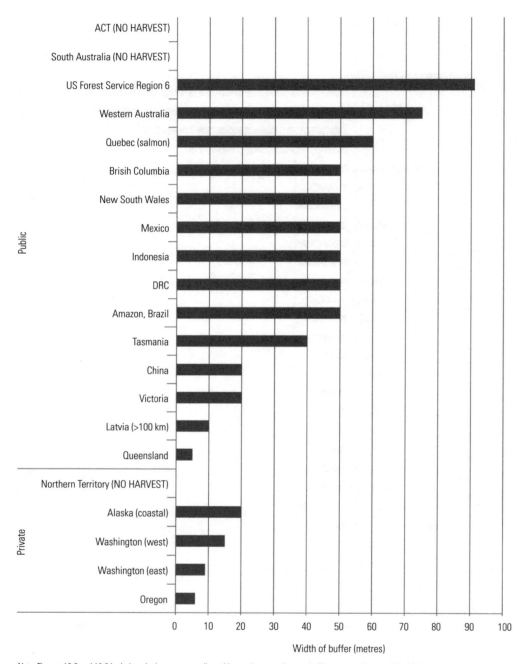

Note: Figures 10.2 and 10.3 include only those case studies with mandatory no-harvest buffer zone requirements for 30-metre wide rivers.
Source: See corresponding regional chapters for source information

Figure 10.2 *Riparian no-harvest zones: 30-metre wide stream (and/or fourth order,[3] 30 per cent slope, low soil permeability, moderate erodibility)*

Figure 10.1 illustrates the largest buffer zone width requirements for each case study jurisdiction. The types of streams to which this applies varies, but in general these buffer widths apply to most of the largest rivers in the respective regions.

The largest buffer zones are found in Brazil, Russia, Chile and Latvia. The 500-metre Brazilian no-harvest buffers, however, apply only to very large-sized rivers greater than 600 metres in width. Likewise, the 500-metre special management zones in Russia apply to rivers greater than 500km in length. The Chilean buffers apply to mountain springs, rather than streams, but – as noted in Chapter 7 – are reportedly not enforced. In riparian buffer zones that are special management zones, rather than no-harvest zones, there are substantial differences in activities allowed within these zones. For example, in some cases such as Latvia, restrictions may be limited to those on chemical use, while in others such as Tasmania, they may apply to machinery access and/or the removal of streamside vegetation, including restrictions on the proportion of trees that can be harvested.

If we compare buffer requirements on rivers that are 30 metres in width, different results emerge (Figure 10.2).

In this case, no-harvest zones apply to fewer – only 15 – of the case study jurisdictions. The largest buffer width is that required by the US Forest Service, 91 metres. The contrast between requirements on different land tenures is particularly evident in this case; buffer widths on other tenures in the US Pacific Northwest range from 6 to 20 metres. At the global scale, the prevalence of a 50-metre no-harvest zone, across a range of ecologically and geographically very diverse jurisdictions, is striking. This requirement applies, for example, to public lands in British Columbia and New South Wales and to both public and private lands in Brazil. This phenomenon is further explored later in this book (see 'Emerging patterns', Chapter 10 and 'Why 50 metres', Chapter 11).

Figure 10.3 compares no-harvest zone requirements for small sized, non-fish bearing, perennial streams of one metre in width. Only nine of the case study jurisdictions specify no-harvest zones for this stream type.

The DRC, Indonesia and Mexico have the largest no-harvest zones of 50m for this one metre stream type. It is also notable that these countries' policies make no, or little, distinction between stream sizes in the specification of riparian buffer widths. Of those jurisdictions where buffer width requirements vary with stream size for this stream class, the USFS and Brazilian Amazon have the largest buffer requirements; those for Washington state, Latvia and the state of New South Wales are half or less of these.

In general, the case study developed countries have enacted fewer nominal requirements for riparian buffer zones than the developing country cases. In Germany, Finland, Sweden, New Zealand and South Africa, special management zones are mandatory for 'natural streams', but there are no standardized requirements governing buffer zone sizes. There are no mandatory provincial requirements for streamside buffer zones on Quebec private forestlands, nor are buffer zones required on private lands in the US southeastern states, Portugal and Japanese private, non-protection forests.[5] In the US southeast, however, all case study states have developed best management practices with recommended buffer strip sizes.

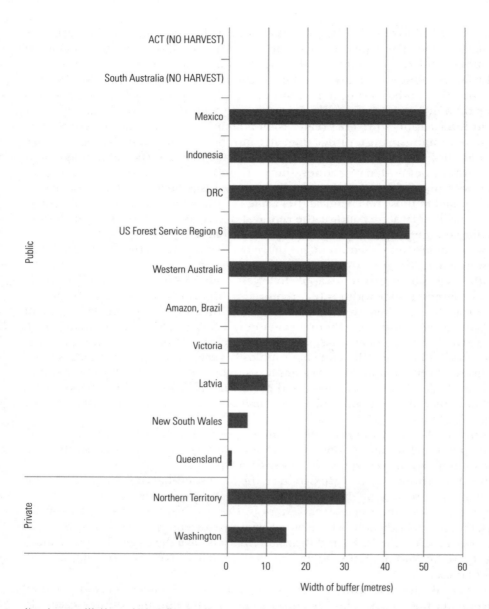

Notes: In western Washington, riparian buffers along Np streams (in other words non-fish-bearing perennial streams whose rate of flow is less than or equal to 0.57 cubic metres per second (streams that do not qualify as Washington 'shoreline')) must measure a minimum of 15 metres in width. The proportion of the stream for which the no-harvest rule applies depends on the distance of the Np stream from shoreline and/or fish-bearing streams.
In eastern Washington, 15-metre no-harvest zones are required along a portion of Np streams when clearcutting is used within the riparian special management zone.
Source: See corresponding regional chapters for source information

Figure 10.3 *Riparian no-harvest zones: one-metre wide stream (and/or first order,[4]*
perennial, no fish present, high soil permeability, moderate erodibility)

While their implementation is 'voluntary', they do have quasi-legal standing as potential proof of due diligence in terms of compliance with the US Clean Water Act.

Policy approach
The policy approach to riparian zone management for the case study jurisdictions is summarized in Tables 10.6 and 10.7. Sixty-two per cent of the 45 case study jurisdictions have established mandatory, substantive (in other words standardized threshold) requirements for riparian buffer zones. No-harvest buffer zones are found in 51 per cent of the cases. 'Mixed' rules, requiring special management buffer zones but without specified width requirements, amount to 9 per cent of the cases. Mandatory procedural rules characterize policies in 7 per cent of the cases; 20 per cent have 'contingent' voluntary rules, in that they may have legal standing in association with water quality laws, and 2 per cent (one case) has voluntary guidelines with no legal standing.

Roads (Indicators: Culvert size at stream crossings, road abandonment)

Table 10.3 lists each jurisdiction by its policy approach to culvert sizes at stream crossings and to road decommissioning. The most consistently prescriptive approaches to these two criteria were found in three Canadian provinces, two Pacific Coast western US states and two Australian states. European case studies were generally 'mixed'; as with the case of Japan, this might reflect relatively stable road networks and the prevalence of privately owned forestlands. Developing country case studies commonly took a procedural approach, or – as in the cases of Chile and the DRC – lacked any specific policies for these criteria. Many of the US southeastern states have developed guidelines for culvert sizing but, as with riparian buffer widths, apply this through a largely voluntary best management practices approach.

Policy approach
The policy approach to roads for the case study jurisdictions is sumaried in Tables 10.6 and 10.7. Sixteen per cent of the case study jurisdictions have enacted prescriptive requirements for both culvert sizes and road decommissioning. Forty-four per cent of the cases take a mixed approach, involving some form of mandatory requirement that is not consistently prescriptive for both culverts and decommissioning. A prescriptive approach to culvert sizing (18 jurisdictions) is more common than to road decommissioning (seven jurisdictions). Seven per cent of case study jurisdictions have enacted purely procedural requirements. Voluntary substantive policies in areas where Clean Water Acts apply (as discussed in Chapter 3) account for 22 per cent of cases, while 2 per cent have purely voluntary guidelines and 9 per cent contain no specific rules or guidelines.

Clearcutting (Indicator: Clearcut size limits (temperate and boreal forests) and cutting rules (tropical forests))

Figure 10.4 illustrates the clearcut size limits in all temperate and boreal jurisdictions for which limits have been established. The US southeastern states and some

Table 10.3 *Policy approach towards culvert size at stream crossings and road decommissioning*

Mandatory substantive rules		Procedural rules only	No mandatory requirements
Culvert sizes	Decommissioning requirements		
Alberta	Alberta	Madhya Pradesh	Alabama
BC	BC	Brazilian Amazon	Arkansas
New Brunswick	New Brunswick (public)	Mexico	Georgia
(public and private)	Ontario (mixed)	New Zealand	Louisiana
Ontario (mixed)	Quebec (mixed)	South Africa	Mississippi
Quebec	California	**Total: 5 cases**	Montana
Alaska	Idaho (mixed)		North Carolina
California	Oregon (mixed)		South Carolina
Idaho	USFS (mixed)		Texas
Oregon	Washington		Virginia
USFS	Germany (mixed)		Portugal
Washington	New South Wales (mixed)		Sweden
Finland	China		Poland
Germany (mixed)	Indonesia (procedural)		Chile
Indonesia	New South Wales (mixed)		DRC
Japan (mixed)	Queensland (mixed)		**Total: 15 cases**
Latvia (surfaced roads only)	Tasmania		
(mixed)	Victoria		
Russia	W Australia (mixed)		
New South Wales	**Prescriptive: 8 cases**		
N Territory (mixed)	**Total: 19 cases**		
Queensland			
Tasmania			
Victoria			
W Australia (mixed)			
Prescriptive: 18 cases			
Total: 24 cases			

Notes: * Note our final summary classification of approach to road policies presented in Tables 10.6 and 10.7 pools the results for both culvert sizing and road decommissioning. Only those jurisdictions that take a mandatory substantive approach to both indicators are classified as 'mandatory substantive' in their road policies. Those with 'mandatory substantive' approaches to only one indicator, and/or 'mixed' approaches to either indicator, are classified as 'mixed'.

Western European jurisdictions are notable for their lack of any specific limits. Only modestly more restrictive are the relatively large-size limits in Ontario (260 hectares) and Russian Far East pioneer hardwoods (250 hectares) (depicted in Figure 10.4 as 'Russia 7').

At the other end of the spectrum, no clearcutting on any scale is allowed in native public forests of New South Wales, New Zealand (with exceptions of a maximum 0.5ha allowed for beech forests) or South Africa and clearcutting is restricted to specific forest types in Chile, Victoria and Western Australia. In these jurisdictions, the requirements for other silvicultural systems are generally highly specified.

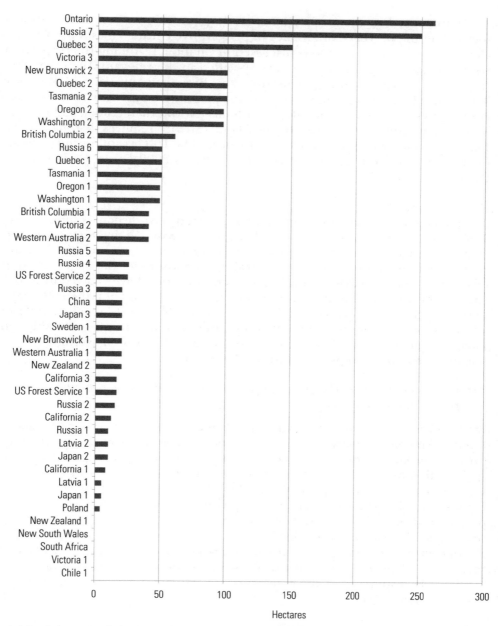

Notes: Includes only those case studies for which mandatory clearcut size limits have been established, including those non-tropical case studies where clearcutting is prohibited in natural forests (e.g. on some public lands in New Zealand and in New South Wales and South Africa). The figure does not include tropical case studies. A number is provided for those cases where more than one size limit has been established, starting with '1' for the smallest limits that apply. See regional chapters for specifics as to when the different sizes apply.

Source: See corresponding regional chapters for source information

Figure 10.4 *Clearcut size limits*

With regard to the tropical case study jurisdictions, our comparison has focused on the presence or absence of prescriptive cutting rules, rather than on clearcut size limits, for reasons discussed in Chapter 1. Among the tropical case study jurisdictions, cutting rules are specified in the relevant Australian states (Northern Territory and Queensland), the Brazilian Amazon and Indonesia. These vary from requirements for harvest planning on an individual tree basis (Brazil), to standard minimum diameter limits (Northern Territory, Indonesia), to more site-specific requirements (Queensland). Mexico requires that management plans describe proposed harvesting practices. The DRC is still in the process of developing technical guidelines to accompany its 2002 Forest Code (Debroux 2004).

Policy approach
The policy approach to cutting rules for the case study jurisdictions is summarized in Tables 10.6 and 10.7. Twenty-two out of forty-five case studies (in other words 49 per cent) have established maximum clearcut sizes and/or mandatory substantive cutting rules that apply to the entirety of natural forestlands under the major forestland ownership type. Two case studies (developing countries) have enacted procedural rules governing harvesting patterns (4 per cent). Four countries (the DRC, Japan, Latvia and Sweden) (9 per cent of the cases) have established clearcut limits affecting a portion of the natural forestlands under the dominant forest ownership type. Finally, 16 jurisdictions (36 per cent) have no mandatory rules governing clearcutting or, in the case of tropical countries, other cutting rules.

Reforestation (Indicator: Requirements for reforestation, including specified time frames and stocking levels)

Prescriptive reforestation policies that specify the required time frames and stocking levels (stems per hectare) for forest regeneration in natural forests are found throughout most of our case studies. In general, precise prescriptions involving specification of stocking levels are most commonly associated with clearcutting, whereas natural regeneration is expected to play a greater role in selective harvesting systems. Nevertheless, even in some jurisdictions where selective harvesting is practised, such as in the Australian states and New Zealand, both time frames and stocking levels have been prescribed.

In contrast, reforestation requirements are not specified on private lands in the southeastern United States, in Portugal and in non-protection forests in Japan.

Table 10.4 summarizes the policy frameworks governing reforestation in each case study jurisdiction.

Policy approach
The policy approach to reforestation for the case study jurisdictions is sumarized in Tables 10.6 and 10.7. Fifty-one per cent of the forty-five case study jurisdictions have established required time frames and stocking levels for reforestation, 11 per cent require reforestation but do not have mandatory time frames and/or stocking levels that apply to all forests, 13 per cent have established procedural reforestation rules,

Table 10.4 *Reforestation requirements*

Mandatory substantive reforestation rules	Procedural reforestation rules	Voluntary guidelines	No rules/guidelines
Alberta * ⇑	Louisiana (commercial forestlands only)	New Brunswick (private)	DRC
British Columbia * ⇑	Chile	Alabama	**Total: 1 case**
New Brunswick (public) * ⇑	Mexico	Arkansas	
Ontario * ⇑	Indonesia	Georgia	
Quebec * ⇑	Madhya Pradesh	Montana	
Alaska *	South Africa	Mississippi	
California * ⇑	**Total: 6 cases**	North Carolina	
Idaho * ⇑		South Carolina	
Oregon * ⇑		Texas	
Virginia (some loblolly/ white pine forests) (mixed)		Portugal	
Washington * ⇑		Japan (non-protection forests)	
USFS * ⇑		**Total: 1 cases**	
Finland * ⇑			
Germany * ⇑			
Sweden * ⇑			
China * ⇑			
Japan (protection forests)			
Poland * ⇑			
Latvia * ⇑			
Russia * ⇑			
Brazilian Amazon * ⇑			
New South Wales * ⇑			
Northern Territory ⇑			
Queensland ⇑			
Tasmania * ⇑			
Victoria * ⇑			
Western Australia * ⇑			
New Zealand * ⇑			
Total: 28			

(23 with time frames and stocking levels for all forest types)

Mandatory Standardized Timeframes *
Mandatory Standardized Stocking Levels ⇑

22 per cent have voluntary guidelines and the DRC has yet to develop rules as directed by its 2002 Forest Code.

Annual allowable cut (AAC) (Indicator: Cut limits based on sustained yield)

Of the five standardized forest practice policy criteria we assess, AAC is that which is not specified by the largest number of jurisdictions (18 cases), across several world regions (Table 10.5). Seventeen jurisdictions in various regions of the world establish

Table 10.5 *AAC requirements*

Stringent (AAC capped by an even-flow policy)	Mixed (AAC capped by sustained yield, but not even flow)	Procedural (AAC required but not capped by sustained yield)	No rules (No mandatory requirements to cap AAC)
USFS	Alberta	BC	New Brunswick (private)
Latvia	New Brunswick (public)	Ontario	Alabama
Poland	Quebec	Bavaria	Alaska
Russia	California	China	Arkansas
New South Wales	Indonesia	Madhya Pradesh	Georgia
Queensland	Japan	DRC	Idaho
Tasmania	Brazilian Amazon	**Total cases: 6**	Louisiana
Victoria	Chile		Montana
Western Australia	Mexico		Mississippi
New Zealand	N Territory		North Carolina
Total cases: 10	South Africa		Oregon
	Total cases: 11		South Carolina
			Texas
			Virginia
			Washington
			Finland
			Portugal
			Sweden
			Total cases: 18

annual cut levels based on a variety of factors including more general concepts of sustained yield and/or a range of economic, social and environmental considerations. The most prescriptive approach is taken by the USFS, Latvia, Poland, Russia, five Australian states and New Zealand, each of which employ a mandatory substantive approach to capping Annual allowable cut at the equivalent of non-declining even flow.

Policy approach

The policy approach to AAC for the case study jurisdictions is sumarized in Tables 10.6 and 10.7. Twenty-two per cent of the jurisdictions we assess have mandatory substantive AAC requirements, in other words, cap AAC at non-declining even flow; 20 per cent have mixed requirements, i.e. cap harvests at long-term sustained yield; 13 per cent have procedural requirements, involving various calculations not necessarily capped by sustained yield; and 40 per cent, all on private lands, have no requirements regarding AAC.

Table 10.6 *Percentage of case study jurisdictions by policy approach*

	Mandatory substantive	Mixed	Procedural	Voluntary contingent	Voluntary	No rules
1) Riparian zones	62%	9%	7%	20%	2%	0%
2) Roads	16%	44%	7%	22%	2%	9%
3) Cutting rules	49%	9%	4%	0%	0%	36%
4) Reforestation	51%	11%	13%	0%	22%	2%
5) AAC	22%	20%	13%	0%	0%	40%

Notes: "Mixed" = mandatory substantive policies, but with thresholds not specified. "Voluntary contingent" = voluntary policies that may be used as proof of due diligence in compliance with mandatory regulations (e.g. clean water Acts).
The above figures do not include Australian Capital Territory (ACT) and South Australia where commercial timber harvest is prohibited in all native forests.

Summary of regulatory approaches

This section summarizes the regulatory approach for each of the above five policy criteria, for the 45 case study jurisdictions and ownerships selected on the basis explained above.

We have scored jurisdictions in terms of the policy stringency for each of the five policy criteria, using an arbitrary scoring system that differentiates according to classification described in Chapter 1: whether the rules enacted are 'voluntary' (discretionary) versus 'mandatory' (non-discretionary) and 'substantive' (performance-based) versus 'procedural' (systems-based). The 'mixed' policy category applies to those mandatory substantive rules that are limited to statements of general objectives (for example, requiring that harvested areas be reforested) as opposed to specific policy prescriptions (for example, within one year of harvesting, stands must be restocked to a level of 1000 seedlings per hectare). Specific policy prescriptions that apply to only a part of the forested land base (e.g. riparian buffer requirements in Japan) are also considered 'mixed'. Finally, 'contingent voluntary' policies are those that are legally recognized as proof of due diligence in meeting legal requirements, such as the prevention of non-point source water pollution under the US Clean Water Act.

Our scoring system assigns two 'points' for mandatory substantive policies that include specific policy prescriptions, one 'point' for 'mixed' rules, one 'point' for mandatory procedural policies and half a 'point' for 'contingent voluntary' policies. The aggregate score, which we have called 'level of prescriptiveness' is simply the sum of scores for each of the five policy criteria for that jurisdiction. This scoring system was designed exclusively for the purpose of comparing relative policy prescriptiveness. The score does not imply policy superiority, nor does it address the issues of policy enforcement and effectiveness, which we discuss later in this and the subsequent chapter.

Table 10.7 presents the results for each policy criteria for all case study jurisdictions, ranked by aggregate prescriptiveness score.

Emerging patterns

The substantial variation in policy approach across case study jurisdictions is a striking feature of these results, with aggregate levels of prescriptiveness ranging from the maximum (ten) to the minimum (zero) possible. Policy settings for public lands in the US, parts of Canada, Australia and Russia are highly and consistently prescriptive. In contrast, very non-prescriptive approaches typify private lands in the US southeast, parts of Canada and Portugal.

If we consider all land ownership types, rather than just the leading ownership covered in Table 10.7, the contrast between public and private lands in many case study jurisdictions is marked – for example, in New Brunswick and New South Wales public forests are each scored at nine, whereas their private forests are scored at three and zero, respectively. For the 19 case study jurisdictions in which we have reviewed both public and private tenures,[6] some 70 per cent had different levels of prescriptiveness for the two tenures; the average score for public lands in these jurisdictions was eight and that for private lands, four. Only in 30 per cent of case studies are public and private land treated equivalently in terms of levels of prescriptiveness – the Australian states of Queensland (eight) and the Northern Territory (seven), the Brazilian Amazon (eight), California (nine), Latvia (nine) and Mexico (six).[7]

There is also a marked difference in the average level of prescriptiveness between 'developed' and 'developing' case study countries. Table 10.8 summarizes these differences by showing average prescriptiveness scores by both level of development and land ownership type.

Across all case study jurisdictions, the average prescriptiveness score for public lands is twice that for private lands. Overall, the developing case study country jurisdictions are around 20 per cent more prescriptive than are developed countries. However, this average masks differences in levels of prescriptiveness on public and private land in developing and developed case study countries. Developed countries take a more consistently prescriptive approach on public lands and a less prescriptive approach on private lands than do developing countries.

Turning to democratic, rather than economic, development, we find no clear relationship between estimated levels of democracy and prescriptiveness. Table 10.9 presents case study countries categorized according to the public land prescriptiveness score and the *Economist*'s democracy index (see Chapter 2). It illustrates that there is a wide range of forest practice prescriptiveness for any level of democratic development and, correspondingly, that countries with very different levels of democracy maintain forest practices systems with similar levels of prescriptiveness.

If, instead of looking variables such as 'development' or 'democracy' individually, we explore the overall distribution of countries at different levels of prescription and consider the broader contextual analysis of preceding chapters, further patterns emerge. Of the countries with the most prescriptive forest practices, one group (Australia, Brazil, New Zealand, western Canada and US public lands) has experienced high levels of public debate and/or international pressure about forest management. A second group (China, Latvia, Poland, Russia) has, or has only recently emerged from, a strongly 'top-down' political system. Countries with

Table 10.7 *Summary of jurisdictional approaches to all five forest practice criteria*

Level of Prescription (1–10) Case Study	1) Riparian	2) Roads	3) Clearcuts	4) Reforestation	5) AAC
10 Tasmania (Public)					
10 Victoria (Public)					
9 Alberta (Public)					
9 British Columbia (Public)					
9 California (Private)					
9 Latvia (Public)					
9 New Brunswick (Public)					
9 New South Wales (Public)					
9 Poland (Public)					
9 Russia (Public)					
9 USFS Region 6* (Public)					
9 Western Australia (Public)					
8 Brazilian Amazon (Public)					
8 China (Public)					
8 New Zealand (Private)					
8 Ontario (Public)					
8 Quebec (Public)					
8 Queensland (Public)					
8 Washington (Private)					
7 Indonesia (Public)					
7 Northern Territory (Private)					
7 Oregon (Private)					
6 Chile (Private)					
6 Mexico (Public)					
5 Idaho (Private)					
5 Japan (Private)					
5 Madhya Pradesh (Public)					
5 South Africa (Public)					
5 Sweden (Private)					
4 Alaska (Private)					
4 Bavaria (Private)					
4 DRC (Public)					
4 Finland (Private)					
3 New Brunswick (Private)					
2.5 Montana (Private)*					
2 Louisiana (Private)*					
2 Virginia (Private)*					
1 Alabama (Private)*					
1 Arkansas (Private)*					
1 Georgia (Private)*					
1 Mississippi (Private)*					
1 North Carolina (Private)*					
1 South Carolina (Private)*					
1 Texas (Private)*					
0 Portugal (Private)					

Legend:
- Mandatory prescriptive
- Mandatory mixed
- Mandatory procedural
- Voluntary
- No policy

Note: Native forest harvest is prohibited on: ACT public lands (there are no ACT private native forests); SA public/private lands, NZ public lands and Chilean public lands.

* USFS lands account for the following percentages of forest cover in the US western case study states: 75 per cent ID; 27 per cent MT; 9 per cent AK; 48 per cent OR; 37 per cent WA; 43 per cent CA. Harvest = 9 per cent ID; 4 per cent MT; 9 per cent AK; 5 per cent OR; 1 per cent WA; 8 per cent CA..

moderate (four–seven) levels of forest policy prescriptiveness also fall into two groups: the rich democracies of Finland, Germany, Japan and Sweden, which have long histories of forest management, and a diverse group of less developed countries. This latter group includes those that take a relatively procedural approach (e.g. India, Mexico and South Africa), harvest minimally in natural forest (Chile, South Africa), have never fully developed their forestry regulations (DRC) and/or are undergoing a complex process of decentralization (Indonesia). Very non-prescriptive forest practice regulations typify private lands in three rich democracies – Canada, Portugal and the US.

Clearly there are numerous possible patterns to explore regarding the relationship between context and policy prescriptiveness. But what of the actual content of policy prescriptions: might our data reveal patterns there as well? The above comparison of riparian threshold requirements, illustrated in Figures 10.1–10.3, suggests there are. For example, developing countries make up only one-quarter of all case study jurisdictions. Nevertheless, the top four largest buffer widths, and six out of the top ten, are located in developing countries, including countries in tropical, temperate and boreal regions. Likewise, the three largest no-harvest zones for 1-metre no-fish streams have been established in developing countries.

Another pattern among riparian thresholds, as noted above, is the prevalence of 50-metre threshold sizes across a wide range of ecosystem conditions. Fifty-metre buffer requirements are found across regions and stream types, from the DRC and Indonesia to New South Wales and Quebec. We discuss possible reasons for this apparent policy convergence in Chapter 11.

The final trend we note relates to changes in forest policies *over time*. While the assessment of policy change was not the focus of our standardized comparison, our contextual analysis of policy development, together with previous parallel studies of policy settings (Cashore and McDermott, 2004; McDermott and Cashore, 2007), do suggest a possible broad longitudinal pattern worthy of further research. That pattern is of a general, cumulative increase in the number of jurisdictions with forest practice codes and accompanying regulations and/or best practice policies. However, this overall tendency towards increased prescriptiveness was not without variation. For example, a number of our case study countries have taken steps to decentralize forest authority (e.g. Brazil, Indonesia, Mexico and Russia) and/or move towards

Table 10.8 *Average level of prescriptiveness by level of development and land ownership type across five environmental practices measures (riparian buffers, road building, clearcutting, reforestation, annual allowable cut)*

Level of Development[*]	Public	Private	Total
Developed countries	8.8	3.4	6.1
Developing countries	6.8	6.0	6.7
Total	7.9	3.5	5.6

Note: Scale 0–10, 10 = most prescriptive, 0 = least prescriptive.
[*] As discussed in Chapter 2, 'developed' refers to countries that had a per capita GDP of US $10,000 or more in 2004, and an HDI ranking in the top 33 countries worldwide.

Table 10.9 *Democracy Index and Policy Prescriptiveness**

		Prescriptiveness score			
		< 3	4–5	6–7	8–10
Democracy Index	< 3		DRC		China
	4–5				
	6–7		India Mexico South Africa	*Chile* Indonesia	**Brazil** **Latvia** Poland Russia
	8–10	*Canada* *Portugal*	*Finland* *Germany* *Japan* *Sweden* *US*	*Australia* *Finland* *Germany* *Japan*	Australia Canada *New Zealand* US

Note: *Average prescriptiveness score for public and private lands. (Italics denote policies on *private* lands, regular font denotes policies on public lands, and bold denotes both public and private lands.)

more 'results-based' rather than prescriptive approaches (e.g. British Columbia). However, these policy changes do not appear, at least as of yet, to have in practice decreased specific threshold requirements for our policy variables. Furthermore, there has been notable persistence in these requirements (e.g. riparian policies in Chile's 1931 Forest Law) and/or increases (e.g. the expansion of riparian zone buffers in Washington and Oregon, in response to endangered salmon listings) suggesting that quantitative policy prescriptions tend to persist or become more demanding over time, even with fluctuations in the stringency of the larger framework of environmental protection and enforcement.

The exploration of *why* all of the above patterns may have emerged, and what they may mean for future policy development, is covered in Chapter 11.

CASE STUDY POLICY APPROACHES TO PLANTATION MANAGEMENT

The definition, role and importance of 'planted forests' vary considerably among the case study countries. Nevertheless, there is a clear trend in many of the case study jurisdictions, as there is globally, towards increased emphasis on wood production from intensively managed plantation forests (Kanowski and Murray, 2008). This trend has significant implications for wood production, for some of the environmental services delivered by forests and for forest practices systems. This trend raises the issue of whether intensively managed plantation forests warrant different forest practice policies to those that apply to natural forests.

Definitions of plantations typically exclude large areas of planted forest in many jurisdictions, in Europe and elsewhere, that are classified as 'semi-natural planted'

Table 10.10 *Comparison of policies governing natural forests versus plantations in countries where productive plantations produce a large percentage of the wood supply*

	Total plantation area (1000 hectares)	Separate policies for NF versus tree plantations	Policies relating to plantations are primarily voluntary	Limits on forest conversion to tree plantations*	Regional/state-level prohibitions on natural forest logging
China (public)	31,369	Yes	No	Yes	Yes
USA (SE, private)	17,061	No	Yes	No	No
Russia (public)	16,962	No	No	No	No
Brazil (private)	5,384	Yes	No	Yes	No
Indonesia (public)	3,399	Yes	No	Yes	No
Chile (private)	2,661	Yes	Yes	Yes	Very little NF logging
New Zealand (both)	1,852	Yes	Yes	Yes	Yes
Australia (both)	1,766	Yes	No	Yes	Yes
South Africa (private)	1,426	Yes	Yes	Yes	Very little NF logging
Portugal (private)	1,234	No	Yes	No	No

Note: * This refers to conversion that does not change the land classification from forest to non-forest. In some US states, for example, short-rotation tree crops are considered agriculture and are subject to agricultural regulations.

production forests (FAO, 2006b). Many of these forests are also managed relatively intensively for wood production and other values and many of the issues which apply to plantation forests may also apply to them.

As outlined in Chapters 1 and 2, the largest productive plantation areas are found in China, the USA, Russia, Brazil, Chile, Indonesia, New Zealand, Australia, South Africa and Portugal. In Chile, New Zealand, South Africa and Portugal, nearly all (more than 95 per cent) industrial wood production is now sourced from plantations; amongst the other countries in the top ten, the proportion varies from around 70 per cent for Australia, to nominal for Russia. We have compared the forest practice policies that apply to 'plantations' with those applied to other 'forests' in these ten case study jurisdictions for which plantation production is most important (Table 10.10).

In most cases, forest practice requirements for plantation management are distinct from those for natural forest management and allow for considerably more management discretion. The exceptions to this generalization are the US, Russia and Portugal, where no policy distinction is made between plantations and natural forests. However, forest practice rules in the US southeast, where most US plantations are located, and those in Portugal, which has little native forest production, are in general highly discretionary. Plantations in Russia, as noted in Chapter 6, would more likely be classified as 'semi-natural planted forests' since they consist of mostly native species and are less intensively managed than those in warmer, faster-growing climates.

If we compare the level of policy discretion governing plantations across jurisdictions, irrespective of forest policies governing natural forest management, substantial variation is evident. In about half the case study countries – China, Brazil,

Indonesia, Russia and some Australian states – relatively prescriptive forest practice policies apply to plantations. For the other half – the other Australian states, Chile, New Zealand, Portugal, South Africa and the US southeastern states – plantation policies are voluntary; although in some cases (e.g. Chile), civil society pressure may ensure that voluntary requirements are observed by large plantation companies and in others (e.g. South Africa), certification may play a role equivalent to a more prescriptive forest practices code.

There are constraints on the conversion of native forests to plantations in most of the plantation case study jurisdictions, with the exceptions of Portugal, Russia and the US southeast. However, the impacts of conversion policies vary with the type of tree species used and management intensity. In Russia and the US southeast, the majority of species planted are native to the region. However, plantation management is much more intensive in the US southeast than Russia, with regard to such factors as chemical use and mechanized site preparation, tending and harvest practices.

There does not appear to be a simple relationship between the relative importance of plantations in a jurisdiction's wood production and the level of policy prescriptiveness applying to plantation forests. For example, plantation practices are relatively prescribed in the Australian Capital Territory but voluntary in South Australia, the two Australian states in which no harvesting of native forests is allowed. Plantation polices are prescriptive in China, where the intent is to meet the majority of domestic needs through plantation production, but voluntary in Chile, New Zealand, Portugal and South Africa, all of which have only very limited native forest harvesting and very significant plantation production.

ENFORCEMENT AND COMPLIANCE

As noted in Chapter 1, our assessment of forest practice enforcement and compliance arrangements in each of the case study jurisdictions is limited to an overview of the system for oversight of forest practices, including any formal auditing requirements. Given the complexity of these arrangements, a more complete treatment was beyond the scope of this book; we discuss related issues later in this chapter and in Chapter 11.

Categories of enforcement and compliance arrangements

A number of categories of enforcement and compliance arrangements are evident from our case studies:

*   One category is that in which there are no formal forest practice requirements, or where compliance with standards is voluntary. Around one-quarter of our case study jurisdictions are in this category. In such cases, typified by the US southeastern states, enforcement and compliance monitoring are by definition very limited. They occur entirely on a voluntary basis, other than where forestry

activities impinge on other legal requirements, such as the maintenance of water quality or the protection of endangered species. However, as we discussed in Chapter 3, some landowners in these jurisdictions may voluntarily adopt auditing regimes, some of which may be quite comprehensive.

• A second category is that which applies to public lands in the eastern Canadian provinces, the Eastern and Western European countries and those in Asia and Africa. In these cases, the forest agency responsible for forest management is also responsible for enforcement and compliance. Around one-quarter of our case study jurisdictions are in this category.

• A third category is that in which there is third-party oversight of forest practices by a government agency. On public lands, third-party oversight necessarily involves an agency other than that responsible for forest management, such as an environmental protection agency. For private lands, it may involve a forestry or environmental agency. In a few cases, such as Japan, this responsibility may fall to municipalities. This form of enforcement and compliance monitoring applies to the Latin American and most of the Oceanic case study jurisdictions, to both federal and private forests in the western US and to those European countries (Finland, Latvia and Sweden) in which public forestry agencies oversee private forest management. Nearly half our case study jurisdictions are in this category.

• A fourth category is that in which dedicated, independent forest practices agencies have been established specifically to oversee and report compliance with the forest practices system. This is the case for two of our case study jurisdictions, the Australian state of Tasmania and the Canadian province of British Columbia. In both cases, the forest practices agencies are responsible for oversight of forest practices across all tenures.

These results suggest that, while institutional arrangements for formal oversight of forest practices vary, some form of oversight is the norm globally; the exceptions amongst our case studies are concentrated in the US southeastern states. The institutional arrangements also reflect a trend towards the use of independent third-party monitors.

Enforcement arrangements and regulatory contexts

The character and structure of forest practice enforcement also varies considerably across case study jurisdictions, reflecting in part the variation in public policy and regulatory norms between case study countries. At one extreme are the highly regulated requirements for, and the highly litigious context of, management of US federal forests, where the ability of interest groups to sue government agencies for non-compliance has been a major force shaping the conduct of forest practices. At the other extreme are the entirely voluntary forest practices systems that, by definition, do not require enforcement.

In the majority of case study jurisdictions, however, enforcement is based primarily on government agencies acting in first-, second- or third-party roles. In a number of jurisdictions – for example, in Finland, Ontario, Sweden and Tasmania –

the forest practices systems are explicitly or implicitly co-regulatory, emphasizing the role of the overseeing agency in facilitating the improvement of forest practices as well as in assessing compliance. A similar rationale has led British Columbia to adopt a 'results-based' forest practices regime (Chapter 3).

Compliance auditing and reporting

Some form of systematic auditing of forest practices occurs in most of our developed country case study jurisdictions: almost universally in Australasia and Europe, and in many of the Canadian and US jurisdictions. Among non-European developing country case studies, only South Africa, where commercial wood production is almost entirely plantation-based, has a comparable auditing system. Regular formal public reporting of the results of compliance audits occurs in a smaller number of cases, including the Australian states of Tasmania and Victoria, the Canadian provinces of British Columbia and Ontario and some of the European case study countries. The most comprehensive examples of this are the annual reports produced by the forest practices agencies in British Columbia and Tasmania, and by Ontario's Ministry of Natural Resources. The extent of such reporting appears to be increasing, with similar initiatives planned in other Australian states and in South Africa.

Regular enforcement and compliance reporting on a comparable basis allows the assessment of trends over time within jurisdictions; differences in regulations and practices between jurisdictions mean that cross-jurisdictional comparisons are less meaningful. Such a temporal analysis was possible in only a few of the case study jurisdictions. In British Columbia, the number of 'enforcement actions' and 'monetary penalties and tickets' issued changed little over the decade to 2005/6, although the number of more minor 'compliance actions' increased by 30 per cent (BC MFR, 2006, see Indicator 21–3, page 102). In Ontario, rates of compliance assessed by self-reporting were stable at around 96 per cent over the three years to 2005 and rates of compliance assessed by independent audits of a sample of operations were generally stable over the period 2001–2005, with 83 per cent fully complying and 6 per cent generally complying (MNR, 2008: Chapter 7). In Tasmania, the number of enforcement notices issued declined by 30 per cent over the period 2003/4–2007/8 and the proportion of forest operations judged to meet or exceed forest practice requirements increased from 86 per cent to 91 per cent over the same period (Forest Practices Board (Tasmania), 2004; Forest Practices Authority (Tasmania), 2008).

Results such as these, which indicate high rates of compliance and functional enforcement mechanisms, are likely to be typical of our case study countries which rate highly in terms of good governance; broadly, the more economically developed nations. They are not typical of those countries that suffer poor governance. We discuss these issues further below.

Enforcement capacity and forest governance

Enforcement capacity is particularly an issue in countries with poor forest governance – a situation characterized by the World Bank (2008) as lacking 'adherence to the rule of law, transparency and low levels of corruption, inputs of all stake-holders in decision-making, accountability of all officials, low regulatory burden, and political stability'. Many of these countries suffer from high levels of illegal logging and other forest crime, indicative of poor enforcement capacity and forest gover-nance.

Four of our case study countries – Brazil, DRC, Indonesia and Russia – are among those identified as having the highest levels of illegal logging globally (Seneca Creek Associates and Wood Resources International, 2004). In the case of Brazil and Russia, these issues are more problematic in the frontier regions of the country – the Amazon and the Russian Far East, respectively – than elsewhere. As discussed in Chapter 7, Mexico also suffers high levels of forest crime and enforcement capacity is acknowledged to be poor in Chile and India (Chapters 7 and 5, respectively). These countries are also amongst the lowest ranked of our case studies in terms of social indicators such as human development and GDP per capita, and governance indicators such as the Corruption and Environmental Regulatory Regime Indices (Chapter 2). In these countries, broad ranging measures to improve forest governance, and strengthen legal frameworks and institutions relevant to the forest sector, are fundamental to improving legal compliance across the forestry sector, including in respect of forest practices (FAO, 2006a; ITTO, 2006; World Bank, 2008).

Enforcement capacity can also be an issue in richer, more-developed countries. Here, reductions in public sector expenditure and agency staffing may challenge the capacity to adequately implement regulatory regimes. In the best cases, such pressures prompt the redesign of regulatory regimes and the implementation of 'new generation' approaches, as we note in Chapters 1 and 11. However, without such changes, reduced capacity can result in the diminution of the effectiveness of forest practices systems, and the consequent deterioration of forest practices.

FOREST CERTIFICATION

As summarized in Chapter 2 and further elaborated in the empirical chapters, 'non-state market driven' (NSMD) forest certification systems, which develop their own private standards regarding sustainable forest management and which seek to earn forest manager and firm support from market incentives, have developed at very different rates among our case study countries (see Figures 2.18 and 2.19 in Chapter 2). We've included certification in our empirical chapters in recognition of its growing influence on forest practice policies worldwide, both directly through certi-fication standards, and indirectly through its impacts on public policy (Auld et al, 2008).

Our comparative analysis situates our case studies within the context of two widely observed trends regarding certification at the global level (see, for example, Rametsteiner and Simula, 2003; Klooster, 2005; Cashore et al, 2006; Cashore et al, 2007; Auld et al, 2008). Firstly, despite the strong initial focus of many certification supporters on halting tropical forest degradation and promoting community benefit, the vast majority of certified land lies within Europe and North America. Furthermore, a major focus of certification activism to date has been the ongoing competition among ENGO- and industry-backed certification systems for dominance of large-scale, industrial production (Cashore et al, 2004a); most of this is concentrated in the north, as well as in some southern plantation forests.

The literature on certification offers a range of possible explanations for these trends (see, for example, Rametsteiner and Simula, 2003; Klooster, 2005; Cashore et al, 2006; Cashore et al, 2007; Auld et al, 2008). The first focuses on the central issues of civil society involvement and market demand. The majority of high capacity, well-coordinated environmental and social organizations are based in developed countries, e.g. our case studies in North America, Western Europe and Australasia. These countries also contain the largest, most lucrative markets for green products, and the majority of wood products feeding these markets comes from the developed world (McDermott and Cashore, 2008). Thus, to the extent that wood products from all sources are considered equivalent, developed country producers would dominate markets for eco-certified products. Pushing demand for developed country certification further, domestically oriented environmental activists in North America and Europe took early advantage of forest certification in market campaigns to target controversial domestic forest practices.

Among the countries quickest to certify on a large-scale basis were export-oriented producers such as Canada and Sweden. Eastern European emerging countries such as Poland and Latvia were also early adopters of forest certification as a means to gain market access into lucrative Western European markets. Some producer groups were quite resistant to adopting the environmentalist-backed FSC, leading to vigorous debates about the merits of the FSC versus the producer-backed PEFC-endorsed national schemes, most of which were initially located in Europe, Canada and the United States. These debates also served to fuel ongoing interest in forest certification itself, with the FSC and competitor programmes 'competing' to be seen as the most appropriate and legitimate (Cashore 2002; Cashore et al, 2003).

Second, in many of our developing country case studies, a lack of forest management and marketing capacities has impeded growth in certified area (Rametsteiner and Simula, 2003; Cashore et al, 2005, 2006; Pattberg, 2006; Espach, 2009). There is a general lack of resources among these countries to invest in forest certification, particularly in the face of poor market infrastructure and inadequate forest governance, both of which raise the cost of becoming certified. Among the notable exceptions are high capacity plantation producers in Brazil, Chile and South Africa (Espach, 2006).

As discussed further in Chapter 11, there are a growing number of initiatives that seek to address forest management and capacity constraints through programmes linked to certification, including organizations that provide assistance for a 'step-wise' approach from legal compliance to full certification, e.g. The Forest Trust's work in Africa and Southeast Asia (The Forest Trust, 2009). However, these are still quite limited in comparison to the scale of commercial forest harvesting.

Certification's influence on forest practices depends not only on area certified, but also on a range of other factors, including the nature of the certification standards themselves. A growing body of research on the relationship between certification and public policy indicates a dynamic and complex interaction. For instance, studies conducted in the 1990s and early 2000s in Europe and North America revealed how FSC-style forest certification rules, especially in Canada and the United States where standards are developed in sub-national regions, were often much more prescriptive, and contained higher thresholds, than government regulations (Cashore and Lawson, 2003; Cashore et al, 2004a, 2004b). At the same time, competition among the FSC and rival industry-initiated certification programmes was found to have generated two important dynamics. First, the rival certification programmes relied much more heavily on government regulations as a benchmark for sustainable forest management, demonstrating a clear relationship between public policy and private authority. Second, the competition placed pressure on the FSC to revise and harmonize its rules closer to public policy requirements, with its 'core audiences' of social and environmental activists acting as 'brake' on just how far, and when, this convergence could proceed.

Subsequent research conducted in the late 2000s indicates the importance of assessing changes over time as well as across regions (McDermott et al, 2009). McDermott et al (2009) found that prescriptive approaches within the FSC diverged considerably across the nine FSC regions in the US and that these differences mirrored public policy approaches within those regions. Hence, FSC standards were more prescriptive, and had higher thresholds, in the US Pacific Northwest and British Columbia, while FSC standards were much less prescriptive, with lower thresholds, in the US southeast.

Extending this analysis, of the relationship between public and private standards, to all the case studies represented in this book proved beyond the scope of our study. But the analyses above point to the critical importance of undertaking explanatory research across time and at multiple scales (in other words, international, national and sub-national), given the highly dynamic relationship between public policy and private authority. In Chapter 11 we discuss the implications of these trends, both positive and negative, for fostering improved environmental forest performance worldwide.

CONCLUSION

As outlined above, several key findings have emerged through this research. The first is that policies not only vary considerably between countries, but often *within* countries as well, sometimes belying generalizations about a country's 'regulatory style'. A number of contextual factors contribute to this intra-country diversity. These include socio-political variations between national and sub-national *jurisdictions*; variations among government agencies; differences among *land ownership types*, and differences across *forest management issues* (e.g. riparian buffer zone management versus road building). In general, forest practices on private lands are much less prescriptively regulated than those on public lands; although in nearly one-third of our case studies forest practice prescriptiveness was equivalent on public and private lands.

Nevertheless, important trends do emerge. The differences in policy prescriptiveness among our case study countries are striking. The highest levels of prescriptiveness are found in public lands in the US, parts of Australia and Canada and Russia. The most flexible approaches are found on private lands in the US southeast, Portugal and parts of Australia and Canada. There is no apparent relationship between the level of democratic development and the level of forest practice prescriptiveness. There is also enormous variation in environmental threshold requirements, with the most extensive thresholds found in developing countries. How we might use these results, and how they might help inform future research efforts, will be explored in our concluding Chapter 11.

Forest practice requirements for productive plantations in the ten most important plantation countries are differentiated from those for natural forests in all except three cases; in two cases where they are not, both plantation and natural forest policies are voluntary. Relatively prescriptive plantation practice policies apply in about half the sampled countries and, in some of the others, forest certification requirements may emulate many of the requirements of more prescriptive forest practice systems. In some instances, most notably in Oceania and China, plantation expansion has been accompanied by major reduction in natural forest harvest. In others, such as the US southeast, tree plantations have continued to replace natural forests.

There is considerable variation amongst case study countries in capacity to enforce forest practice requirements and in levels of compliance. In general terms, there is a division between the rich developed nations and those that are poor and rank poorly against social and governance indicators. About half the case study jurisdictions have established third-party arrangements for oversight of forest practices requirements. We discuss the challenges of improving enforcement and compliance further in Chapter 11.

Forest certification is strongest among the industrial developed countries, that is, within jurisdictions with strong forest governance. There are many reasons for this, as well as for variation in the growth of certification among countries in the developed world. While certified area continues to expand worldwide, it is likely to continue to do so unevenly, with uncertain effects on the promotion of sustainable forestry among lower performing regions and firms.

Chapter 11 reflects on the patterns and trends identified in this chapter, and the ways in which the domestic policies we uncovered might be explained from a policy perspective. We are particularly interested in evolutionary and dynamic processes that lead to an overall improvement of environmental performance through global competition, cooperation and/or cross-stakeholder, international policy learning over time.

NOTES

1   As discussed in Chapter 1 and elsewhere in this book, the authors do not take a position on which policy approach is 'better' or 'worse'. The prescriptiveness 'scores' presented in this chapter should therefore be understood only as a relative measure of prescriptiveness, not an evaluative judgement. The evaluation of policy effectiveness would require further research that examined on-the-ground environmental outcomes.
2   Data for about 29 per cent of the world's forests in the MODIS Continuous Vegetation Data Fields Dataset (MODIS05 VCF) (Hansen et al, 2006) were not of sufficient quality to identify forest type.
3   Corresponding to fourth order streams in Strahler's (1952) classification.
4   Corresponding to first order streams in Strahler's (1952) classification.
5   As stated earlier in this book, this study does not examine municipal laws, or other laws enacted by local governments.
6   Considering USFS lands as the public land category for the US western states of Alaska, California, Idaho, Montana, Oregon and Washington.
7   Considering Mexico's communal lands as equivalent to public lands for this purpose.

REFERENCES

Auld, Graeme, Lars H. Gulbrandsen and Constance L. McDermott (2008) 'Certification schemes and the impacts on forests and forestry', in G. Matson (ed) *Annual Review of Environment and Resources*, Palo Alto, CA: Annual Reviews

BC MFR (2006) 'The state of British Columbia's forests 2006', British Columbia Ministry of Forests and Range 2006, www.for.gov.bc.ca/hfp/sof/2006/, accessed May 2009

Cashore, Benjamin (2002) 'Legitimacy and the privatization of environmental governance: How non state market-driven (NSMD) governance systems gain rule making authority', *Governance*, 15, 4, October, pp503–529

Cashore, Benjamin and James Lawson (2003) 'Comparing forest certification in the US northeast and the Canadian maritimes', *Canadian–American Public Policy*, 53, pp1–50

Cashore, Benjamin and Constance L. McDermott (2004) 'Global environmental forest policy: Canada as a constant case comparison of select forest practice regulations', Victoria: International Forest Resources

Cashore, Benjamin, Graeme Auld and Deanna Newsom (2003) 'Forest certification (Eco-labeling) programs and their policy-making authority: Explaining divergence among North American and European case studies', *Forest Policy and Economics*, 5, 3, pp225–247

Cashore, Benjamin, Graeme Auld and Deanna Newsom (2004a) *Governing Through Markets: Forest Certification and the Emergence of Non-State Authority*, New Haven, CT: Yale University Press

Cashore, Benjamin, Graeme Auld and Deanna Newsom (2004b) 'The United States' race to certify sustainable forestry: Non-state environmental governance and the competition for policy-making authority', *Business and Politics*, 5, 3

Cashore, Benjamin, Fred Gale, Errol Meidinger and Deanna Newsom (eds) (2005) *Confronting Sustainability: Forest Certification in Developing and Transitioning Societies*, New Haven, CT: Yale School of Forestry and Environmental Studies Publication Series

Cashore, B., F. Gale, E. Meidinger and D. Newsom (2006) 'Forest certification in developing and transitioning countries: Part of a sustainable future?', *Environment*, 48, 9, pp6–25

Cashore, Benjamin, Graeme Auld, Steven Bernstein and Constance McDermott (2007) 'Can non-state governance "ratchet up" global environmental standards? Lessons from the forest sector', *Review of European Community and International Environmental Law*, 16, 2

CBD (2004) COP 7, Decision VII/30. 'Strategic plan: Future evaluation of progress', Seventh Meeting of the Conference of the Parties (COP) to the Convention on Biological Diversity (CBD), Kuala Lumpur, Malaysia, 9–20 February 2004

CBD (2006) COP 8, Decision XIII/15, 'Framework for monitoring implementation of the achievement of the 2010 target and integration of targets into the thematic programmes of work', Eighth Meeting of the Conference of the Parties (COP) to the Convention on Biological Diversity (CBD), Curitiba, Brazil, 20–31 March 2006

CBD (2008) COP 9 Decision IX/5, 'Forest biodiversity', Ninth Meeting of the Conference of the Parties (COP) to the Convention on Biological Diversity (CBD), Bonn, Germany, 19–30 May 2008

Debroux, Laurent (2004) Email communication with Laurent Debroux, Senior Natural Resource Management Specialist, World Bank, 28 April 2004

Espach, R. (2006) 'When is sustainable forestry sustainable? The Forest Stewardship Council in Argentina and Brazil', *Global Environmental Politics*, 6, 2, pp55–84

Espach, Ralph (2009) *Private Environmental Regimes in Developing Countries: Globally Sown, Locally Grown*, Manchester: Palgrave Macmillan

FAO (2006a) 'Global forest resources assessment 2005: Progress towards sustainable forest management', in *FAO Forestry Paper 147*, Rome: Food and Agricultural Organization of the United Nations

FAO (2006b) 'Global planted forests thematic study: Results and analysis', A. Del Lungo, J. Ball and J. Carle (eds) *Responsible Management of Planted Forests: Voluntary Guidelines*, Rome: United Nations Food and Agriculture Organization

Forest Practices Authority (Tasmania) (2008) 'Annual report on forest practices 2007–08', Forest Practices Authority, www.fpa.gov.au > Publications > Annual Reports, accessed December 2007

Forest Practices Board (Tasmania) (2004) 'Annual report on forest practices 2003/2004', Forest Practices Authority, www.fpa.gov.au > Publications > Annual Reports, accessed December 2007

Hansen, M. C., R. S. DeFries, J. R. Townshend, M. Carroll, C. Dimiceli and R. Sohlberg (2006) 'Vegetation continuous fields MOD44B, 2005 percent tree cover, collection 4', College Park, MD: University of Maryland

ITTO (2006) 'Status of tropical forest management 2005', ITTO Technical Series No. 24, Yokohama: International Tropical Timber Organization

Kanowski, P. J. and H. Murray (2008) 'TFD review: Intensively-managed planted forests', http://research.yale.edu/gisf/tfd/ifm.html > TFD IMPF Review, accessed November 2009

Klooster, D. (2005) 'Environmental certification of forests: The evolution of environmental governance in a commodity network', *Journal of Rural Studies*, 21, 4, pp403–417

McDermott, Constance and Benjamin Cashore (2007) 'A global comparison of forest practice policies using Tasmania as a constant case', New Haven, CT: Yale Program on Forest Policy and Governance, Global Institute of Sustainable Forestry

McDermott, Constance and Benjamin Cashore (2008) 'Forestry driver mapping project: Global and US trade report', New Haven, CT: Yale Program on Forest Policy and Governance

McDermott, Constance L., Benjamin Cashore and Peter Kanowski (2009) 'Setting the bar: An international comparison of public and private policy specifications and implications for explaining policy trends', *Journal of Integrative Environmental Sciences*, 6, 3, pp1–21

MNR (2008) 'Annual report on forest management: For the year April 1, 2005 to March 31, 2006', Minister of Natural Resources of the Province of Ontario

Montreal Process Implementation Group (2008) 'Australia's state of the forests report', Bureau of Rural Sciences, http://adl.brs.gov.au/forestsaustralia/, accessed September 2008

Pattberg, P. (2006) 'Private governance and the south: Lessons from global forest politics', *Third World Quarterly*, 27, 4, pp579–593

Rametsteiner, Ewald and Markku Simula (2003) 'Forest certification – An instrument to promote sustainable forest management?', *Journal of Environmental Management*, 67, 1, pp8–98

Ridder, Ralph M. (2007) 'Global forest resources assessment 2010: Options and recommendations for a global and remote sensing survey of forests', Rome: Forest Resources Assessment Programme, Food and Agriculture Organization of the United Nations

Schmitt, C. B., A. Belokurov, C. Besançon, L. Boisrobert, N. D. Burgess, A. Campbell, L. Coad, L. Fish, D. Gliddon, K. Humphries, V. Kapos, C. Loucks, I. Lysenko, L. Miles, C. Mills, S. Minnemeyer, T. Pistorius, C. Ravilious, M. Steininger and G. Winkel (2008) 'Global ecological forest classification and forest protected area gap analysis: Analyses and recommendations in view of the 10% target for forest protection under the Convention on Biological Diversity (CBD)', Freiburg: Freiburg University

Schmitt, Christine B., Neil D. Burgess, Lauren Coad, Alexander Belokuruv, Charles Besançon, Lauriane Boisrobert, Alison Campbell, Lucy Fish, Derek Gliddon, Kate Humphries, Valerie Kapos, Colby Loucks, Igor Lysenko, Lera Miles, Craig Mills, Susan Minnemeyer, Till Pistorius, Corinna Ravilious, Marc Steininger and Georg Winkel (2009) 'Global analysis of the protection status of the world's forests', *Biological Conservation*, 142, pp2122–2130

Seneca Creek Associates and Wood Resources International (2004) 'Summary: "Illegal" logging and global wood markets: The competitive impacts on the U.S. wood products industry', report prepared for the American Forest and Paper Association, www.illegal-logging.info/uploads/afandpa.pdf, accessed November 2009

Strahler, A. N. (1952) 'Hypsometric (area-altitude) analysis of erosional topography', *Geological Society of America Bulletin*, 63, pp1117–1142

The Forest Trust (2009) 'TFT projects', www.tropicalforesttrust.com/projects.php, accessed August 2009

WDPA Consortium (2009) 'Protected areas management effectiveness information module', World Database on Protected Areas (WDPA) Consortium, www.wdpa.org/ME/Default.aspx, accessed May 2009

World Bank (2008) 'Forest source book', Washington, DC: World Bank

Three Puzzles, a Conundrum and a Question: Towards a Dynamic and Problem-Focused Policy Research Agenda

INTRODUCTION

Our research on environmental policy specifications regulating commercial forestry operations around the world has opened up what was hitherto treated, by most comparative and international relations scholars, as a 'black box'. The sheer complexity of regulations in some jurisdictions explains, in part, why such comparative work has been absent or limited. Our classification framework directly addressed these challenges by focusing on key structural attributes of policy requirements on the one hand and their actual thresholds on the other. And by focusing inductively on any policy arena that carried the force of the state (including legislation, regulations and agency policy directives), we differ from comparative work that focuses solely on distinguishing legislative and judicial approaches (Howlett, 1994) or differences in national and sub-national styles of regulations (Eisner, 1993; Hoberg, 1997). Our approach illustrates to students of the policy process in general, and of environmental regulations specifically, that there exist identifiable patterns of policy approaches that both unite and distinguish specific cases in ways that were previously largely hidden. Uncovering these trends allows us to identify critical research questions and preliminary hypotheses with which to guide future research.

This chapter focuses on three puzzles, and a conundrum, that emerge from our data as a means to explore trends in policy development more generally. The first puzzle concerns why precisely the same riparian buffer zone regulations have been applied within jurisdictions that manage very different forest types. The second puzzle is why clear differences exist in regulatory 'prescriptiveness' between public and private landownership within developed countries and why such differences are much less within developing countries. The third puzzle is why many jurisdictions with low capacity for enforcement have adopted highly prescriptive approaches and high threshold requirements.

The conundrum concerns environmental regulation more generally: on the one hand, there is widespread civil society demand for prescriptive regulations to ensure high environmental performance from forest managers who are linked to consumers through complex and diverse global supply chains. Calls for greater prescriptiveness stem in part from the recognition that without precise, standardized requirements, it will be difficult for stakeholders and customers in distant markets to have any assurance of the level of environmental practice followed. On the other hand, there is increasing recognition by many of the same actors and practitioners that locally based decision-making is needed if forest management is to be appropriately tailored to current and (changing) local environmental and social conditions. Yet the greater the prescriptiveness at the national, state or provincial levels, the less the room for local, field-based discretion. How to overcome this conundrum is arguably, in the global era, one of the greatest challenges facing sustainable forest management.

Attention to changes in forest policy over time, and their impacts on *behavioural* change, is critical for answering these puzzles and finding a solution to the conundrum. Our analysis has been limited to policy specifications at one point in time, but we know (as the contextual analysis in each of our case study chapters has demonstrated) that policies continue to change and evolve. Understanding whether, when and why existing policy specifications might change is an obvious and desirable next step in work on this topic.

Analysis of policy change, following systematic policy classifications, sets the stage for research on the *effectiveness* of existing policies in changing *behaviour* and achieving environmental outcomes. As discussed in Chapter 1, policy effectiveness is clearly of central importance to all those concerned with conservation and sustainable management of the world's forests, including policy-makers and scholars. While we have not, in this work, systematically assessed the impacts of policy decisions on behaviour and, ultimately, effectiveness of forest practice policies, our analyses have established part of the platform necessary for doing so. Our work also provides insights into how and why policies develop and change over time and what factors might lead to overall, lasting improvements in forestry performance worldwide.

This chapter therefore considers key factors that might impact policy durability and dynamism. To what extent do policy changes reflect scientific and/or other forms of learning regarding environmental conditions? Alternatively, to what extent do domestic 'structural' features, e.g. land tenure or institutional capacity, shape or constrain policy choices – and what does this mean about the likelihood that environmental standards will be either raised or lowered over time? What is the influence of international forces such as inter-governmental discourse and market pressures? Finally, how might improved understanding of policy change and durability inform strategic interventions aimed at increasing global environmental forest performance?

The remainder of this chapter develops and illustrates our questions in five analytical steps. Following this introduction, we examine the three puzzles and develop, deductively and inductively, hypotheses to assist future research on historical trends and future policy decisions. The next section sheds light on the conundrum between demands for 'high prescriptiveness' on the one hand and local and contextual forest management on the other. Identifying what appears to be an overall tendency towards increasing prescriptiveness, and the concurrent growth in market-driven forest certification as a new arena for policy setting on the other hand, we turn to the notion of the 'paradox of trust' as important for understanding the development of both government regulations and certification standards. If accurate, this explanation carries with it additional implications regarding policy prescriptions often missed by scholars of regulation: that higher regulatory arenas may not necessarily be due solely to stronger commitments to environmental protection and sustainability *per se*, but also to the lack of legitimacy and trust within global, domestic and local political communities. Such a lack of trust conflicts with demands for more flexible, adaptive governance that many consider a core component of sustainable forestry.

The last section then discusses strategic interventions that might foster an overall global-scale improvement in environmental forest practices over time. We conclude by reflecting on the need for more research that expressly links policy with on-the-ground environmental effectiveness and sustainable forest management.

THE THREE PUZZLES

The three puzzles we discuss were identified, inductively, on the basis of patterns observed across our 20 case study countries and 45 jurisdictions. A series of hypotheses are presented to address these puzzles and to identify strategic areas for future research that may help to explain more systematically policy convergence and divergence at both national and sub-national levels.

Why 50-metre buffer zones?

Our data on five forest practice criteria revealed a wide range of policy prescriptions across the 45 case study jurisdictions. Within this diversity, however, some notable patterns emerged. For example, as illustrated in Chapter 10, we found repeated use of 50-metre riparian buffer zones within a substantial subset of countries (see Figures 10.1–10.3, Chapter 10). Such diverse jurisdictions as the DRC, Indonesia, Mexico and the US Forest Service established this *same* threshold for streams of various sizes. Our first puzzle, therefore, concerns why 50 metres appears as such a common threshold size and whether this phenomenon might offer clues regarding how policies are formed and why they endure.

Of the five forest practice criteria assessed, riparian zone policies appear to be the most consistently prescriptive. McDermott et al (2009) suggest this may reflect some form of consensus, among scientists and/or policy-makers, that riparian protection is *important*. However, a review of riparian literature from different countries reveals no consensus on the precise buffer zone sizes or 'thresholds' that should be prescribed. Furthermore, we found no agreement among natural scientists that any single standard size, such as 50 metres, should be applied to case studies representing such a wide diversity of environments, or – conversely – that very different requirements should be developed in apparently similar environments (McDermott et al, 2009).

A focus on the role of external forces in shaping policy responses offers clues to this puzzle of policy convergence (Bennett and Howlett, 1992). DiMaggio and Powell's (1991) seminal work, among others, provides direction on possible explanations for such a phenomenon.

DiMaggio and Powell identify three processes – mimetic, coercive and normative – that might explain motivations for adopting similar policy settings (in this case, 50-metre buffer zones). *Mimetic* refers to a situation in which one organization simply 'copies' another, as a result of uncertainty. Hence, it may be that a 50-metre buffer zone was first established in one country, for either scientific or other reasons, and was then copied by other governments simply because they were uncertain about what else to do. *Coercive* isomorphism refers to the process through which an organization is forced to adopt a particular policy. For example, a donor agency, such as the World Bank, may commit structural adjustment funds on the condition that a country will improve its forest management (Brown et al, 2008). *Normative* isomorphism refers to homogeneity resulting from the 'professionalization' of institutions and associated policies, reflecting the education and discourse of the recognized 'expert' community. The concept can also be applied to populist or political 'norms', for which it operates similarly.

Drawing on this literature, we can identify hypotheses that individually, or collectively, might help explain the policy convergence behind the prevalence of 50-metre buffer zones in forest practice requirements.

Hypotheses (H1) accounting for 50-metre buffer zones policy convergence

Mimetic
H1.1: Convergence of the 50-metre buffer zone is explained by policy mimicry when lack of certainty exists.

Coercive
H1.2: Convergence of the 50-metre buffer zone is explained by other parties creating (positive or negative) incentives to adopt the same buffer size.

Normative
H1.3: Convergence of the 50-metre buffer zone is explained by the emergence of an international norm of 50-metre buffer zones. This norm may originate from within an international professional community and/or from other sources, such as NGOs, the business community or government networks.

Further research on the above question and its associated hypotheses is critical for understanding existing and future drivers of domestic forest policy. That is, understanding whether a policy approach was taken for mimetic, normative or coercive reasons may not just explain the conditions under which a particular policy might

emerge, but also whether it is likely to endure. In the case of these three classifications, normative can be seen as relatively durable as norms tend to change more
slowly; mimetic is arguably more susceptible to change, since there is usually a
variety of available policies to copy, and the persistence of settings of coercive origins
depends on the strength and durability of the incentives or pressure.

However, it is ultimately effectiveness, not policy durability in itself, that is the
commonly stated goal of environmental forest policies. Durability *per se* could also be
a sign of *ineffectiveness*. This could be illustrated by the case of some of the longest-
standing riparian zone policies identified in our case studies: Chile's buffer zone
requirements of 200–400 metres which were established in 1931. Such policies may
endure simply because they are ignored, or are of little consequence in practice, and
thus there is no need to invest resources in changing them.

Any mimetic, coercive or normative explanations for institutional support of a
given policy could be seen as working counter to the policy responsiveness and flexibility needed to address on-the-ground realities, unless the definition of 'normative'
is expressly tied to end goals and values. The systematic application of DiMaggio
and Powell's isomorphism framework to our specific policy settings could serve to
operationalize such abstract concepts as 'normative' isomorphism, for example, by
distinguishing between professional standardization versus shared values regarding
desired environmental outcomes.

Earlier in this book (see Chapter 1), we identified many of the common goals of
riparian zones, e.g. the protection of water quality and yield and protection of riparian habitats. We also noted the considerable scientific research that might inform
policies on differing ways to achieve those goals, depending on a range of environmental variables. From the point of view of strategic intervention, future research is
needed on how to better integrate field-based learning into buffer zone prescriptions
so that policy design is linked as least as strongly to effectiveness as it is to institutional features such as isomorphism. This would require understanding not only
environmental impacts, but also the social dynamics of policy-making, including
international pressures and influence, as factors that help or hinder effective and
durable policy change and implementation.

Why public/private differences in developed countries?

A second trend our analysis discovered was the strong difference in policy prescriptiveness governing public and private land management in developed countries
(Table 10.8, Chapter 10). This is revealing in that it confronts assumptions, common
in much of the literature on public policy and political culture, that nations can be
distinguished according to their approach to regulation. According to such assumptions some nations, such as the US, are classified as much more prone to neoliberal
market solutions; while others, such as the European nations – especially those who
come out of the 'corporatist tradition' – are classified as much more likely to encourage government intervention. Instead, what seems to be a stronger predictor of
policy prescriptiveness – at least in the case of forest practice requirements – is
whether the land is publicly, rather than privately owned.

The literature on the regulation of public and private land bears out these findings. A wide-ranging literature on environmental protection and public policy has found that – at least in North America – private property rights, including the requirement to compensate forest owners once a regulation has been deemed by the courts to infringe upon such rights, make it much more difficult for governments to regulate private rather than public forests (Flick et al, 1995; Cashore, 1999; Zhang, 2000). This literature, by focusing not only on timber sustainability, but also environmental sustainability, cautions against prescriptions to privatize, where possible, natural resources as a means to address resource depletion owing to a 'tragedy of the commons' (Hardin, 1968).

Two hypotheses emerge from the literature on private property rights that may explain divergence across public and private forestlands.

Hypotheses (H2) for differences in public and private forestland regulation in developed countries

Pressures for divergence

- **H2.1:** Private property rights make it much more difficult and/or costly for governments to regulate privately held land than public land. This places pressure on policy-makers to develop less prescriptive regulatory approaches on private lands.
- **H2.2:** Governments in developed countries respond to pressure from environmental activists and the community for high forest management standards by developing high levels of policy prescriptiveness, and high performance thresholds, for public forestlands.

Our study suggests that each of these hypotheses may apply in specific cases. However, in some cases where they apply jointly in a particular jurisdiction, a 'spillover' may occur where higher public land regulations lead to similarly high private land regulations, while at other times the two may diverge rapidly. The case of the US Pacific Northwest illustrates this well. As Cashore and Howlett (2007) and Hoberg (1993) noted, forest practice regulations governing forest management on public and private forestlands in the US Pacific Northwest were very similar in the 1990s. Given that Washington, Oregon and California all contain large areas of both public and private forests, this similarity could be attributed to an earlier 'spillover' of public to private regulation. However, in the 1990s rules governing federal lands were changed dramatically in response to the listing of the old-growth-dependent northern spotted owl (*Strix occidentalis caurina*) as a threatened species (Sher, 1993). This listing resulted in a major reduction in harvesting on federal lands with no similar corresponding trend on private lands. Similarly, the relatively large no-harvest riparian zones on US Forest Service lands were an indirect product of the spotted owl controversies, emerging in the process of developing an action plan to resolve federal forest conflicts (FEMAT, 1993). Although buffer sizes on private lands have subsequently increased, in response to the listing of salmon as endangered species, they have done so only to a lesser extent.

Two somewhat different questions emerge from the case of the US Pacific Northwest. One has to do with policy divergence across federal and private lands within this region (Hoberg, 1993; Cashore and Howlett, 2007), owing to higher relative thresholds and the removal of much of the remaining publicly owned old-growth forest from the commercial land base. But when these public and private land regulations are viewed alongside our broader global comparison, they seem – in contrast – much more similar. Both have high policy prescriptiveness and high thresholds. Hence, a scholar comparing federal versus private forestland management in the US Pacific Northwest is likely to focus on policy divergence (in other words, why higher thresholds on federal lands?), but a scholar focusing on global-scale comparisons is likely to focus on why there is relative policy convergence in these same cases. Both questions are legitimate, and point to the need to be very careful as to what, precisely, is the overall policy approach in a given jurisdiction and how it compares within and across countries. Whereas the previous two hypotheses were developed to explain differences across public and private forest management, it is equally important – given the much higher degrees of policy prescriptiveness in Oregon and Washington private forestlands compared to most other US private forestland management, in particular that in the US southeast – to reflect on why and how such higher prescriptions on private lands emerge.

Pressures for convergence

- **H2.3:** Where there are substantial public and private lands within a given jurisdiction, greater regulation of public lands may, over time, result in increased pressures from civil society and environmental groups for greater regulation on private lands.
- **H2.4:** Where there are substantial increases in public land regulation in a given jurisdiction, industry and landowners' associates may proactively support regulations in private lands as a means to control, and limit, further government oversight.

The two hypotheses above are important for understanding pressures on private forestland regulation as emanating from public policy interventions on government owned land. The first hypothesis has to do with a 'spillover effect', in that societal awareness of environmentally informed stewardship on public lands should, or could, also apply on private lands. At the same time, it may well be that apparent 'convergence' of forest policies on private lands towards a more prescriptive approach (H2.4) may be owing to strategic interventions of industry interests to limit, or direct, the societal pressures towards recognition of the economic and social benefits that harvesting of private forests provides. This hypothesis draws on Cashore (1997, 1999), who found that the Forest Practices Act changes of the 1970s, which dramatically increased government regulations of private forest management in Oregon, were initiated by this state's peak timber industry association as a means to 'head off' concerns that the federal government might move in to regulate private forestry – which the industry felt would bias environmental concerns over economic and social values. Further exploration of these hypotheses is important, not only for understanding current policy differences, but also in anticipating the potential for policy change, and the implications for policy divergence and convergence.

Why higher average prescriptions in developing countries?

A third puzzle we identified is that of higher levels of prescriptiveness, on average, in developing than in developed countries. This data, at first glance, would appear contradict a body of literature emerging from Inglehart's post-modern (though widely critiqued) thesis that citizens in developed countries are more likely to support environmental protection because their material needs have been satisfied. According to this thesis, developing countries have relatively little motivation to pursue environmental conservation because there is not the same degree of civil interest in these matters (Inglehart 1977, 1995). While the high level of prescriptiveness we found among developing countries may appear to contradict this assertion, a more complex picture begins to emerge if we place these policies in a broader governance context.

One possible explanation lies in the quest for government legitimacy, particularly in newly emerging democracies with unstable political histories and relatively high levels of corruption. For example, Rose-Ackerman's research on Eastern European economies notes how a lack of trust in post-socialist governments has created citizen demand for centralized, authoritarian rule to control distrusted government actors. Her survey research revealed that distrustful citizens favoured stricter laws and penalties as a means of social control over institutional reform that might change behavioural incentives (Rose-Ackerman, 2001). Relating these ideas more broadly to our case studies, a similar logic may help to explain why we found highly prescriptive policies not only in Eastern Europe, but also in non-European countries that are becoming more democratic after a history of authoritarian regimes (e.g. Brazil and Indonesia).

Coupled with the challenge of legitimacy, it is also widely understood that developing countries often lack capacity to enforce their environmental regulations (see, for example, Chapter 2, Figure 2.5 Environmental regime index). As we discussed in Chapter 10, that countries might have prescriptive laws 'on the books' does not mean that they are enforced and have impact in the field. Victor's research on international treaties found that the more shallow an international treaty, the more likely it is that countries will agree to it (Victor, 1999). Correspondingly, it is possible that developing countries were willing to enact more prescriptive regulations, with relatively high threshold requirements, because of the low likelihood of actual enforcement.

These two sets of theories, one focused on legitimacy and the other on a lack of enforcement, suggest the following two hypotheses.

Hypotheses (H3) for higher average prescriptions in developing countries

- **H3.1:** Prescriptive policies are more likely to be enacted in emerging democratic regimes in developing countries as a means to increase government legitimacy.

- **H3.2:** Developing countries create highly prescriptive rules in response to various pressures (e.g. in response to donor agency pressure) because they expect they will not be enforced.

Greater attention to the motivations for prescriptive policies in developing countries will be of critical importance in further analyses of environmental policies. Even if a country does not expect its environmental forest practices policies to be followed, this is very different from no policy existing. If a policy exists, an international agency can offer to assist in implementation and enforcement (that is, build state capacity) of its laws rather than challenging state sovereignty through coercive demands for new policies and approaches.

Of course, if policies were designed primarily to gain legitimacy rather than protect the environment, then improved enforcement may create pressure for policy change. As observed in the next section, however, historical trends suggest that environmental threshold requirements, once established, appear relatively 'sticky' and resistant to change. The effect of this stickiness, furthermore, is complex and variable. The following section introduces a 'trust paradox' involving reflexive social dynamics among stakeholders, policy-makers and policy implementers. It considers how this paradox may contribute to prescriptive requirements, which in turn may inhibit incentives for on-the-ground cooperation with these requirements, leading to further demands for prescription.

POLICY CHANGE, PRESCRIPTIVENESS AND THE TRUST PARADOX

As discussed in Chapter 10 (see 'Emerging patterns'), a general trend across our case study countries has been an overall increase in the scope and level of environmental regulation, and/or best practice policies, applying to our select forest policy criteria. While neoliberal reforms in some countries may have led to slippage of certain environmental policies (Liverman and Vilas, 2006), there was little evidence of slippage in the environmental threshold requirements covered by our study. Instead, there was a notable expansion of the use of such thresholds across case studies and apparent durability of these thresholds once established. This durability is particularly notable if we consider that some jurisdictions have undergone major institutional restructuring and/or multiple Forest Codes, yet the threshold requirements have effectively remained in place (e.g. in BC and Russia), or increased (e.g. in Washington and Oregon).

What kinds of factors might explain the increasing prevalence, and relative durability, of policy prescriptions and what might this ultimately imply about the environmental impacts of forest management?

One set of factors is addressed in our hypothesized explanations for the three puzzles above. Another set may be associated with the broad, global-scale trend away from traditional, 'informal' modes of forest production towards formalized regulation. In countries where forest governance has been based on informal or traditional modes of social organization and natural resource management, the implementation of formal regulatory control has typically displaced, rather than complemented and reinforced, these modes. In these cases, the introduction of state-based regulatory structures is often driven by increasing interest in commercial exploitation of forest resources and thereby disempowers other users and interests. In such cases, increased regulation may be as much a sign of greater environmental threats as it is of environmental progress. Thus it cannot be assumed that regions with higher levels of regulation are actually performing better than those with lesser levels. Furthermore, due to economies of scale and differing levels of social influence and capacity, increased regulation may favour intensive forest producers over traditional users, thereby displacing traditional users to marginal lands and possibly increasing overall environmental impacts (Liverman and Vilas, 2006; Robledo et al, 2008; Unruh, 2008; RRI and ITTO, 2009).

A third set of factors that may play a role in shaping a policy approach relates to trust among forest stakeholders and in forest governance. In situations characterized by low levels of trust between key actors, highly prescriptive performance requirements may be necessary to provide assurance that distrusted forest producers will meet high environmental standards. However, where there is uncertainty as to the precise environmental prescriptions needed to achieve desired environmental outcomes, variability in environmental and social conditions, and/or high levels of stakeholder distrust, prescriptiveness may inhibit adaptive management, social learning and the building of trust (Shapiro, 1987; Sitkin and Roth, 1993; Murnighan et al, 2007), all of which can contribute to improving environmental performance (Holling and Meffe, 1996; McDermott, 2003; Folke, 2006; Blackmore, 2007; Ison et al, 2007). Trust levels may vary, however, depending on the type of forest producer and distrust-driven policies may be more effective with some forest producers than with others (McDermott, 2003). Thus, the effectiveness of prescriptive versus other types of policy approaches may vary – in different circumstances and for different modes of forest production.

Forest certification and shifting modes of production

Perhaps nowhere are the dilemmas regarding different modes of production and the prescriptiveness of regulations as evident as in the dynamic and expanding arena of forest certification. As discussed in Chapter 1, forest certification emerged as a new form of non-state governance designed to provide market incentives for forest practices that meet a set of agreed upon environmental and social standards for responsible forestry.

Many of the early supporters of forest certification aimed to promote local, indigenous and/or community based management as an alternative to, and/or replacement for, large-scale industrial production (McDermott and Hoberg, 2003). At the same time, concern about the impacts of the global wood products trade fuelled international market campaigns focused on pressuring large-scale retailers to commit to buying only certified forest products (Cashore et al, 2004). These large retailers, in turn, created demand for certified wood from large forest suppliers, thereby contributing to certification's disproportionate growth among multi-national forest companies serving North American and European markets (see Figure 2.17 in Chapter 2 for the global distribution of forest certification) (Rametsteiner and Simula, 2003).

The entrance of large-scale industrial forestry into the forest certification arena, however, also served to shift the dynamics of stakeholder trust in the system. As highlighted in a study of the FSC in British Columbia, high levels of distrust in the forest industry, as well as in foreign certifying bodies, generated pressure for highly prescriptive forest certification standards that would control industry and auditor behaviour. In other words, building sufficient trust in the certification *system* required removing discretion – and hence much of the need for trust – in *individual* firms (McDermott, 2003; Power, 2003).

Theorists have observed the emergence of a 'paradox of trust' in the use of formal rules to address individual trustworthiness (Murnighan et al, 2007), particularly where there are high levels of distrust and perceived value differences (Shapiro, 1987; Sitkin and Roth, 1993). It is arguably impossible in a third-party system, such as forest certification, to create standards and procedures that are both fully comprehensive and entirely non-discretionary, thereby removing all need for trust in individual actors (including all auditors and producers involved). The removal of a certain degree of discretion, meanwhile, creates social distance (and possibly resistance), limits incentives to exceed minimal requirements and prevents opportunities for voluntary displays of trustworthiness – thereby contributing to a perceived dependence on formal rules. Where there are value-based conflicts, furthermore, rule formalization may inhibit the creation of shared meaning and values (Shapiro, 1987; Shapiro, 1997; McDermott, 2003).

In the context of forest certification, this 'paradox of trust' is compounded when high levels of prescription with extensive documentation requirements disproportionately burden the most trusted, small-scale producers, thereby favouring the participation of large-scale, distrusted actors. To the degree that the system excludes the most trusted actors, this further perpetuates a 'spiral of distrust' (McDermott, 2003). Meanwhile, global market competition is such that if one set of distrusted actors is forced to abide by prescriptive standards, they may demand that all other producers, trustworthy or otherwise, be subject to the same prescriptive rules.

With all of the above explanatory pathways in mind, the following section looks at their implications for future directions in environmental forest policy.

STRATEGIC INTERVENTION AND THE CALIFORNIA EFFECT: HOW MIGHT MARKET-BASED APPROACHES INTERACT WITH GOVERNMENT POLICY TO FACILITATE INCREASED ENVIRONMENTAL FORESTRY PERFORMANCE?

As described in Chapter 1, forest certification emerged in part out of frustration over the lack of governmental progress in tackling tropical deforestation and degradation. It thus represents a new market-based mechanism for writing forest practice rules that lies outside direct government control. While certification and state-based governance may be institutionally distinct, the influence of certification on market dynamics, and its interaction with government policy, have led scholars to speculate what the effects of certification might be in shaping what international policy theorists call 'regulatory competition'.

Political scientist David Vogel, a scholar of 'regulatory competition', has described how market competition may result in either a lowering or raising of environmental policies. The former, referred to as the 'Delaware effect', occurs when firms are attracted to regions (e.g. US states such as Delaware) with lax, or largely unenforced, environmental or social regulations. In the context of forest policy, this phenomenon might address not only global forest loss and degradation in the tropics, but also relatively lax regulations in some developed jurisdictions. On the other hand, increased environmental requirements, known as the 'California effect', occur when firms seek access to lucrative markets that require adherence to high environmental standards (e.g. US states such as California). The initial establishment of these standards is often sparked, Vogel notes, by an active civil society. Once such standards are instituted, however, they give firms operating under them the incentive to support similar regulation of their competitors (Vogel, 1995, 2005).

The question is whether certification or other market-based strategies could play a role in creating a global 'California effect' (DeSombre, 2000), and if so, how might strategists go about fostering such a phenomenon? We argue that a focus on strategic behaviour reveals a potential two-pronged approach. The first prong would focus on rewarding 'best practices' by making the 'bar' of performance more transparent. That is, the focus is placed on recognizing where firms are held to higher governmental and/or certification requirements and this recognition is used to increase pressure to get their competitors to 'come up to' their standards. This means that instead of using certification merely to increase rules on firms who operate under already relatively progressive public policy regulations, certification would distinguish among regions in terms of both the standards used and their implementation, thereby creating competitive pressure to raise standards and/or improve implementation in regions with relatively lax performance. This would address the accusation, for example, that certification is favouring those regions with the least restrictive requirements.

Efforts to recognize higher performance might benefit not only from more precise product labelling that distinguishes, and therefore rewards, high forest practice and/or certification standards, but also from capturing other market incentives such as payments for ecosystem services and evolving markets for carbon sequestration to counteract global warming. For example, larger riparian buffer zones may help to regulate water quality and quantity as well as sequester forest carbon. The more avenues available to recognize the ecosystem services provided by the effective implementation of high forest practice and/or certification standards, the better the incentive for all countries to raise their levels of performance.

While most supporters of forest certification have focused on certifying exemplary models of forestry, even more challenging has been how to preclude the most destructive practices from ever getting market access. Ironically, this seems in part caused by a focus on rewarding the top that then does not distinguish the middle from the very worst. This highlights the importance of the second prong, which draws on the logic of the California effect to weed out the 'worst' practices by forbidding access to lucrative markets.

The growing number of initiatives against 'illegal logging' are pursuing precisely such an approach. These efforts, furthermore, have managed to win relatively widespread support across previously warring factions of environmental groups and industry and between developed and developing country governments. Virtually all forest product producers in Western Europe and North America, and some in Eastern Europe, Russia and in the tropics have a strategic self-interest to participate in a more exclusive market for verified legal wood. Everything else being equal, market discrimination against illegal wood would provide new market access and/or increased prices for wood harvested legally. Indeed, the American Forest and Paper Association recently estimated that, as a result of the proliferation of illegal logging, the price of wood products in world markets is deflated by an average of 5 to 10 per cent (Seneca Creek Associates and Wood Resources International, 2004). Governments are interested in capturing lost revenue from illegal forest trade, estimated to be as high as US$15 billion (World Bank, 2004), and in gaining outside support to enforce their own, domestically generated forestry laws.

Reflecting this broad base of support, the various illegal logging initiatives that have emerged range from inter-governmental, to domestic, to non-governmental. Key inter-governmental processes are regional Forest Law Enforcement and Governance (FLEG) and the EU Forest Law Enforcement, Governance and Trade (FLEGT) initiatives (Chatham House, 2009). Domestically, there are a growing number of government procurement policies that give preference to legal and/or sustainable wood (CPET, 2009), as well as new trade prohibitions on the import of illegally produced wood (e.g. the US Lacey Act 2008) (US CBP, 2009). Non-governmental efforts include the development of voluntary standards (e.g. those of SGS and SmartWood) that verify the origin and legality of wood products, as well as 'stepwise' approaches among these and other actors. The latter include, for example, The Forest Trust and Global Forest and Trade Network initiatives that support companies in moving up a performance 'ladder' from legality verification to sustainable forest management certification (Metafore, 2009).

While the groundswell of support for illegal logging initiatives and related market mechanisms has indeed triggered notable action, it is nevertheless important to take a critical look at the alignment of interests and consider who benefits and who loses and if there might also be unintended effects. For example, if US southeastern timber companies are aligning to denounce wood production in tropical countries on the basis of 'illegality', the question arises over just what constitutes 'legal' production in these two regions and who has actually set higher standards? As explained in the contextual analysis of Chapters 4 (Latin America) and 3 (Canada and the US), it is very challenging to obtain permits for forest management in the Amazon, while some US southeastern states do not require any form of harvest permit. Granted some might argue that the high biodiversity and frontier nature of Amazonian forests justifies more stringent regulation, but the US southeast also contains areas of biodiversity 'hotspots' and better articulating the debate in this way can lead to more probing, and equitable, global dialogue.

Another challenge with addressing legality alone is that prosecution for illegal trade could trigger countries to *lower* their environmental requirements to prevent future interruptions to trade. This has indeed already occurred, for example, in a US case involving prosecution under the Lacey Act for trade in illegally harvested Honduran lobsters. The lobsters were found to be below Honduras' legal size limit; in response, the Honduran government subsequently lowered its size limit (Ortiz, 2005). This renders all the more critical the combined efforts of excluding the poorest performers from the market while also recognizing and rewarding efforts to achieve higher standards and performance.

A two-pronged strategy offers 'something for everyone' in the debate over global forest performance. It allows, on the one hand, developed country firms to demand that all producers be held to basic requirements of the law. On the other hand, at the same time it allows developing country firms to highlight instances where 'legal' practices and/or certification standards in their respective countries may involve higher performance requirements than those found in developed countries. Distinguishing among forest policies and certification standards on the basis of the level of environmental requirements also enables 'policy entrepreneurs' in places like the US southeast to develop higher standards in order to distinguish themselves in the marketplace. In contrast, if all forest practice policies and certification standards are presumed equal, 'competition' favours the lower cost producers and provides no reward for higher performance.

CONCLUSION

One of the most important questions of our times is to understand whether, when and how public policy might be developed in a way that nurtures environmental stewardship of the world's resources. Increasing economic globalization, consumption and population growth all combine to create enormous pressures on the world's forests. Just how governments are able to address these pressures, and the consequences of their policy choices on other governments, across time and ultimately on the natural environment, are among the most important and thorny questions for those who care about environmental stewardship in the global era.

Our book has helped lay the groundwork necessary to answer such questions, through the development and empirical application of a precise classification system with which to compare environmental policy in general, and forest practice settings in particular, across jurisdictions and countries. While our work cannot, and should not, replace much more nuanced and historically grounded scholarship within particular countries, our framework has permitted a comparison of policy choices across a range of cases, revealing policies that were previously viewed only in national or regional isolation and/or remained hidden across a range of complex planning statutes and regulations.

And by focusing on five critical questions of importance for promoting sustainable forestry, our inductive effort to classify government policies – wherever they reside – has revealed the strength of a problem-focused regulatory analysis that addresses key 'on-the-ground' challenges, such as riparian zone protection.

Our focus on clearly identifying and classifying existing regulatory policies, and examining the patterns that emerge from this assessment, is intended not only to inform practitioners and scholars about current policy, but also to set the stage for future analysis that systematically addresses when, why and where the policies identified actually achieve durable environmental stewardship objectives. To be sure, other policy interventions beyond the scope of this book, such as tax incentives, landowner education, subsidies and trade policies, will be critical components of such an approach. Yet by focusing on forest regulations – arguably the most preferred approach by governments in addressing environmental sustainability in commercial forests – we have uncovered important results that challenge some existing studies of global business regulation.

While our results are consistent with the observation of Braithwaite and Drahos (2000) that a 'neoliberal fairy tale' has led many scholars incorrectly to assume that business regulations are on the decline, our data challenges their claims in two important ways. First, the existence of agencies designed to regulate environmental impacts is insufficient for understanding the nature of the environmental requirements those agencies espouse. Second, our results identify the need for policy analysis to 'bring the state back in'. It is not only that global business regulations have developed beyond the state; there are many prescriptive approaches employed by a range of countries that defy existing understandings and characterizations of state retrenchment.

This concluding chapter has focused attention on the need for future research *explaining* the policy patterns we have identified and the changes that have or might result. Such a step is a critical, but often overlooked, precondition for providing governments, businesses and non-governmental organizations with sound policy advice. In other words, further explanatory research is important not only for students of forest policy and governance, but also for practitioners who seek to develop policies that encourage environmental stewardship and effectiveness.

For these reasons, our approach and results are directly relevant to existing domestic and global efforts to promote sustainable forest management. We have discussed elsewhere in this book the need to understand domestic policy dynamics in the context of forest law enforcement and governance (FLEG) and trade (FLEGT) processes, which are focused explicitly on domestic-level forest policy. At the same time, our results are also important for informing other, more cross-sectoral international policy dialogues and deliberations, including those focused on the design of forest policies to limit carbon emissions through 'reduced emissions from deforestation and degradation' (REDD) efforts (Karsenty, 2008). While increasing loss and degradation of the world's forests has not yet resulted in progress towards international agreements about forests (Humphreys, 2006), growing concern over climate change, and carbon emissions from forest loss, have led to a renewed international emphasis on conservation and sustainable management of the world's forests, offering new hope and opportunities for concerted global action on promoting, and implementing, sustainable forest policy.

Whether this hope will be realized, and hence lead to outcomes different from previous international efforts (Levin et al, 2008), depends in large part on just what types of environmental stewardship will be required of REDD supported projects. The empirical approach in our book may offer forest and climate policy stakeholders and actors – governments, international agencies, non-governmental organizations, businesses and communities – a dispassionate way to measure and assess commitments to sustainable forest management, something virtually all REDD negotiations agree is critical for any project linking forest stewardship and climate change (Humphreys, 2008).

As with previous attempts for coordinated international action for forest conservation and sustainable forest management, any international REDD regime will need to address the multiple interests in forests, at local, national and global levels; the need to sustain and improve livelihoods of the poor while also sustaining forests; and the challenges of policy design and implementation that deliver the outcomes which decision- and policy-makers are seeking. National and sub-national policies will continue to be fundamental to realizing policy objectives agreed internationally. The goals of any REDD regime will still only be realized by effective policy implementation in the same forests and jurisdictions that have been the subject of the forest practice policies we have sampled, informed by the lessons which can be drawn from this and similar studies.

What is clear is that if progress is to be made in linking policy development to behavioural change and, ultimately, to maintenance and improvement of environmental quality, a range of key stakeholders and policy actors – from government agencies to businesses to environmental groups – must engage in problem-focused 'policy learning'. Over 20 years of research and analysis by Sabatier and his colleagues (Sabatier, 1998, 1999; Sabatier and Jenkins-Smith, 1999; Leach and Sabatier, 2005) has found that while multi-stakeholder policy learning can rarely change deep seated value differences, it can indeed lead to broad acceptance for particular means-focused policy interventions, especially when they uncover 'win-win' arrangements that simultaneously champion some degree of environmental, economic, and social goals.

A focus on policy learning moves actors away from a focus on their pre-established and perceived self-interest (which is often expressed in terms of environment versus development goals) towards a collective effort that identifies means to address ends (Elliott, 2000). Interventions that foster policy learning, such as the joint development of certification in Indonesia and Sweden (Elliott, 2000), and field-based approaches that integrate local knowledge, multi-stakeholder involvement and scientific research (Holling and Meffe, 1996; Glasbergen, 1996; Ison et al, 2007; Cadman, 2009), are known to be important for building greater awareness, consensus and trust, and enabling adaptive management. Similarly, comparative assessments such as we undertook in this study, are prerequisites for fostering greater consensus and support among a range of stakeholders (Elliott and Schlaepfer, 2001).

For these reasons, we hope that the comparative application of our classification framework may help foster future policy learning processes within domestic and international efforts for forest conservation and sustainable forest management, paving the way for innovative and effective solutions.

More systematic and focused attention to specific forest practice policy specifications, the social dynamics of the policy-making and the effectiveness of policy are critical to better link policy approaches with the on-the-ground impacts in ways that facilitate enduring and effective conservation and sustainable management of the world's forests.

Such an effort will also pave the way for arguably the most important research of all: understanding better the relationships among policy settings, behavioural change and improvements in environmental and social sustainability. By opening up the 'black box' of global forest policy specifications, our book justifies attention to, and paves the way for, a more complete understanding of the causal mechanisms through which policy might be made more effective.

REFERENCES

Bennett, Colin J. and Michael Howlett (1992) 'The lessons of learning: Reconciling theories of policy learning and policy change', *Policy Sciences*, 25, pp275–294

Blackmore, Chris (2007) 'What kinds of knowledge, knowing and learning are required for addressing resource dilemmas?: A theoretical overview', *Environmental Science and Policy*, 10, pp512–525

Braithwaite, John, and Peter Drahos (2000) *Global Business Regulation*, Cambridge: Cambridge University Press

Brown, David, Kate Schreckenberg, Neil Bird, Paolo Cerutti, Filippo Del Gatto, Chimere Diaw, Tim Fomete, Cecilia Luttrell, Guillermo Navarro, Rob Oberndorf, Hans Theil and Adrian Wells (2008) 'Legal timber: Verification and governance in the forest sector', London: Overseas Development Institute (ODI)

Cadman, Timothy Mark (2009) *Quality, Legitimacy and Global Governance: A Comparative Analysis of Four Forest Institutions*, PhD thesis, University of Tasmania,

Cashore, Benjamin (1997) 'Governing forestry: Environmental group influence in British Columbia and the US Pacific Northwest', PhD, Political Science, Toronto: University of Toronto

Cashore, Benjamin (1999) 'Chapter three: US Pacific Northwest', in B. Wilson, K. V. Kooten, I. Vertinsky and L. Arthur (eds) *Forest Policy: International Case Studies*, Wallingford: UK, CABI Publications

Cashore, Benjamin and Michael Howlett (2007) 'Punctuating which equilibrium? Understanding thermostatic policy dynamics in Pacific Northwest forestry', *American Journal of Political Science*, 51, 3, pp532–551

Cashore, Benjamin, Graeme Auld and Deanna Newsom (2004) *Governing Through Markets: Forest Certification and the Emergence of Non-State Authority*, New Haven, CT: Yale University Press

Chatham House (2009) 'Illegal logging', www.illegal-logging.info/, accessed August 2009

CPET (2009) 'National policies: Government timber procurement policies around the world', Central Point of Expertise on Timber (CPET) available from www.proforest.net/cpet/international-context/international-policies-1/, accessed August 2009

DeSombre, Elizabeth R. (2000) *Domestic Sources of International Environmental Policy: Industry, Environmentalists, and U.S. Power*, Cambridge, MA: MIT Press

DiMaggio, Paul J. and Walter W. Powell (1991) 'The iron cage revisited: Institutional isomorphism and collective rationality', in W. W. Powell and P. J. DiMaggio (eds) *The New Institutionalism in Organizational Analysis*, Chicago: The University of Chicago Press

Eisner, Marc Allen (1993) *Regulatory Politics in Transition*, Baltimore, MD: The Johns Hopkins University Press

Elliott, Christopher (2000) *Forest Certification: A Policy Perspective*, Bogor: Center for International Forestry Research (CIFOR)

Elliott, C. and R. Schlaepfer (2001) 'Understanding forest certification using the Advocacy Coalition Framework', *Forest Policy and Economics*, 2, 3–4, pp257–266

FEMAT (1993) 'Forest ecosystem management: An ecological, economic and social assessment', Portland, OR: Forest Ecosystem Management Assessment Team; Departments of Agriculture, Commerce and Interior

Flick, Warren A., Allen Barnes and Robert A. Tufts (1995) 'Public purpose and private property: The evolution of regulatory taking', *Journal of Forestry*, 93, 6, June, pp21–24

Folke, Carle (2006) 'Resilience: The emergence of a perspective for social-ecological systems analyses', *Global Environmental Change*, 16, pp253–267

Glasbergen, Pieter (1996) 'Learning to manage the environment', in W. M. Lafferty and J. Meadowcroft (eds) *Democracy and the Environment: Problems and Prospects*, Cheltenham, UK and Brookfield, MA: Edward Elgar

Hardin, Garrett (1968) 'The tragedy of the commons', *Science*, 162, 3859, pp1243–1248

Hoberg, George (1993) 'Regulating forestry: A comparison of institutions and policies in British Columbia and the US Pacific Northwest', Vancouver: Forest Economics and Policy Analysis Research Unit, University of British Columbia

Hoberg, George (1997) 'Governing the environment; Comparing policy in Canada and the United States', in K. Banting, G. Hoberg and R. Simeon (eds) *Degrees of Freedom: Canada and the United States in a Changing Global Context*, Montreal and Kingston: McGill-Queens

Holling, C. S. and Gary K. Meffe (1996) 'Command and control and the pathology of natural resource management', *Conservation Biology*, 10, 2, pp328–337

Howlett, Michael (1994) 'The judicialization of Canadian environmental policy, 1980–1990: A test of the Canada-United States convergence thesis', *Canadian Journal of Political Science*, XXVII, 1, pp99–127

Humphreys, David (2006) *Logjam: Deforestation and the Crisis of Global Governance*, London: Earthscan

Humphreys, David (2008) 'The politics of "avoided deforestation": Historical context and contemporary issues', *International Forestry Review*, 10, 13, pp433–442

Inglehart, Ronald (1977) *The Silent Revolution: Changing Values and Political Styles among Western Publics*, Princeton, NJ: Princeton University Press

Inglehart, Ronald (1995) 'Public support for environmental protection: Objective and subjective values in 43 societies', *Political Science & Politics*, 28, 1, pp57–73

Ison, Ray, Niels Rölling and Drennan Watson (2007) 'Challenges to science and society in the sustainable management and use of water: Investigating social learning', *Environmental Science and Policy*, 10, pp499–511

Karsenty, Alain (2008) 'The architecture of proposed REDD schemes after Bali: Facing critical choices', *International Forestry Review*, 10, 13, pp443–457

Leach, William and Paul A. Sabatier (2005) 'To trust an adversary: Integrating rational and psychological models of collaborative policymaking', *American Political Science Review*, 99, 4

Levin, Kelly, Constance McDermott and Benjamin Cashore (2008) 'The climate regime as global forest governance: Can reduced emissions from deforestation and forest degradation (REDD) initiatives pass a "dual effectiveness test"?', *International Forestry Review*, 10, 3, pp538–549

Liverman, Diana M. and Silvina Vilas (2006) 'Neoliberalism and the environment in Latin America', *Annual Review of Environment and Resources*, 31, 1, pp327–363

McDermott, Constance L. (2003) 'Personal trust and trust in abstract systems: A study of forest stewardship council-accredited certification in British Columbia', PhD, Vancouver: Department of Forest Resources Management, Faculty of Forestry, University of British Columbia

McDermott, Constance L. and George Hoberg (2003) 'From state to market: Forestry certification in the U.S. and Canada', in B. Schindler, T. Beckley and C. Finley (eds) *Two Paths Toward Sustainable Forests: Public Values in Canada and the United States*, Corvallis, OR: Oregon State University Press

McDermott, Constance L., Benjamin Cashore and Peter Kanowski (2009) 'Setting the bar: An international comparison of public and private policy specifications and implications for explaining policy trends', *Journal of Integrative Environmental Sciences*, 6, 3, pp1–21

Metafore (2009) *Forest Certification Resource Center: Stepwise Approach*, www.metafore.org/index.php?p=Stepwise+approach&s=174, accessed August 2009

Murnighan, J. Keith, Deepak Malhotra and Mark Weber (2007) 'Chapter 12: Paradoxes of trust: Empirical and theoretical departures from a traditional model', in R. M. Kramer and K. C. Cook (eds) *Trust and Distrust in Organizations: Dilemmas and Approaches*, New York: Russell Sage

Ortiz, Paul A. (2005) 'An overview of the U.S. Lacey Act Amendments of 1981 and a proposal for a model Port State Fisheries Enforcement Act', report prepared for the ministerially led task force on illegal, unreported and unregulated fishing on the high seas, Senior Enforcement Attorney, US National Oceanic and Atmospheric Administration, November 2005, www.high-seas.org/docs/Lacey_Act_Paper.pdf, accessed November 2009

Power, Michael (2003) 'Evaluating the audit explosion', *Law & Policy*, 25, 3, pp185–202

Rametsteiner, Ewald and Markku Simula (2003) 'Forest certification – An instrument to promote sustainable forest management?', *Journal of Environmental Management* 67, 1, pp87–98

Robledo, Carmenza, Jürgen Blaser, Sarah Byrne and Kaspar Schmidt (2008) *Climate Change and Governance in the Forest Sector: An Overview of the Issues on Forests and Climate Change with Specific Consideration of Sector Governance, Tenure, and Access for Local Stakeholders*, Washington, DC: Rights and Resources Initiative

Rose-Ackerman, Susan (2001) 'Trust and honesty in post-socialist societies', *Kyklos*, 54, pp415–444

RRI and ITTO (2009) 'Tropical forest tenure assessment: Trends, challenges and opportunities', Rights and Resources Initiative (RRI) and the International Tropical Timber Organization (ITTO), prepared for the International Conference on Forest Tenure, Governance and Enterprise: New Opportunities for Central and West Africa, 25–29 May 2009, Hôtel Mont Fébé, Yaoundé, Cameroon

Sabatier, Paul A. (1998) 'Top-down and bottom-up approaches to implementation research: A critical analysis and suggested synthesis', *Journal of Public Policy*, 6, 1, pp21–48

Sabatier, Paul (ed) (1999) *An Advocacy Coalition Lens on Environmental Policy*, Cambridge, MA: MIT Press

Sabatier, Paul A. and Hank C. Jenkins-Smith (1999) 'The advocacy coalition framework: An assessment', in Paul Sabatier (ed) *Theories of the Policy Process*, Boulder, CO: Westview

Seneca Creek Associates and Wood Resources International (2004) 'Summary: "Illegal" logging and global wood markets: The competitive impacts on the U.S. wood products industry', report prepared for the American Forest and Paper Association, www.illegal-logging.info/uploads/afandpa.pdf, accessed November 2009

Shapiro, Arthur M. (1997) 'Science or spin: Harvesting Chile's Lenga forests', *New Leader*, 10 March, pp14–15

Shapiro, Susan P. (1987) 'The social control of impersonal trust', *American Journal of Sociology*, 93, 3, pp623–658

Sher, Victor M. (1993) 'Travels with Strix: The spotted owl's journey through the federal courts', *The Public Land Law Review*, 14, pp41–79

Sitkin, S. B. and N. L. Roth (1993) 'Explaining limited effectiveness of legalistic "remedies" for trust/distrust', *Organizational Dynamics*, 16, 3, pp73–79

Unruh, J. D. (2008) 'Carbon sequestration in Africa: The land tenure problem', *Global Environmental Change – Human and Policy Dimensions*, 18, 4, pp700–707

US CBP (2009) 'Guidance on the Lacey Act Declaration', US Department of Homeland Security, Customs and Border Patrol, www.cbp.gov/xp/cgov/trade/trade_programs/entry_summary/laws/food_energy/amended_lacey_act/guidance_lacey_act.xml, accessed November 2009

Victor, David G. (1999) 'Enforcing international law: Implications for an effective global warming regime', *Duke Environmental Law and Policy Forum*, 10, 1, pp47–184

Vogel, David (1995) *Trading Up: Consumer and Environmental Regulation in a Global Economy*, Cambridge, MA: Harvard University Press

Vogel, David (2005) *The Market for Virtue*, Washington, DC: Brookings Institution Press

World Bank (2004) 'Sustaining forests: A development strategy', Washington, DC: World Bank

Zhang, Daowei (2000) 'Endangered species and timber harvesting: The case of Red-Cockaded Woodpeckers', Auburn, AL: Forest Policy Center, School of Forestry and Wildlife Sciences, Auburn University

Index